A Smith

Living Under the Shadow

One World Archaeology Series
Sponsored by the World Archaeological Congress
Series Editors: Joan Gero, Mark Leone, and Robin Torrence

One World Archaeology volumes contain carefully edited selections of the exemplary papers presented at the World Archaeology Congress (WAC), held every four years, and intercongress meetings. WAC gives place to considerations of power and politics in framing archaeological questions and results. The organisation also gives place and privilege to minorities who have often been silenced or regarded as beyond capable of making main line contributions to the field. All royalties from the series are used to help the wider work of the organisation. The series is published by Left Coast Press, Inc. beginning with volume 48.

Living Under The Shadow
Cultural Impacts of Volcanic Eruptions

Edited by
John Grattan and Robin Torrence

Left Coast Press Inc.

Walnut Creek, California

Left Coast Press Inc.

LEFT COAST PRESS, INC.
1630 North Main Street, #400
Walnut Creek, CA 94596
http://www.LCoastPress.com

Cataloging-in-Publication Library of Congress Data

Living under the shadow: cultural impacts of volcanic eruptions/
John Grattan and Robin Torrence, editors.
p. cm. — (One world archaeology series; 53)
Includes bibliographical references and index.

ISBN-13: 978-1-59874-268-8 (hardcover: alk. paper)

1. Archaeology and natural disasters. 2. Volcanoes—Social aspects—History. 3. Social change—History. 4. Human ecology—History.
5. Human beings—Effect of environment on—
History. I. Grattan, John. II. Torrence, Robin.
CC77.N36L585 2007
930.1028—dc22
2007018705

Printed in the United States of America

∞™ The paper used in this publication meets the minimum requirements of American National Standard for Information Sciences—Permanence of Paper for Printed Library Materials, ANSI/NISO Z39.48–1992.

07 08 09 10 11 5 4 3 2 1

Contents

List of Illustrations

Figures

Tables

CHAPTER 1

Beyond Gloom and Doom: The Long-Term Consequences of Volcanic Disasters

John Grattan and Robin Torrence

MODERN BIASES

Natural disasters are a major concern for modern society because their incidence and resulting mortalities are steadily rising (Leroy 2006: 4; Tobin and Montz 1997: 1–2). Despite speculation about the possible role of climatic change in increasing the incidence of hazardous events, the increasing vulnerability of societies probably plays a more important role in this trend. As global population continues its rapid growth, more and more people are exposed to environmental forcing agents (Figure 1.1), especially because vulnerable groups have naturally expanded into or been pushed into hazardous zones (Sheets, Chapter 4; Small and Naumann 2001). Given the powerful effects of disasters on modern societies, it is not surprising that recent experiences have had a major influence on the way the general public, and archaeologists, conceive past events.

The most obvious result has been a boom in archaeological research focused on the effects of ancient catastrophes on cultural change (eg, Bawden and Reycraft 2000; Hoffman and Oliver-Smith 2002; McCoy and Heiken 2000; McGuire *et al* 2000; Oliver-Smith and Hoffman 1999), including our previous book *Natural Disasters and Cultural Change* (Torrence and Grattan 2002a). As we pointed out in the book's introduction, however, a great deal of the literature examining the role of natural disasters in human history is sensationalist and based on unproven correlation. A more useful approach utilises sound analyses that clearly and unambiguously demonstrate causation. Well-researched, balanced case studies make the most important contribution to our understanding of how human societies have lived and coped with environmental perturbations (Torrence and Grattan 2002b). One of the aims of this book, therefore, is to present studies that take a penetrating and critical view of whether and how disasters have shaped the past.

Hard-hitting empirical studies have already produced some surprises. As illustrated in Torrence and Grattan (2002a), popularist

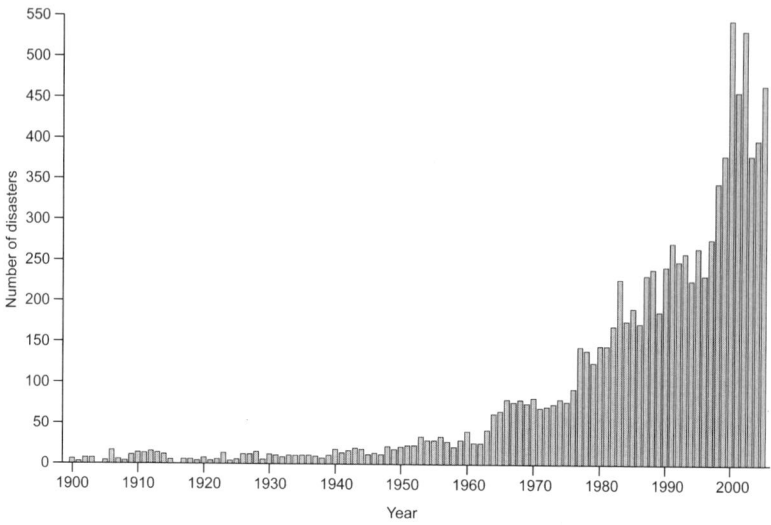

Figure 1.1 Natural disasters reported 1900–2005. Data from the International Disaster Database www.em-dat.net.

treatments of ancient disasters and their obsession with death, destruction, and flight have been grossly overstated. In contrast, many of the book's case studies showed that human societies have been incredibly resilient and have recovered remarkably well despite having experienced wide-scale destruction and/or significant mortality (cf Moseley 2002: 194, 211). The few examples of collapse were limited to disasters triggered by environmental agents of very high magnitude or involved societies that were unstable or already in decline. The new case studies presented here significantly expand knowledge about the varied ways that human societies have coped with and responded to environmental disasters. The global and temporal coverage of the research reported is impressive, comprising studies from North and Central America, Europe, Asia, and the wider Pacific region, and these range in time from the late Pleistocene (Middle Palaeolithic) up to the modern day. These historical studies are also relevant to modern hazard management because they provide records for a far wider range of events and responses than have been recorded in written records (cf Leroy 2006; Mastrolorenzo *et al* 2006; Siebe *et al* 1996).

A second way that modern experiences of natural disasters have influenced scholarship is through shifting the primary focus away from the strength of the environmental forcing mechanism as the primary agent of change to the vulnerability of past societies (Torrence and Grattan 2002b: 5). Disasters are now seen as social rather than simply environmental phenomena (eg, Blaikie *et al* 1994; Oliver-Smith 2002;

Shimoyama 2002a). Although current thinking is clearly an improvement over the previous separation of environmental and social science approaches to disasters, the new emphasis on vulnerability has meant that people are mainly conceived of as helpless 'victims'. Whereas incorporating humans into disasters is obviously essential, theoretical conceptions about the range of potential responses have been grossly limited. The most important goal of this book, therefore, is to move beyond the gloom and doom image of civilisations collapsing or humans cowering in the face of disaster and take a broader and more considered approach informed by comprehensive, solid data.

When a much wider range of potential outcomes is considered, one discovers a complex mix of responses to disasters. Most significantly, creative as well as destructive effects on human societies are commonly observed. The key to this new approach is the consideration of vulnerability, disaster, and recovery as stages in a process that unfolds over a reasonably long period of time (cf Oliver-Smith and Hoffman 2002: 12–13). Scholars who have examined disasters over the long term have found their effects can linger on and/or be used and reinterpreted by subsequent societies (eg, Blong 1982; Cronin, Gaylord *et al* 2004; Cronin, Petterson *et al* 2004; Dillian 2004; Driessen 2002; Elson *et al* 2002; Gaillard *et al* 2005; Galipaud 2002; Hoffman 1999: 306, 2002; Lowe, Newnham, and McCraw 2002; Lowe, Newnham, McFadgen *et al* 2002; Plunket and Uruñuela 1998a, 1998b, and Chapters 6, 9, 10, 12, 13). Oral history and myth have been important in keeping the memory of disasters alive, acting as invaluable sources of information about how to handle similar risks in the future and providing raw material for re-interpretation. Humans have not just responded to their environment, even when coping with its catastrophic consequences. Instead, people often move beyond the devastating initial impacts to incorporate their experiences in a myriad of creative and fascinating ways. As illustrated by the studies in this book, this broader and more complex view of how humans and disasters have interacted has yielded tantalising results and has opened up new avenues of research.

WHY VOLCANOES?

By limiting the focus of this book to volcanoes – a single, although admittedly variable, type of environmental hazard – the chapters achieve a focused comparative framework. Volcanic disasters are, however, rather special, and it is worth noting the particular features that set them apart from other types. Ironically, archaeology often benefits from volcanic disasters because the eruptive products can create conditions for excellent preservation of ancient dwelling places, villages, and whole cities as well as extensive cultural landscapes. Furthermore, the widespread air-fall tephras often form well-dated chronostratigraphic units, which, when spread over large areas, enable correlation of cultural responses among widely spaced regions (eg, Cronin and Neall 2000; Dugmore *et al* 1995; Lowe *et al* 2000; Machida 2002; Machida and Sugiyama 2002; Newnham *et al* 1998; Sheets 1983, 2002; Sheets and McKee 1994; Shimoyama 2002b; Torrence *et al* 2000; Zeidler and Isaacson 2003, and Chapters 2–7 and 11).

Apart from very rare events such as giant meteors, volcanoes are among the most hazardous of all natural disasters because they are unpredictable, sudden, and often catastrophic. It is worth examining these features in more detail.

Unpredictable

Volcanoes are especially dangerous hazards because, without extensive modern scientific research and monitoring, they are highly unpredictable. Seemingly innocuous mountains can suddenly explode with very little warning. There are many notable examples of populations being taken by surprise by large eruptions, the most recent being the major eruption of Pinatubo, the Philippines, in 1991 (Gaillard *et al* 2005, and Chapter 11). The eruption of Lamington in Papua New Guinea in 1951 caused high mortality and enormous distress to local populations and colonial authorities alike because no one had even recognised that the mountain was a volcano (Blong 1982: 51). The cinder-cone eruption of Parícutin, described by Elson *et al* (Chapter 6) also seemed to appear out of nowhere, and in 1783, most educated Europeans thought the sulphur gases they could smell had been released from the earth by an earthquake rather than from an Iceland volcano (Grattan *et al*, Chapter 8). Since the length of time between eruptions might be in the order of thousands of years, it can be difficult for human groups to maintain memories of previous eruptions from a particular volcano. In contrast, groups resident in highly active volcanic environments often pass on stories about volcanic disasters and these may be useful in familiarising the population with volcanism in general and specific attributes in particular (eg, Chester and Duncan, Chapter 10; Cronin and Cashman, Chapter 9; Elson *et al*, Chapter 6; Lowe *et al* 2002).

Large Scale

As demonstrated in the chapters in this book, the scale of impacts from volcanic events is quite variable, but many volcanoes have created total destruction over huge areas. The well-studied eruptions of Tambora in 1815 and Laki in 1783 are often cited as major volcanic eruptions that influenced climate and environment, but what is surprising when viewed from a global perspective is that large volcanic eruptions of a similar scale are not rare events. A brief consideration of the record of volcanic sulphate contained in the GISP ice-core record (Figure 1.2) suggests that there have been numerous eruptions in the past six thousand years that dwarf the eruptions of Tambora and Laki. It is instructive to remember that these took place during the time when agriculture and complex urban living were developed. The provenance of most of these large volcanic events is unknown, as, for example, the recently discovered signal for an eruption recorded ~1256 AD. Although this is the greatest eruption of the past 12,000 years, its source is not known and we are only now beginning to assess its impacts, which could have been considerable. The GISP sulphate record makes it clear that there have been numerous volcanic events of considerable magnitude and capable of perturbing climate on a global scale, but they have yet to attract the attention of archaeologists or palaeoenvironmentalists. Furthermore, there is the risk to modern global civilisation of a super volcanic eruption. Sparks *et al* (2005) estimated that the recurrence interval for an 'entry level' super eruption, releasing 200–300 km^3 magma into the atmosphere, could be as high as one every 10,000 years; thus there is a one in 100 probability that such an eruption could occur in the 21st century!

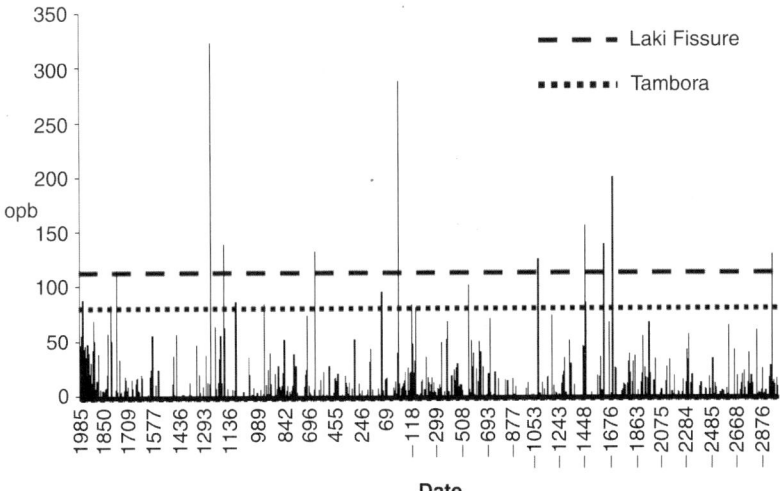

Figure 1.2 Volcanic sulphate in the GISP2 ice core. Input of the Laki fissure and Tambora eruptions highlighted, as these were recent significant volcanic eruptions.

Longevity and Lag Effects

Unlike some natural hazards, volcanic impacts can begin very suddenly but continue for long periods. Sometimes the eruptions last for years, if not decades. For instance, the town of Rabaul in Papua New Guinea has continued to be showered with rocks, ash, and foul gases since the eruption began in 1994, and some countries, such as Ecuador, endure frequent volcanic eruptions and live with a considerable degree of risk (Lane *et al* 2004). Even if volcanoes are not in eruption, they can remain dangerous for thousands of years. When geologists use the term 'active volcano', they mean that there is a potential for further eruptions even if the volcano has remained quiet for decades or centuries. Highly active volcanoes such as Etna in Italy, Tungurahua in Ecuador, and Ngaurahoe in New Zealand may experience periods with frequent eruptions punctuated by relatively short intervals of quiescence lasting a few years or decades. Large and dangerous volcanoes such as Tambora in Indonesia, Vesuvius in Italy, and Santorini in Greece, although currently quiet, are all still active and pose a threat for the future. In addition, supervolcanoes such as Yellowstone, Toba, Taupo, and Rotorua, to name but a few, are still considered 'active' in geological terms and will erupt one day, posing a challenge to our interconnected and interdependent world (Engvild 2003; Sparks *et al* 2005).

Even after short events, the effects of the erupted products (eg, lavas, pyroclastic flows, and air-fall tephras) can continue for centuries because they form a sterile blanket over the landscape with great variation in the rates of natural breakdown and colonisation by plants (eg, VanderHoek and Nelson, Chapter 7). Furthermore, the secondary effects in the form of major landslides and highly mobile lahars can create serious new hazards that may last over centuries (eg, Blong 1984: 285–291; Crittenden and Rodolfo 2002; Gaillard, Chapter 11). Finally, there can be lags in the effects. For example, dams created by flows can eventually break, resulting in extensive damage many years after the initial eruption (Gaillard *et al*, Chapter 11).

Range of Impacts

The wide range of volcanic impacts makes them attractive for research on disasters. Volcanoes produce hot materials near the source but also cool air-fall tephras at great distances. The variability in the sizes of materials produced is also impressive: from house-sized blocks to fine ashes to gases. In addition to the obvious erupted products from flows and air falls, there are more insidious impacts involving noxious and lethal gases, which, in some cases, can spread over large regions (eg, Grattan *et al* 2002; Grattan *et al*, Chapter 8). In a world increasingly sensitive to air quality, the risk posed by the potential addition to serious urban smog of millions of tonnes of volcanic sulphur and other volatile gases should not be underestimated.

The numerous potential consequences of volcanic events enable comparison amongst phenomena of difference scales; these range from the global impacts of climate change to the immediately local impacts of lavas. This variability is expressed in the chapters in this book, with contributions focusing on short-term, local events such as Loma Caldera (Sheets, Chapter 4), Sunset Crater in Arizona (Elson *et al*, Chapter 6) and Glass Mountain in California (Dillian, Chapter 12) to tephras spread across large regions (Torrence and Doelman, Chapter 3; Sheets, Chapter 4) up to the scale of the Alaskan Peninsula (VanderHoek and Nelson, Chapter 7) or tephras and gases over Europe (Fedele *et al*, Chapter 2; Grattan *et al*, Chapter 8) and even with potentially global effects as in the case of the Toba eruption (Oppenheimer 2002; Rampino and Self 1993; Sparks *et al* 2005).

Methodological Difficulties

Having stressed why volcanic hazards provide good opportunities for disaster research, it is also worth reflecting briefly on some of the difficulties faced by researchers. The first and most important is dating. Short-term events, such as a volcanic eruption, are notoriously hard to pin-point precisely with techniques such as radiocarbon. Accurately dating prehistoric volcanic activity can be very difficult if it cannot be identified in records that have a yearly resolution (eg, ice cores, coral reefs, speleotherms, tree-rings, or laminated lake sediments) (Leroy 2006: 6–7). Second, it can be quite problematic to link a volcanic event with a short-term duration to a cultural response in the absence of highly precise dates, as perhaps best exemplified by the controversy over whether the Santorini volcano was directly responsible for changes in ancient Cretan society (eg, Manning and Sewell 2002). The lack of good dating can be an especially significant obstacle to reconstruction in terms of the time scales relevant to human behaviour (eg, a few years or a generation). Precise dating is also required for examining the scale of impact as measured by length of abandonment or in studying the nature of recovery and recolonisation (eg, Chapters 3, 4, 6, 7).

Making correlations between a volcanic event and a cultural or biological response is generally clear when there is a well-studied and well-dated stratigraphic sequence and when archaeological remains are buried in or under diagnostic tephra (eg, Chapters 2–6, 11), but when one looks outside the area of direct impact, it can be difficult to make a precise link between a specific volcanic event and a particular cultural outcome. Great improvements are being made in geochemical methods used to characterise particular tephras, however, and as demonstrated by Grattan *et al* (Chapter 8), other disciplines such as history can provide useful data for studying particular eruptions and their effects, even when the environmental agents were virtually invisible, as in the case of noxious gases. Certainly, the study of volcanic disasters would benefit from the development of more precise methods for dating and correlation, in particular, but the chapters in this book demonstrate that these obstacles can often be overcome by combining techniques and data from a range of disciplines and data sources.

FOCUS ON CREATIVITY

Despite our brief but worrying review of volcanic hazards, it is important to acknowledge that so far humanity has survived a wide range of extreme risks. The species thrived following the eruption of the last supervolcano, Toba, around 74,000 years ago. As the chapters in this book demonstrate, the focus of attention should move away from failure and collapse in the face of geological disasters to human resilience and creativity (Grattan 2006). This book also represents a significant shift away from the simplistic black and white question of whether disasters have played an important role in human history. The authors accept that volcanic events have certainly been a potent force, but unlike much previous discussion in which volcanic impacts have only been equated with collapse or decimation, the new approach showcased here examines a much broader range of potential relationships between volcanoes and societies.

To begin with, there is no doubt that major volcanic eruptions can have catastrophic consequences that lead to extinctions and subsequent biological and cultural changes. The very high magnitude eruptions discussed in Chapters 2–5, 7, and 11 obliterated populations and cultural forms over large areas. Eventually these were replaced by others, surprisingly not all of which were markedly different from those they had replaced. The events described by Fedele *et al* (Chapter 2) and Gonzalez *et al* (Chapter 5) were particularly significant in the long term because one or more species was rendered extinct as a consequence of these large events. Furthermore, VanderHoek and Nelson (Chapter 7) show how a major volcanic eruption can devastate resources necessary for humans, especially in an environment characterised by low primary productivity, and their findings illustrate the considerable time that may be needed for recovery. As Sheets (Chapter 4) reminds us, however, many human societies are remarkably resilient and resist the effects of volcanic events, despite periods of forced abandonment from their territories.

The destructive effects of volcanic catastrophes are well known and have been catalogued in basic texts (eg, Blong 1984; Bryant 1991; Sheets and Grayson 1979: 2–4; Sigurdsson 2000), but what has not been widely appreciated are the positive effects that volcanic events can have on the environment and on cultural process. Despite the seemingly devastating nature of many events, volcanic hazards as a class should not be viewed in a purely negative light. As many of the chapters show, destruction of life and environment can also provide new opportunities for alternative lifeways and/or form the platform for new forms of life and cultural innovations. For example, it may be useful to conceive of volcanic hazards as an intense form of natural selection leading to evolutionary change as well as extinction (Torrence and Doelman, Chapter 3). The wide range of creative forces of volcanic hazards discussed in the chapters can be usefully grouped into a number of themes that we now briefly review. These include physical enhancements to the environment, flexibility, and creativity.

Physical Enhancements

Volcanoes often produce products that are useful and desirable for human communities (eg, pumice for polishing and grinding, volcanic glasses for stone tools and ornaments, and lavas for building stone). An obvious beneficial trait of volcanoes that has been long recognised is the addition of mineral-rich tephras, many of which are highly porous. These can greatly increase the productivity of agricultural lands, as in the many examples presented in Blong (1984: 348–50; cf Sheets and Grayson 1979: 2). Neall (2006) notes that volcanic soils are ideal for vegetables and especially root crops because they can expand freely in the well-drained subsoil. Over the short term, thin layers or dustings of air-fall tephra can have dramatic effects. Lentfer and Boyd (2001: 50) report that in some areas following the 1994 Rabaul eruption, 'people interviewed were pleased with their gardens, noting higher crop yields, improved vigour and healthier foliage'. Thicker layers may need a longer period of weathering to break down the lithic fragments and release the useful elements, but in the meantime these layers may restrict the growth of competing weeds. Air falls may also remove or reduce insect pests. Interestingly, tephras can act as a layer of mulch that creates an organically rich layer of rotting vegetation that promotes growth (eg, Lentfer and Boyd 2001: 50). Elson *et al* (Chapter 6) provide an intriguing example of how cinders from an eruption of Sunset Crater in Arizona created a mulch that has had very long-term beneficial effects on maize agriculture through reduction of evaporation and regulation of soil temperature. In fact, groups displaced by heavy tephra falls in one region actively immigrated into other afflicted areas because the cinder mulch created new arable lands.

Flexibility

Many of the chapters demonstrate the social and residential flexibility of cultural groups in dealing with volcanic disasters. People often abandon areas and recolonise but show remarkably few cultural changes. It is worth noting that many societies have lived in risky environments for a very long time and have had considerable experience in dealing with disasters. This is especially the case with groups resident in volcanically active regions. Adaptation developed over many centuries may help explain why many cultures appear to be resilient, as in the case of the groups described by Torrence and Doelman (Chapter 3) and Sheets (Chapter 4). As a possible type of adaptation, Torrence and Doelman emphasise that the maintenance of systems of social exchange can create links between what may become an impacted region and another area that might provide temporary refuge in times of crisis.

Creativity

Social groups impacted by volcanic disasters should not just be considered as victims of violent environmental factors. Many societies are extremely resilient to disruption (eg, Sheets, Chapter 4), actively resist change (eg, Crittenden and Rodolfo 2002), and/or use these experiences to shape aspects of their culture (eg, Torrence and Doelman, Chapter 3). For many cultures, environmental disasters are not conceived as natural events but as the result of social processes (eg, Galipaud 2002). Elson *et al* (Chapter 6), Cronin and Cashman (Chapter 9), and Chester and Duncan (Chapter 10) describe a wide range of societies that have incorporated volcanic events into belief systems, story telling, myths, and oral histories. Holmberg (2005 and Chapter 13) takes this approach further to discuss how groups have incorporated their volcanic landscape into their culture in creative ways. She shows that the volcanic environment is a part of daily life as expressed through art. Her chapter reminds us that the importance of volcanoes is not limited to the short period when they are erupting. Their influence and impact can be very pervasive and long lasting, and above all engender creative responses. In Chapter 12, Dillian reconstructs the ideological importance of obsidian among groups who might have witnessed the eruption in which it was formed. This is a powerful example of both the large impact of volcanism on culture and the creative ways in which people have incorporated volcanic events into their lives.

The creative use and reworking of previous disasters in myth and ritual can also become valuable assets because these experiences, observations, and knowledge may assist descendants who face similar situations (Cronin and Cashman, Chapter 9). The experiences of the ancestors may help people avoid hazards in the first place and may provide guidance about successful ways to find shelter and refuge during a volcanic event. Merely having knowledge of these unusual and very frightening phenomena can also reduce panic. The power of prior knowledge has been greatly emphasised in recent cases, such as the Pinatubo eruption, where modernisation has led to the loss of oral history and its useful knowledge (Davies 2002; Gaillard *et al*, Chapter 11).

A NEW DISCIPLINE

Leroy (2006) has observed that a new field of study, which she calls 'the science of environmental catastrophes', has emerged and has an important role to play in helping populations adjust to the growing environmental threats in the modern world. To be effective, this type of research must cross-disciplinary boundaries separating anthropology, archaeology, sociology, geology, biology, and practical risk assessment and ensure close collaboration among researchers and practitioners in these diverse fields. Highly sophisticated studies along these lines are becoming increasingly in demand because effective modern hazard management depends on the quality of the knowledge about the history of previous volcanic events and their impact on human communities (eg, Arce *et al* 2003; Berger 2006; Gaillard *et al*, Chapter 11; Mastrolorenzo *et al* 2006; Siebe *et al* 1996).

The chapters in this book showcase the importance and resulting value of cross-disciplinary collaborations in the study of natural disasters, in general, and volcanic events, in particular. Currently, scholars studying volcanic hazards must incorporate ideas, methods, and data from both the traditional 'hard' sciences and the social sciences and humanities. As illustrated here, people trained in geology are now actively learning about oral history from modern traditional communities (eg, Chester and Duncan, Chapter 10; Cronin and Cashman, Chapter 9); geologists have become historians (eg, Grattan *et al*, Chapter 8); and archaeologists have spearheaded research in volcanology to understand the conditions experienced by ancient peoples (eg, Gonzalez and Huddart, Chapter 5; Holmberg, Chapter 13; Sheets, Chapter 4; Torrence and Doelman, Chapter 3; VanderHoek and Nelson, Chapter 7).

LESSONS FOR THE MODERN WORLD

Interdisciplinary research like that reported here has an important role to play in directing risk avoidance in the modern world. The studies have identified specific problems in particular countries as well as important general principles about the value of behaviours that were successful in the past and the maintenance of traditional knowledge. Several authors in this book warn modern governments and planners that they are putting their citizens at extreme risk of death and loss of property from volcanic catastrophes by not paying enough attention to the history of a particular volcano and its past impacts.

These warnings are also pertinent to many areas of the world where there are active volcanoes. Sheets (Chapter 3, El Salvador), Gonzalez and Huddart (Chapter 5, Mexico), and Gaillard *et al* (Chapter 11, Philippines) note that people are being either pushed into or allowed to move into and develop areas that have been prone to volcanic activity in the past or, in the case of Pinatubo, are likely to be impacted in the very near future. Gaillard *et al* (Chapter 11) predict a further series of eruptions of Pinatubo relatively soon, based on a combination of geological, historical, oral history, and archaeological data that show the pattern of past activity is a series of closely spaced events and not a single, isolated eruption.

Along these lines, the findings of Grattan *et al* (Chapter 8) are also important because they illustrate that volcanic products can have dire consequences in terms of mortality and cultural disruption at surprisingly large distances from the volcanic centre where they were produced. Consequently, knowledge about areas that are likely to produce gaseous emissions, together with their potential patterns of dispersion, should be a high priority for research and planning.

Archaeological studies have demonstrated the importance of social factors in assisting recovery from volcanic disasters. Sheets's descriptions (Chapter 4) of variations in the degree to which prehistoric groups in Central America have coped with volcanoes should alert disaster managers that not all societies have the same potential for survival and that the risks may actually be higher for those with a significant dependence on infrastructure. He also points out that the nature of internal conflict can lower the resilience of societies in the face of volcanic disasters. Torrence and Doelman (Chapter 3) also argue that the organisation of society is a key element in mitigating the impact of volcanic disasters. They stress the role of social networks in providing safe refuges and recovery assistance over the longer term. In many cases, social links stretch over considerable distances and provide accommodation far from the affected area. Following their observations, it would be useful for managers to plan evacuation routes and places for recovery utilising knowledge of current social links between groups at risks and those in safe areas. In many cases, supporting and even augmenting traditional social ties might be beneficial.

Several chapters argue that active hazard management can take advantage of the keen observations and long memories about their environments common among traditional societies living in risky areas of the world. Cronin and Cashman (Chapter 8; cf Cronin, Gaylord *et al* 2004; Cronin, Petterson *et al* 2004) use a broad range of case studies from modern societies in the Pacific region and Central America to demonstrate that groups living among volcanoes have useful knowledge both about the history and character of past events as well as about current hazards, although these valuable observations may be expressed in oral history and beliefs rather than as scientific facts. In addition to providing geologists with essential information for predicting future volcanic activity, traditional knowledge and beliefs help maintain awareness that particular areas are potentially dangerous, even if eruptions have not occurred within living memory (eg, Holmberg, Chapter 13), and, in the case of many Polynesian societies, actively prevent people from going near risky places. Much of the mythological and ritual behaviour and beliefs recorded by Chester and Duncan (Chapter 10) and reconstructed from archaeology by Elson *et al* (Chapter 6) may have played these roles in the past (cf Lowe *et al* 2002). The loss of traditional knowledge about past disasters has already been shown to have dire consequences, as in the case of the 1998 Papua New Guinea tsunami (Davies 2002). By tapping into these rich sources of information and reinforcing sensible prohibitions, scientists can work with communities to put their culture to good use to avoid future disasters.

Risk managers have generally limited their concerns to the immediate consequences of the hazards, particularly deaths and physical damage, but archaeologists and other scholars have demonstrated that a better understanding of the social effects of catastrophes demands a longer term perspective. As pointed out by Oliver-Smith and Hoffman (2002: 12), 'whether rapid or slow in onset, disasters and the vulnerability leading to them unfold over time, often considerable amounts of time' and so 'disasters have pasts, presents and futures'. This is particularly true of the period during which people are coping with the impacts and either putting their society back into place or changing it (cf Hoffman 1999). The case studies clearly illustrate that the story of any volcanic disaster does not end shortly after the damage has been inflicted on human groups, but continues for a very long time. For example, new ideas about the sacred and profane worlds and their relationships following the experience of a cataclysmic event like as a large volcanic eruption, such as those described by Elson *et al* (Chapter 6) and Dillian (Chapter 12), do not appear overnight. Cultural groups are likely to explore and reconfigure their experiences with volcanic activity again and again, as described by Holmberg (Chapter 13).

Not only does the long life history of disasters make it difficult for archaeologists to be certain whether a particular volcanic event caused a specific cultural change, it also creates problems for social planners working within limited budgets and governmental time frames. This lesson is important for modern disaster managers, because not only can they utilise experiences and knowledge that have been preserved in living groups, but they can also learn from detailed environmental and archaeological studies of past events. Most importantly, they need to realise that because all future disasters will also have long and complex histories, management should plan for relatively long time scales rather than just the immediate clean up.

Comprehension of a disaster's full set of effects is often improved through comparison with a previous event. Not only can managers gain a better understanding of the potential variation of environmental forcing agents through historical studies, but they can also learn about how particular cultures have reacted in the past and discern the influence of cultural beliefs and various social structures on risk avoidance and patterns of recovery. Berger (2006) notes that historical studies are important because they illustrate what sorts of hazard mitigation worked in the past and which ones were unsuccessful. As demonstrated by Gaillard *et al* (Chapter 11) and evidenced by a growing number of historical studies, increasingly modern planners are looking to studies of the past to help them prepare for the future. Archaeology is beginning to play quite an important role in providing the data that risk managers require (eg, Cronin and Neall 2000; Gaillard *et al* 2005; Mastrolorenzo *et al* 2006; Siebe *et al* 1996).

A RICHER PICTURE

Our hope is that this book will be a valuable contribution to the emerging science of environmental catastrophes and that the paradigms presented will aid the study of past crises and planning for future events. The case studies confirm that in many respects the past is certainly a key to the future. The interdisciplinary global research showcased here has used archaeology, oral tradition, recorded literature, earth science, and geology to elucidate environmental and cultural responses to past disasters. In a world where volcanoes may suddenly erupt into the present, the key to understanding and anticipating the threat may lie in the past. A major Indian Ocean tsunami was in the written record just 124 years ago, following the eruption of Krakatau – if only we had paid more attention to the past!

The myriad case studies in this book show that by adopting wider perspectives and combining forces to enhance databases, archaeologists and environmental and social scientists can mobilise historical studies together with the information contained in oral history and ideologies to achieve a rich understanding of how human societies have experienced, coped with, and used volcanic events in creative ways. Based on the chapters presented here, we see the future of disaster research as increasingly based on very broad-scale, multidisciplinary research that acknowledges both the creative and destructive potentials of large magnitude environmental events. The new picture that is emerging is coloured not only by death, destruction, and collapse, but also by recovery, regeneration, and creation of new environments and cultural patterns. Volcanic disasters have been very powerful factors in human history, but the consequences have not all been negative.

ACKNOWLEDGMENTS

This book is derived from a session of the same name at WAC5, Washington, DC, in 2003. We are very grateful to the conference organisers for providing a full day in which we could learn from each other's different perspectives and case studies and discuss and debate the impacts of volcanic disasters on human history. We also thank all the session participants for the lively and stimulating discussion and enthusiasm that encouraged us to take on another edited book. Thanks, too, to the authors who later joined the project for producing thought-provoking papers in a timely fashion. We appreciate the useful suggestions from our referees and we acknowledge our series editor, Mark Leone, for his patience and support.

REFERENCES

Arce, J, Macías, J and Vázquez-Selem, L (2003) 'The 10.5 ka Plinian eruption of Nevado de Toluca volcano, Mexico: stratigraphy and hazard implications', *Bulletin of the Geological Society of America* 115, 230–48

Bawden, G and Reycraft, R (eds) (2000) *Environmental Disaster and the Archaeology of Human Response*, Anthropological Papers No 7, Albuquerque: Maxwell Museum of Anthropology

Berger, AR (2006) 'Abrupt geological changes: causes, effects, and public issues', *Quaternary International* 151, 3–9

Blaikie, P, Cannon, T, Davis, I and Wisner, B (1994) *At Risk: Natural Hazards, People's Vulnerability, and Disasters*, London: Routledge

Blong, R (1982) *The Time of Darkness: Local Legends and Volcanic Reality in Papua New Guinea*, Seattle: University of Washington Press

Blong, R (1984) *Volcanic Hazards: A Sourcebook on the Effects of Eruptions*, Sydney: Academic Press

Bryant, E (1991) *Natural Hazards*, Cambridge: Cambridge University Press

Crittenden, KS and Rodolfo KS (2002) 'Bacolor town and Pinatubo volcano, Philippines: coping with recurrent lahar disaster', in Torrence, R and Grattan, J (eds), *Natural Disasters and Cultural Change*, pp 43–65, London: Routledge

Cronin, SJ, Gaylord, DR, Charley, D, Wallez, S, Alloway, B and Esau, J (2004) 'Participatory methods of incorporating scientific with traditional knowledge for volcanic hazard management on Ambae Island, Vanuatu', *Bulletin of Volcanology* 66, 652–68

Cronin, SJ and Neall, V (2000) 'Impacts of volcanism on pre-European inhabitants of Taveuni, Fiji', *Bulletin of Volcanology* 62, 199–213

Cronin, SJ, Petterson, MG, Taylor, PW and Biliki, R (2004) 'Maximising multi-stakeholder participation in government and community volcanic hazard management programs, a case study from Savo, Solomon Islands', *Natural Hazards* 33, 105–36

Davies, H (2002) 'Tsunamis and the coastal communities of Papua New Guinea', in Torrence, R and Grattan, J (eds), *Natural Disasters and Cultural Change*, pp 28–42, London: Routledge

Dillian, C (2004) 'Sourcing belief: using obsidian sourcing to understand prehistoric ideology in northeastern California, U.S.A.', *Mediterranean Archaeology and Archaeometry* 4, 33–52

Driessen, J (2002) 'Towards an archaeology of crisis: defining the long-term impact of the Bronze Age Santorini eruption', in Torrence, R and Grattan, J (eds), *Natural Disasters and Cultural Change*, pp 250–63, London: Routledge

Dugmore, A, Larsen, G and Newton, A (1995) 'Seven tephra isochrones in Scotland', *The Holocene* 5, 257–66

Elson, M, Ort, M, Hesse J and Duffield, W (2002) 'Lava, corn, and ritual in the northern Southwest', *American Antiquity* 67, 119–35

Engvild, KC (2003) 'A review of the risks of sudden global cooling and its effects on agriculture', *Agriculture and Forest Meteorology* 115, 127–37

Gaillard, J, Delfin, F, Dizon, E, Larkin, J, Paz, V, Ramos, E, Remotigue, C, Rodolfo, K, Siringan, F, Soria, J and Umbak, J (2005) 'Dimension anthropique de l'éruption du Mont Pinatubo, Philippines, entre 800 et 500 ans BP', *L'anthropologie* 109, 249–66

Galipaud, J-C (2002) 'Under the volcano: Ni-Vanuatu and their environment', in Torrence, R and Grattan, J (eds), *Natural Disasters and Cultural Change*, pp 162–71, London: Routledge

Grattan, J (2006) 'Volcanic eruptions and archaeology: cultural catastrophe or stimulus?', *Quaternary International* 151, 10–11

Grattan, J Brayshay, M, and Schüttenhelm (2002) 'The end is nigh? Social and environmental responses to volcanic gas pollution', in Torrence, R and Grattan, J (eds), *Natural Disasters and Cultural Change*, pp 87–106, London: Routledge

Hoffman, S (1999) 'After Atlas shrugs: cultural change or persistence after a disaster', in Oliver-Smith, A and Hoffman, S (eds), *The Angry Earth: Disaster in Anthropological Perspective*, pp 302–26, London: Routledge

Hoffman, S (2002) 'The monster and the mother: the symbolism of disaster', in Hoffman, S and Oliver-Smith, A (eds), *Catastrophe and Culture*, pp 113–42, Sante Fe, NM: School of American Research

Hoffman, S and Oliver-Smith, A (eds) (2002) *Catastrophe and Culture*, Sante Fe, NM: School of American Research

Holmberg, K (2005) 'The voices of stones: unthinkable materiality in the volcanic context of Western Panamá', in Meskell, L (ed), *Archaeologies of Materiality*, pp 190–211, London: Blackwell

Lane, LR, Tobin, GA and Whiteford, LM (2004) 'Volcanic hazard or economic destitution: hard choices in Baños. Ecuador', *Environmental Hazards: Global Environmental Change*, Part B, 5, 23–34

Lentfer, C and Boyd, W (2001) *Maunten Paia: Volcanoes, People and Environment: The 1994 Rabaul Eruption*, Lismore, Australia: Southern Cross University Press

Leroy, S (2006) 'From natural hazard to environmental catastrophe: past and present', *Quaternary International*, 158, 4–12

Lowe, D, Newnham, R, McFadgen, B and Higham, T (2000) 'Tephras and New Zealand archaeology', *Journal of Archaeological Science* 27, 859–70

Lowe, D, Newnham, R and McCraw, J (2002) 'Volcanism and early Maori society in New Zealand', in Torrence, R and Grattan, J (eds), *Natural Disasters and Cultural Change*, pp 126–61, London: Routledge

Machida, H (2002) 'Quaternary volcanoes and widespread tephras of the world', available online at http://ns.airies.or.jp/publication/ger/pdf/06-2-02.pdf

Machida, H and Sugiyama, S (2002) 'The impact of the Kikai-Akahoya explosive eruptions on human societies', in Torrence, R and Grattan, J (eds), *Natural Disasters and Cultural Change*, pp 313–25, London: Routledge

Manning, SW and Sewell, DA (2002) 'Volcanoes and history: a significant relationship? The case of Santorini', in Torrence, R and Grattan, J (eds), *Natural Disasters and Cultural Change*, pp 264–91, London: Routledge

Mastrolorenzo, G, Petrone, P, Pappalardo, L and Sheridan, M (2006) 'The Avellino 3780-yr-BP catastrophe as a worst-case scenario for a future eruption at Vesuvius', *Proceedings of the National Academy of Sciences* 103, 4366–70

McCoy, F and Heiken, G (eds) (2000) *Volcanic Hazards and Disasters in Human Antiquity*, Geological Society of America Special Paper 345, Denver: Geological Society of America

McGuire, B, Griffiths, D and Stewart, I (eds) (2000) *The Archaeology of Geological Catastrophes*, Geological Society Special Publications 171, London: Geological Society of London

Moseley, M (2002) 'Modeling protracted drought, collateral natural disaster, and human responses in the Andes', in Hoffman, S and Oliver-Smith, A (eds), *Catastrophe and Culture*, pp 187–212, Sante Fe, NM: School of American Research

Neall, V (2006) 'Volcanic soils', in Verheye, W (ed), *Land Cover and Land Use*, in *Encyclopedia of Life Support Systems* (EOLSS), developed under the auspices of the UNESCO, EOLSS Publishers, Oxford, UK, available online at http://www.eolss.net

Newnham, R, Lowe, D, McGlone, M, Wilmshurst, J and Higham, T (1998) 'The Kaharoa Tephra as a critical datum for earliest human impact in northern New Zealand', *Journal of Archaeological Science* 25, 533–44

Oliver-Smith, A (2002) 'Theorizing disasters: nature, power, and culture', in Hoffman, S and Oliver-Smith (eds), *Catastrophe and Culture*, pp 23–48, Sante Fe, NM: School of American Research

Oliver-Smith, A and Hoffman, S (eds) (1999) *The Angry Earth: Disaster in Anthropological Perspective*, New York: Routledge

Oliver-Smith, A and Hoffman, S (2002) 'Introduction: why anthropologists should study disasters', in Hoffman, S and Oliver-Smith (eds), *Catastrophe and Culture*, pp 3–20, Sante Fe, NM: School of American Research

Oppenheimer, C (2002) 'Limited global change due to the largest known Quaternary eruption, Toba ~74 kyr BP?', *Quaternary Science Reviews* 21, 1593–1609

Plunket, P and Uruñuela, G (1998a) 'The impact of Popocatépetl volcano on Preclassic settlement in central Mexico', *Quaternaire* 9, 53–59

Plunket, P and Uruñuela, G (1998b) 'Preclassic household patterns preserved under volcanic ash at Tetimpa, Puebla, Mexico', *Latin American Antiquity* 9, 287–309

Rampino, MR and Self, S (1993) 'Climate-volcanism feedback and the Toba eruption of ~74000 years ago', *Quaternary Research* 40, 269–80

Sheets, P (ed) (1983) *Archaeology and Volcanism in Central America: The Zapotitan Valley of El Salvador*, Austin: University of Texas Press

Sheets, P (ed) (2002) *Before the Volcano Erupted: The Ancient Ceren Village in Central America*, Austin: University of Texas Press

Sheets, P and Grayson, D (1979) 'Introduction', in Sheets, P and Grayson, G (eds), *Volcanic Activity and Human Ecology*, pp 1–8, New York: Academic Press

Sheets, P and McKee, B (eds) (1994) *Archaeology, Volcanism, and Remote Sensing in the Arenal Region, Costa Rica*, Austin: University of Texas

Shimoyama, S (2002a) 'Basic characteristics of disasters', in Torrence, R and Grattan, J (eds), *Natural Disasters and Cultural Change*, pp 19–27, London: Routledge

Shimoyama, S (2002b) 'Volcanic disasters and archaeological sites in Southern Kyushu, Japan', in Torrence, R and Grattan, J (eds), *Natural Disasters and Cultural Change*, pp 326–42, London: Routledge

Siebe, C, Abrams, M, Macías, J and Obenholzner, J (1996) 'Repeated volcanic disasters in prehispanic time at Popocatépetl, central Mexico: past key to the future?', *Geology* 24, 399–402

Sigurdsson, H (ed) (2000) *Encyclopedia of Volcanoes*, San Diego: Academic Press

Small, C, and Naumann, T (2001) 'The global distribution of human population and recent volcanism', *Environmental Hazards* 3, 93–109

Sparks, S, Self, S, Grattan, JP, Oppenheimer, C, Pyle, D and Rymer, H (2005) *Super-eruptions: Global Effects and Future Threats*, report of a Geological Society of London working group, available online at www.geolsoc.org.uk/supereruptions

Tobin, G and Montz, B (1997) *Natural Hazards: Explanation and Integration*, London: The Guilford Press

Torrence, R and Grattan, J (eds) (2002a) *Natural Disasters and Cultural Change*, London: Routledge

Torrence, R and Grattan, J (2002b) 'Trends in the archaeology of disasters', in Torrence, R and Grattan, J (eds), *Natural Disasters and Cultural Change*, pp 1–18, London: Routledge

Torrence, R, Pavlides, C, Jackson, P and Webb, J (2000) 'Volcanic disasters and cultural discontinuities in the Holocene of West New Britain, Papua New Guinea', *Special Memoir, Geological Society of London* 171, 225–44

Zeidler, J and Isaacson, J (2003) 'Settlement process and historical contingency in the western Ecuadorian Formative', in Raymond, J and Burger, R (eds), *Archaeology of Formative Ecuador*, pp 69–123, Washington, DC: Dumbarton Oaks Research Library and Collection

CHAPTER 2

The Campanian Ignimbrite Factor: Towards a Reappraisal of the Middle to Upper Palaeolithic 'Transition'

Francesco G Fedele, Biagio Giaccio, Roberto Isaia,
Giovanni Orsi, Michael Carroll, and Bruno Scaillet

INTRODUCTION

In this chapter, we present results of a detailed investigation of the Campanian Ignimbrite (CI) eruption that took place c 40,000 BP. We endeavour to show that the potential impacts of this underrated eruption, in combination with other environmental factors, demand a reconsideration of the processes and rhythms that took place in western Eurasia at a crucial point in time represented by the so-called Middle to Upper Palaeolithic 'transition'.

The source of the CI eruption was in the Phlegraean Fields Caldera, north-west of Naples in southern Italy (Figure 2.1). Petrologically identified CI ash layers have been reported from inland and marine sequences throughout south-eastern Europe from the Tyrrhenian Sea to the Volga River in Russia. The extensive CI deposits can be recognised as the product of the largest volcanic eruption during the past 200,000 years in the Greater Mediterranean area, defined here as including the Balkans, Anatolia, and the Pontic (Black Sea) regions.

The eruption was large in both intensity and magnitude. The CI sulphate signal has been recognised in the GISP2 ice core, Greenland (Fedele *et al* 2003). This eruption can be dated to c 40,000 BP by means of argon-argon measurements supported by correlations with the Greenland ice record. This temporal placement draws attention to the climatic context of a time period centred on 40,000 BP, midway through the Weichselian (Last Glacial) Interpleniglacial, which corresponds to stage 3 of Late Pleistocene environmental stratigraphy (MIS 3; Figures 2.2 and 2.3). The volcanic catastrophe precisely overlapped with the culmination of the penultimate Bond cycle of Stage 3, a cooling

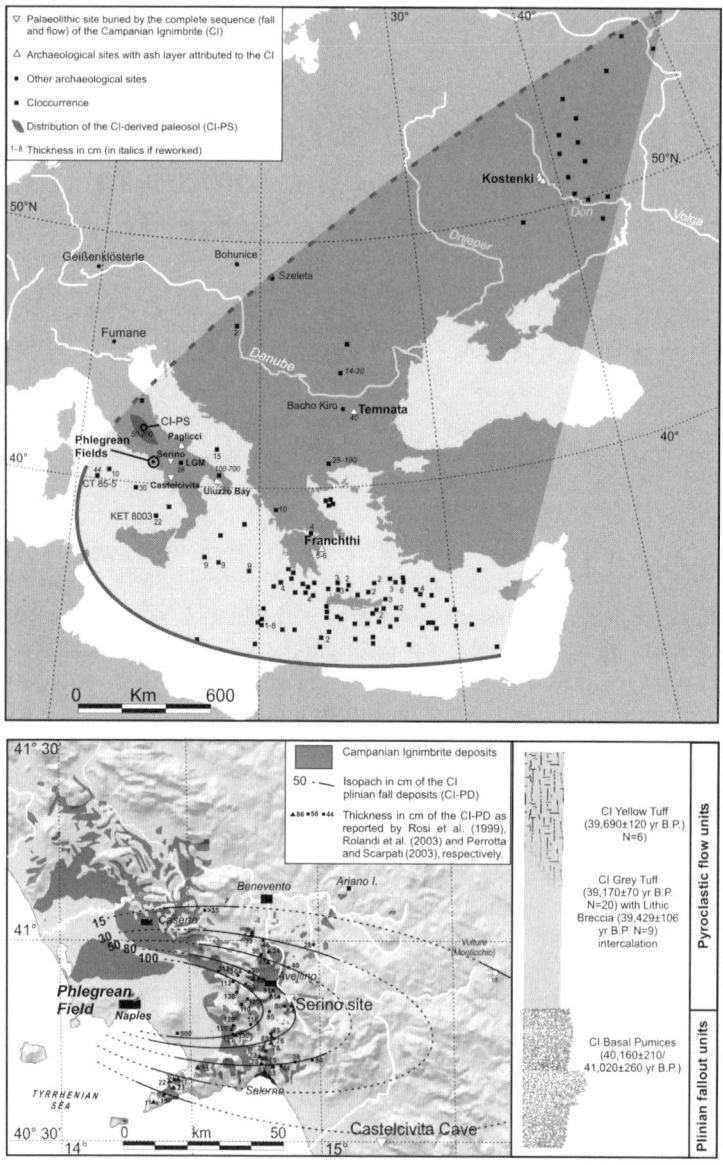

Figure 2.1 The Campanian Ignimbrite deposits: (a) geographic distribution of distal tephra, including main archaeological sites and purely geological occurrences; LGM = Lago Grande di Monticchio (Fedele *et al* 2003 updated); (b) geological map of proximal deposits and superimposed isopachs of the Campanian Ignimbrite plinian fallout (compiled from Perrotta and Scarpati 2003; Rolandi *et al* 2003; Rosi *et al* 1999). Inset: generalised columnar section of the CI in proximal areas, with its distinctive units and related ^{40}Ar/^{39}Ar ages as reported by De Vivo *et al* (2001).

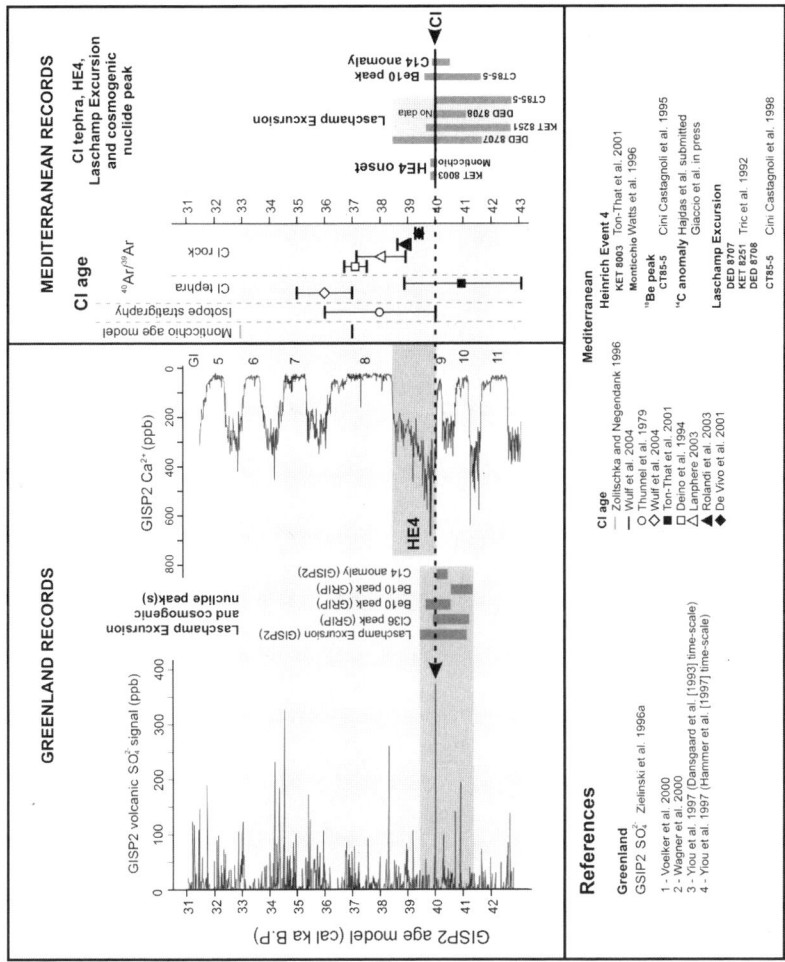

Figure 2.2 The Campanian Ignimbrite and the OIS 3 stratigraphic events.

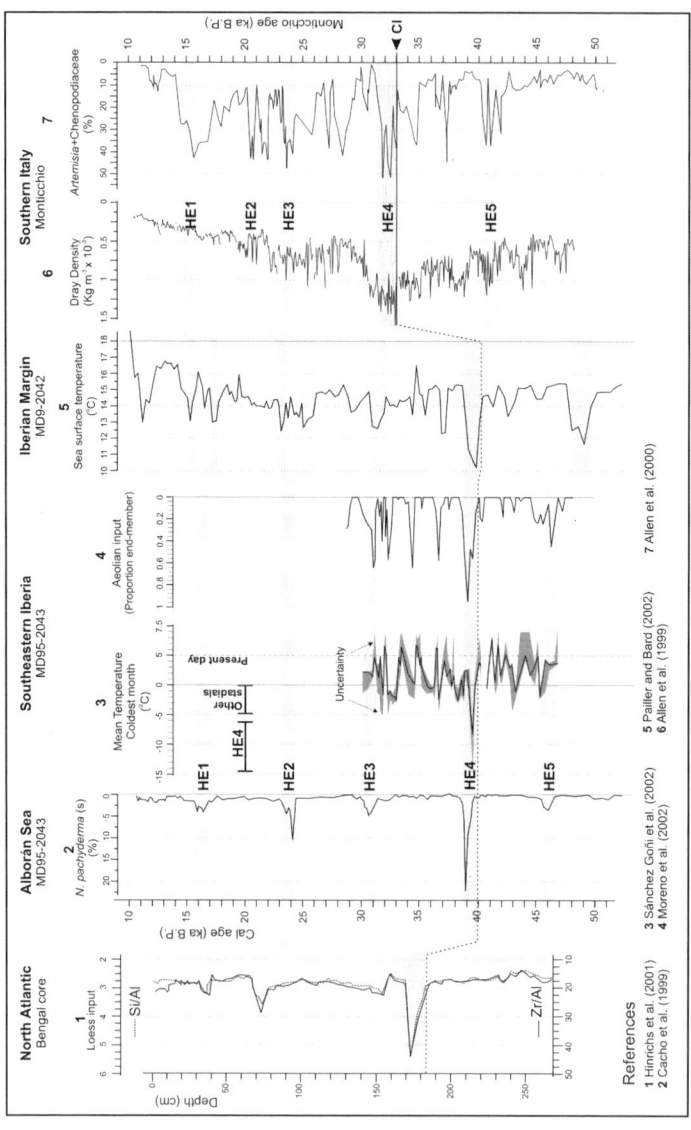

Figure 2.3 Selected palaeoclimatic records displaying the large, anomalous climatic signal associated with HE4 (from Fedele *et al* 2003, updated). In record 2, the abundance data refer to *Neogloboquadrina pachyderma* (sinistral); the timescale to the right, displaying significant chronological discrepancies, exclusively refers to the varve-supported age model for Lago Grande di Monticchio, presently under revision (Wulf *et al* 2004).

trend that began ~45,000 BP. We therefore propose that a positive feedback triggered by an exceptional concurrence of factors – climatic change, glaciation, ocean-ice exchange, and latitude – is likely to have had hemispheric or wider effects. The most pronounced of these was the abrupt cooling coeval with the Heinrich Event 4 (HE4) about 40,000 BP. The scenario of a 'volcanic winter' should therefore be considered.

The archaeological changes immediately following the CI are a sign of conspicuous interference with human populations. The coincidence of a volcanic eruption and Palaeolithic phases had been noted previously (eg, Mussi 2001), but has not been effectively addressed in spite of abundant archaeological evidence in southern Italy and elsewhere. Furthermore, we surmise that the hiatus in human occupation must have had a causal relationship not only with the impact from the blast but with the ecological disruption it generated. It could be argued that only when humans are examined as ecosystem components can the effects of the CI on social and cultural processes – including the Middle to Upper Palaeolithic 'transition' – be understood.

THEORETICAL BACKGROUND

The CI factor as presently appraised may have significantly interfered with cultural processes and trajectories, at least in the Greater Mediterranean. Our major theoretical premise is that the atmospheric impact of the CI was greatly amplified and prolonged through positive feedback, as presented in the model shown in Figure 2.4. We predict a highly contrasted mosaic of local conditions. In the direct-impact zone – ie, the areal extent where pyroclastic cover was sufficiently thick to alter natural life cycles – the eruption may have affected human residents in four principal ways: (1) by disrupting animal populations through the collapse of the herbivore grade in the trophic chain, from suppression or deterioration of pasture; (2) by altering the species composition, growth rhythm, and/or visibility – hence the overall availability – of habitually exploited staple plants; (3) by changing the visibility and/or ease of procurement – hence the overall availability – of familiar rock resources connected with chipped stone technology; and (4) by changing the availability or predictability of water. We predict that at least variations (2) and (3) interacted with lithic and nonlithic technology in significant ways.

We contend that Palaeolithic groups may have been critically impacted in their capacity as upper predators in the trophic chain, not necessarily as direct victims of the eruption or subsequent lowered temperatures. Climate mostly acts on humans through variation in the biome and land (water, soils, ice, landslides, etc), particularly by

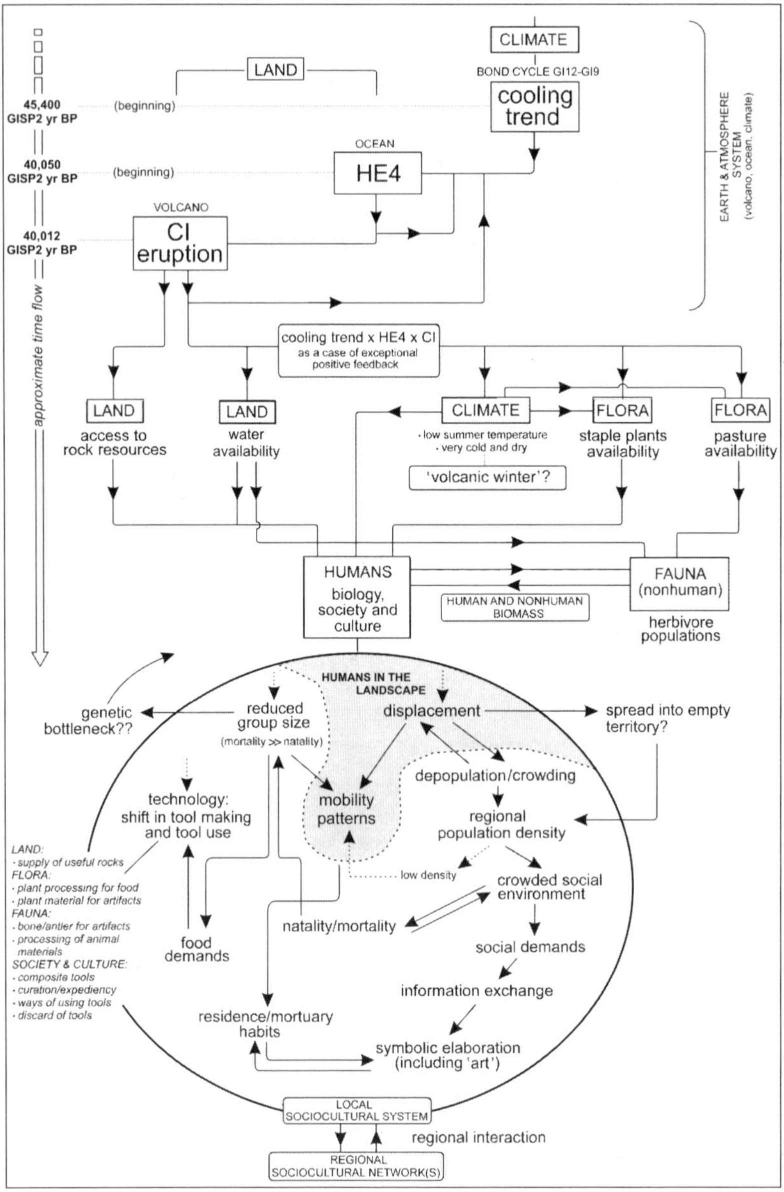

Figure 2.4 A provisional model for the study of the Campanian Ignimbrite impact on human ecosystems, c 40,000 BP.

modulating biodiversity and food availability. In addition to data that bear on resource exploitation, any attempt at modelling a Palaeolithic impact of the CI should consider cultural transmission as well as population density and distribution. Predictably, a result of the environmental alteration caused by the CI was a displacement of human groups.

Data from stratified sites confirm that over a wide but yet undetermined area the CI indirectly forced populations away from traditional and familiar territories. Population displacement entailed two possible outcomes at the regional level: depopulation and crowding. Both, in different ways, may have been responsible for marked changes in within-group and inter-group interaction, leading to innovative social behaviour.

The combination of variables – displacement and trophic alteration of human and nonhuman fauna, availability of critical resources, technology, cultural transmission, social behaviour, and ideology – provide the building bricks for a model. Some of the variables and their states were not novel, of course, but had occurred before. What must have made a difference c 40,000 BP were the abruptness and sharpness of environmental forcing, coupled with the past history and variation of a number of Palaeolithic social systems. In this perspective, the CI-precipitated crisis was not necessarily negative. Rather, it plausibly acted as a filter and catalyst, either constraining or enhancing predispositions and processes.

DATING THE CAMPANIAN IGNIMBRITE ERUPTION

The most recent and accurate age determinations for the CI eruption cluster round 40,000 BP (Figure 2.2), based on ^{40}Ar/^{39}Ar radiometry on several dozens samples from ignimbritic deposits on land and stratified tephra in marine sediments (Fedele *et al* 2003; Giaccio *et al* 2006). The close stratigraphic proximity of the CI eruption to several environmental and geophysical events makes it a unique marker for temporally assessing a series of significant phenomena belonging to the second half of the Interpleniglacial, or MIS 3. These include the onset of HE4 (one of the sudden cooling events connected to Arctic ice discharge), the Laschamp Geomagnetic Excursion, and a distinct peak of ^{10}Be and other cosmogenic nuclides, C^{14} included (see Fedele *et al* 2003; Giaccio *et al* 2006). These relationships have been established in Mediterranean stratigraphic sequences, where the CI products and each of these environmental and geophysical events have been physically repeatedly detected. The uncertainties previously clouding both the CI age and the chronology of MIS 3 have been overcome. The CI is also correlated with the shift from the 'Middle' to the 'Early Upper Palaeolithic' in Europe.

The pollen record of the lacustrine sequence from Lago Grande di Monticchio, southern Italy and the palaeoclimatic data of the Tyrrhenian Sea core KET 8003 both show that the CI tephra coincided with the onset of a cold dry episode corresponding to HE4 (eg, Paterne *et al* 1999; Ton-That *et al* 2001; Watts *et al* 1996). Obviously, Monticchio and KET 8003 cannot contain the typical stratigraphic indicators of Heinrich events commonly found in the North Atlantic, such as layers of ice-rafted debris. However, in light of sufficient evidence indicating a close in-phase relationship of the suborbital climatic oscillations of high-latitude regions with those of the Mediterranean area (eg, Cacho *et al* 1999; Moreno *et al* 2002; Paterne *et al* 1999; Sánchez Goñi, Cacho *et al* 2002; Sánchez Goñi, Turon *et al* 2000; Watts *et al* 1996), the concurrence of the CI eruption and HE4 onset can be accepted.

The close coincidence of the CI tephra with the Laschamp Excursion and the cosmogenic nuclide peak is shown in Figure 2.2, which summarises the Mediterranean palaeomagnetic, cosmogenic, and palaeoclimatic records containing the CI tephra. Laschamp and the nuclide peak are global events, causally interconnected and near-synchronous worldwide (Voelker *et al* 2000; Wagner *et al* 2000), and therefore different from the HE4 signal whose temporal distribution is a debated issue. However, independently of the Mediterranean climatic record and the general debate about the timing of abrupt climatic changes (synchroneity versus time-transgression), Laschamp and the nuclide peak can be shown to have occurred near or across the HE4 onset (Figure 2.2). Consequently, the concurrence of the CI, Laschamp, and ^{10}Be peak provides supporting evidence for the correlation of the CI eruption and the onset of HE4. The unambiguous place of the CI with respect to these stratigraphic events allows the CI volcanic signal to be identified in the GISP2 ice-core (Zielinski *et al* 1996a). Amongst several volcanic sulphate peaks broadly consistent with the radiomet-ric age interval of the CI, only the signal at 40,012 GISP2 yrs BP actu-ally meets the CI stratigraphic constraints (Fedele *et al* 2003; Giaccio *et al* 2006) (Figure 2.2). We take this to be the best available estimate for the true age of the CI eruption.

CHARACTERISTICS OF THE CI EVENT

The Phlegraean Fields (Figure 2.1) was an area of intense explosive vol-canic activity during the Late Quaternary, when the CI was the largest eruption not only for this area but for the Greater Mediterranean region (Barberi *et al* 1978). A summary of the eruption is given in Table 2.1. It was characterised by variable eruptive dynamics that generated different pyroclastic deposits (eg, Civetta *et al* 1997; Fisher *et al* 1993; Orsi *et al* 1996; Ort *et al* 2003; Perrotta and Scarpati 2003; Rolandi *et al* 2003; Rosi *et al* 1996). During the initial stage of the eruption, a plinian column reached a maximum height of 44 km (Rosi *et al* 1996) and dispersed the products towards the east. CI ash layers are ubiquitous in sea cores from the Tyrrhenian Sea ('C-13' tephra; Cini Castagnoli *et al* 1995; Narcisi and Vezzoli 1999; Ton-That *et al* 2001) to the eastern Mediterranean ('Y5' tephra; Thunnel *et al* 1979). They were recognised on land in southern, central, and northern Italy, as well as in Greece, Bulgaria, Ukraine, and

Table 2.1 A summary of information on the Campanian Ignimbrite eruption and its impact (cf Fedele *et al* 2003)

Place of origin	Phlegraean Fields, Campania, S Italy
Best estimated date(s)	40,012 GISP2 yr BP; 39,395 ± 51 yr BP (^{40}Ar/^{39}Ar)
Timing	onset of Heinrich Event 4
Wind direction and prevalent impact	E, NE
Maximum height of Plinian column	c 44 km
Minimum height of co-ignimbrite clouds	c 30–35 km
Volume of extruded magma	c 300 km^3 (dense rock equivalent)
Estimated discharge rate	c 10^{10}–10^{11} kg s^{-1}
Eruptive temperature	c 1000° C
Sulphur injected into the atmosphere[a]	1,17 × 10^{15} g (minimum)
Sulphate signal in Greenland ice-core GISP2	375 ppb SO$_4^{2-}$ (second largest of the whole record)
Minimum area covered by pyroclastic currents	30,000 km^2
Minimum area affected by ash fallout	5,000,000 km^2
Cooling induced[b]	3–4° C
High-latitude (>60° N) amplifying factor for cooling	×4–7 (c 12–20° C)
Sudden abandonment of Palaeolithic sites or locales	yes (S Italy, ?S Balkans)

[a] Extrapolated from Sigurdsson's (1990) curve; cf Figure 2.3.
[b] Based on a very conservative estimate of c 150 km^3 of extruded magma.

Russia (references in Fedele *et al* 2003; Pyle *et al* 2004; Seymour *et al* 2004) (Figure 2.1).

The quantity of sulphur released to the atmosphere by the eruption can be estimated from the mass of magma erupted, together with analytical data concerning sulphur content of degassed matrix glass and sulphur contained in pockets of melt trapped within crystals (hence not degassed during eruption). Combined, these data suggest that the CI eruption injected at least 10^{15} g of sulphur into the atmosphere according to a *minimum* estimate based on only 150 km^3 of erupted magma. This result makes the CI one of the most sulphur-rich eruptions ever identified in the geological record (Scaillet and Pichavant 2003). The calculated mass of sulphur emitted is closely comparable to those of super-eruptions such as Toba and Bishop Tuff.

POTENTIAL CLIMATIC IMPACTS

The impact of the CI eruption on climate needs be assessed on short- and long-term temporal scales. Sulphur is the principal agent in the formation of stratospheric acid aerosols that – enhancing cloud albedo – cause a lowering of the global or hemispheric temperature (Robock 2000). Empirical data show that the cooling degree is a function of the mass of emitted sulphur (Sigurdsson 1990). Direct historical observations are not available for such large eruptions as the CI, but extrapolation of Sigurdsson's empirical curve suggests that the expected global cooling induced by the CI eruption should have been on the order of 3–4° C for at least two–three years. The same value is predicted for Toba, Sumatra, the largest known Late Quaternary volcanic eruption (c 74,000–71,000 yrs BP), for which major effects on global climate (Huang *et al* 2001; Rampino and Self 1992, 1993; Zielinski *et al* 1996b) and living systems have been claimed. Although pertinent criticism has been levelled at Ambrose's (1998; Rampino 2002; Rampino and Ambrose 2000) theory of Toba's impact on hominin evolution (Oppenheimer 2002), the size of the climatic impact can hardly be disputed.

The CI short-term cooling would have been enough to cause severe ecosystem alteration, even in an interglacial climate such as the present one (Rampino 2002). During the Last Glacial, when the climate was highly unstable and led by extremely sensitive threshold mechanisms, the CI possibly prolonged the 'volcanic winter' on a decennial if not centennial time span. High-resolution studies of Late Pleistocene climates show that HE4 was characterised by substantially greater severity than other similar episodes of the Last Glacial (Figure 2.3). The conditions of extreme aridity and cooling particularly affected the Mediterranean and eastern North Atlantic regions. From the indication that the CI eruption was contemporaneous with the onset of HE4, this 'anomaly' can justifiably and parsimoniously be explained in terms of a positive climatic feedback, triggered by the concurrence of the volcanic and climatic factors. Given known Interpleniglacial climate trends, the physics, dynamics, and size of the CI eruption (Fedele *et al* 2003 with references), it would have contributed an additional factor of cooling precisely at the abrupt onset of a markedly cold climatic episode, and that would have shifted the ocean-atmosphere system towards extreme conditions by way of reinforcing circuits or what Renfrew (1972) terms the 'multiplier effect'.

A plausible set of feedback mechanisms can be listed in very simplified form:

Slowdown of thermohyaline circulation induced by HE4, with consequent cooling of the boreal regions ↔
Further global cooling of 3–4° C and possibly >15° C at very high latitudes induced by the CI eruption ↔
Increased and summer-persisting land/ocean snow/ice cover ↔
Enhanced snow/ice albedo with consequent further cooling ↔
Further cooling of the ocean surface and changes in ocean circulation ↔ severe, prolonged 'volcanic winter'.

Such feedback circuits are well suited to explain the peculiar climatic conditions of HE4, which all reliable archives show to have been the coldest and driest of the whole Late Pleistocene (Figure 2.3). Although we are currently unable to quantify the environmental effects of this sudden, possibly amplified and prolonged global cooling at the beginning of a marked stadial phase, the impact on living systems could have been severe, particularly at higher latitudes, where the volcanic-induced cooling was probably amplified by a factor of four to seven (Table 2.1). Continental Europe should have been affected significantly. Even at Mediterranean latitudes, pollen records indicate that HE4 coincided with a swift contraction of forests that were replaced by arid and cold steppes (Sánchez Goñi, Cacho *et al* 2002; Watts *et al* 2000).

Low temperatures and even the extreme aridity of HE4 may not have been limiting factors in themselves (see Vrba *et al* 1995). Rather, it was the exceptional rapidity of environmental change that would have impacted on human groups. The predicted geo-environmental response was such that within a few years many Palaeolithic groups of western Eurasia faced an altered environment in terms of biotic resources and water distribution. We suggest that such a rapid change would have created conditions for selectively transforming human mobility patterns and subsistence strategies. Not every eruption of equal magnitude has the same effects. The effects can be modulated by a great number of factors, amongst which the climatic background of the period is important. Although already large in itself, the effects of the CI eruption were greatly amplified by a special concurrence of environmental conditions. We posit that the Bond cooling trend, the abrupt HE4 event, and the eruption jointly acted as forceful 'catalysts' on selected human processes that were already in motion at about 40,000 BP.

CI TEPHRASTRATIGRAPHY

In the Greater Mediterranean, the CI tephra occurs in a number of stratified archaeological sites where it represents a stratigraphic marker of prime importance (Fedele *et al* 2003; Giaccio *et al* 2006). In all cases, the CI tephra separates the cultural layers containing Mousterian (Middle Palaeolithic) and/or Early Upper Palaeolithic assemblages from the layers in which later and unquestionably defined Upper Palaeolithic industries, often Gravettian, occur (Figure 2.5). Over a large area, the CI tephra – or homotaxial deposits – are regularly followed by culturally sterile layers marking an interruption of human presence (ie, a cultural hiatus).

Several important sites for assessing the culture-stratigraphic position of the CI in Palaeolithic sequences have been identified (Fedele *et al* 2002, 2003 and references cited) (Figure 2.1). At Serino, an open-air campsite in Campania with hearths and other features, a thick pyroclastic deposit immediately overlies a Proto-Aurignacian occupation (Accorsi *et al* 1979); the deposit unambiguously represents the complete CI sequence in the intermediate distal zone. Also in central Campania, Castelcivita cave offers one of the most complete examples of Middle

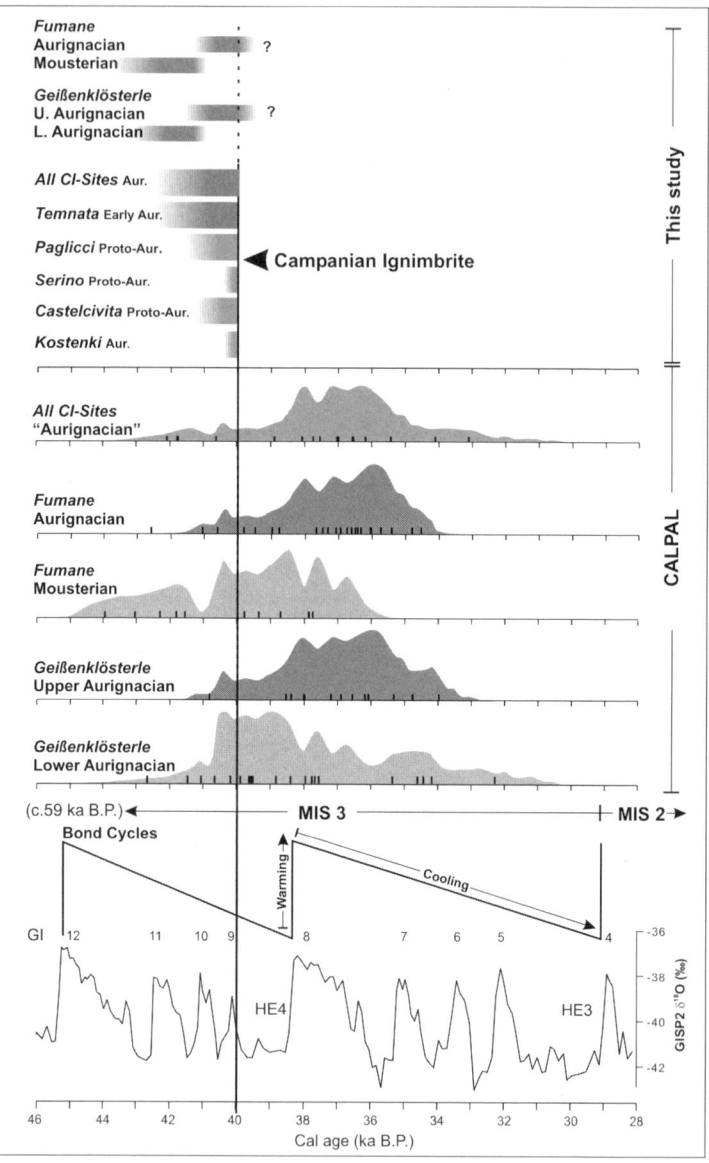

Figure 2.5 Calendar age distribution (above) and paleoclimatic context (below) of Middle/Upper Palaeolithic (M/UP) industries in Europe. Two calendar-age chronologies are compared – the 'CI', derived from the CI-GISP2 correlation (this paper), and the CalPal calibration (Jöris and Weninger 1998, 2000; cf van Andel *et al* 2003). Both chronologies are here applied to sites containing the CI tephra ('CI-sites') and to two key sites without the CI (Fumane and GeBen-klösterle; cf Figure 2.1); the CI-GISP2 dating of these latter sites is based on tephrostratigraphic correlation.

to Upper Palaeolithic transition in peninsular Italy, with its stratified Mousterian, Uluzzian, and Proto-Aurignacian cultural layers at the cave entrance (Gambassini 1997); the CI tephra seals the whole Palaeolithic sequence without any archaeological evidence either within or above. In extreme south-eastern Italy, notable sequences spanning the CI interval and containing a distinctive ash deposit include such caves as Cavallo and Uluzzo C in Uluzzo Bay, Bernardini, Paglicci, and Tana delle Iene (eg, Giaccio and Coppola 2000; Mussi 2001; Palma di Cesnola 2001); here, too, the CI ash seals the last documented industries of the Early Upper Palaeolithic.

In the Aegean-Pontic region, the physical occurrence of the CI has been identified in archaeological sequences from Greece (Franchthi Cave) to Bulgaria (eg, Temnata Cave; Kozłowski *et al* 1992) and farther afield to the Russian Plain, as well as in purely geological stratigraphies (Figure 2.1). Of particular interest is the ongoing excavation of the Kostenki 14/'Markina gora' site, where a short-lived Palaeolithic camp dated to $32,420 \pm 440/420$ C^{14} yrs BP is directly buried below the CI ash (Haesaerts *et al* 2004; Hoffecker 2002; Sinitsyn 2003; Sinitsyn *et al* 2002). The assemblage is rich in bone artefacts and idiosyncratic in lithics; other contemporary assemblages in the Kostenki group either present Middle or 'fully fledged' Upper Palaeolithic affinities (Vishnyatsky and Nehoroshev 2004).

The ecosystem evidence obtained from several Palaeolithic sequences (see Figures 2.1, 2.5) points to landscape conditions before and after the CI tephra in excellent agreement with those inferred from the Monticchio and Greenland records. The time interval occupied by the 'Final' Mousterian and the 'Early Upper Palaeolithic' clearly corresponds in Europe to a rapid progression towards colder and drier habitats (the climatic deterioration is the 'sloping down' of the penultimate Bond cycle in Figure 2.3; c 45,000–38,000 BP). A combined temperature-humidity curve derived from a study of micromammals of Castelcivita cave shows the progression in very great detail and matches the GISP2 palaeotemperature curve perfectly (Giaccio *et al* 2004).

CULTURAL BACKGROUND c 40,000 BP

'Forty thousand years ago, our ancestors wandered into Europe and met another type of human already living there, the brawny, big-brained Neandertals'. This concise statement from a special section on human evolution in *Science* (Gibbons 2001) neatly summarises the mainstream view on the European Late Pleistocene shift. Current debates rage about the nature, timing, and underlying causes of these cultural and anatomical changes. The controversy mainly revolves around continuity or discontinuity, the latter often viewed rather simplistically. Particularly in Europe, there is a peculiar tendency to equate biological and cultural units so that 'Neandertal' is sometimes synonymous with any maker of Middle Palaeolithic tools, or the supposed or expected appearance of Upper Palaeolithic tools (the 'Aurignacian') is equated with 'early modern humans'.

The European Late Pleistocene shift has so far been generally understood in terms of continuity versus discontinuity ('replacement' as in Klein and Edgar 2002 and Mellars 2004 versus Clark 2002 and Wolpoff *et al* 2001; see Balter 2004; Gowlett 2001; van Andel and Davies 2003). In fact, substantial rethinking of pre-existing information, as well as new findings, provide compelling evidence for a less simplistic scenario, patently multivariate and possibly mosaic in nature. Distancing themselves from polarised views, several authors have consistently argued that artefacts, skeletal remains, and even genetics show intriguing variation – if not continuities – between the Middle and the Upper Palaeolithic, and between Neandertals and more modern-looking humans (eg, Ahern *et al* 2004; Brantingham *et al* 2004; Conard 2006; Conard *et al* 2004; Goring-Morris and Belfer-Cohen 2003; INQUA 2003; Trinkaus *et al* 2003).

We believe that the whole debate about replacement or continuity in Europe is irrelevant from a worldwide, cognitive-based perspective of hominin evolution (see Bednarik 2002, 2003; Clark 2003). The peculiar cultural phase during which the CI eruption took place is characterised in Eurasia by a mosaic of variously named stone-tool industries, commonly regarded as innovative when compared to the 'classical' Mousterian technocomplex of the Middle Palaeolithic. Any novelty in tool making perceived as sufficient to herald a new stage (ie, the Upper Palaeolithic) has frequently been called Aurignacian. These evolving or transitional lithic entities acquire archaeological visibility between c 45,000 and 35,000 BP. To avoid preconceived taxonomy and the ambiguities inherent in the word 'transition', this Mousterian- and Aurignacian-like mosaic will be termed the 'Middle/Upper Palaeolithic industrial spectrum' (M/UP) (see Fedele *et al* 2002), of which there are at least two dozen stone-tool configurations or named regional groupings (eg, Gamble 1999). These make for a great deal of variety in the cultural backdrop to the CI eruption, especially if one considers the Near East, Anatolia, and central and north-eastern Asia as well (eg, Brantingham *et al* 2004).

Such a proliferation of archaeological entities has several causes, including the fact that 'most of Palaeolithic archaeology still operates within a techno-cultural taxonomy that was initially developed in Europe over a century ago during the heyday of unilineal evolutionism' (Wynn 2003: 120; see Clark and Willermet 1997 for a critical review). With regard to our model of the CI impact, the proliferation of labels describes to some extent historical reality, namely, a development of regional variants; not only a 'regionalisation', in fact, but also the frequent appearance of evolutionary trends within the Mousterian. Until recently, there has been a tendency to play down the range of variation – material and cognitive – within the later Mousterian. Variation in tool making around and after 40,000 BP appears to be the culmination of a much longer phase of artefactual instability and exploration. Some long-term trends were perhaps set in motion as early as 55,000–60,000 yrs BP (eg, Gamble 1999; Kozłowski 1990). By artefactual 'instability', we refer to a component of variation, cognitively and socially connected with flexibility and experiment (Fedele *et al* 2003). The CI volcanic event inserted itself, more or less dramatically, into this fluctuating spectrum of Palaeolithic developments, and especially the mid-Interpleniglacial mosaic termed the 'M/UP'.

A brief note on timing and chronometry is required. Timing means dates and rhythm, with environmental context as a related concern. In all three domains – chronology, rates of change, and ecology – a consideration of the CI leads to a major reassessment of the timing and context of the European Late Pleistocene shift. No factor is better suited to correlate regionally disparate episodes than a major volcanic event. The CI provides unequalled means to correlate stratigraphic sequences across western Eurasia, and, by further correlation to the Greenland ice-core stratigraphy, affords a unique opportunity to handle – if not actually calibrate – the radiocarbon curve at the limit of its range. This is important because the burgeoning dataset of available radiocarbon dates for Eurasian sequences spanning the M/UP spectrum has encouraged scholars to tackle the problem of the 'transition' through statistical analysis of the dates themselves. But, in fact, these ambitious attempts are questionable because of flawed time-scale control. The CI-GISP2 correlation demonstrates that radiocarbon dates younger than 33,000/32,000 (or even 30,000) C^{14} yrs BP may indeed *predate* HE4 (ie, predate c 40,000 BP in calendar time).

The impressive increase of atmospheric C^{14} concentration that occurred slightly before or across the CI eruption, as an aspect of the cosmogenic nuclide peak mentioned above, is historically very considerable, as shown in Figure 2.5. The dates and rhythm of the Middle to Upper Palaeolithic 'transition' are particularly affected. In Europe at least, the M/UP archaeological spectrum may occupy a time span of no more than two or three thousand years, c 42,500–40,000 yrs BP. The same is true of the apparently long 'Aurignacian' occupations claimed for certain sites. The main implication is that a slow colonisation of Europe by incoming groups – the alleged 'moderns' – is simply not supported by the chronological and tephrostratigraphic evidence. The CI impact should be explored first of all against a background of highly varied, fluctuating human societies of a Middle Palaeolithic continuum.

TOWARDS A MODEL OF THE CI IMPACT

The provisional model presented in Figure 2.4, explicitly based on a human ecosystem framework, represents an attempt at formalising the impact of the CI eruption, a major problem in modelling the CI impact on people is the lack of modern analogues. What we are attempting to study is not just any volcanic impact, but one that occurred at the time of extinct sociocultural systems at a Middle Palaeolithic level of complexity, for which the present-day hunting and gathering groups only provide vague similarities.

Furthermore, the impacts of the CI crisis varied widely according to impact gradient. The interaction between the CI and the Palaeolithic must therefore be explored at different geographic scales: local (ie, central and southern Italy), corresponding to the demonstrable range of main impact; regional, from south-eastern Europe throughout the Greater Mediterranean's core area if not beyond; and hemispheric. Over a sizeable area of the Greater Mediterranean, the CI coincides with an interruption of occupation several millennia long. From this point of view, the CI marks a behavioural watershed. Long series of Mousterian *and* M/UP occupations, implying a certain persistence in circulation habits and regional population, are capped by overlying CI volcanics and immediately followed by site abandonment and prolonged human absence. Archaeological sites disappear all over the direct-impact zone and its immediate surroundings. Sometimes the tephra is actually embedded within freshly abandoned remains (eg, the Serino and Kostenki 14 excavations). When human presence is eventually re-established, occupation generally displays different patterns of land use, if not an altogether novel cultural entity, such as the Gravettian.

Discontinuity of occupation at an archaeological site (as defined by the formation of deposits) provides a fundamental indicator of change in the human ecosystem at large. Conversely, persistence of occupation on the same site is an indicator that human behaviour across the landscape was continuous whatever the variation in other subsystems of culture. Even without prolonged interruption, some kind of settlement change can be observed outside the Greater Mediterranean's core. Although information is biased by the limitation of cave sites, across the region affected by the CI, one observes what can be construed as a population crash, a difference in the kind of specialised use of caves, or both. The most significant changes concern mobility patterns, an all-important aspect of hunter-gatherers' life also relating to social constructs (eg, Kelly 1995).

Eurasian Mousterian technologies, particularly during the mid-Interpleniglacial, display an oscillating reliance on blade-based manufacture and/or miniaturisation of tool-kits, locally supplemented by innovations in hafting and related tool retouch and form (backing or blunting, 'foliate' bifaces etc). From a human ecosystem perspective, they represent adaptations to cope with landscape variation, be it in the way of stress or opportunity, especially if several types were developed out of a need for more efficient composite tools (eg, Bar-Yosef and Kuhn 1999; Kuhn 1995; Shea 1997). A common denominator to all these devices is standardisation of shape, which may have acted as a cost-efficient strategy. Technology becomes more specialised when maximum resources have to be procured in a short period of time (Torrence 2001). A concurrent kind of adaptive response is the shift in raw material procurement or exchange, which shows maximum local variation during the 'Early Aurignacian'.

Such European developments lose most of their novelty if they are compared with similar phenomena at other times and places (Near East, Kenya, the Howieson's Poort industry of South Africa, etc). Blades, miniaturisation, geometrisation, and hafting likely formed a functional package that was probably a response to new demands and opportunities brought about by rapidly expanding, open, cold-dry habitats (ie, environmental 'deterioration'). Specific ecological conditions and flexibility of strategy in coping with an unpredictable, less diverse, and/or sparse food base may explain the greatly expanded use of microblades during the 'Early Aurignacian'.

In our model (Figure 2.4), a particular role is suggested for human 'displacement', which we consider a key factor in the subsystem of society and space dimensions. We propose that by forcing populations away from familiar territory or favouring a change in lifeways without actual migration, the CI may have produced (1) unusual population density in certain regions and (2) a disruption of the cognitive balance of social groups.

First, over a certain period, the environmental alteration may have shifted groups into less-affected regions at the periphery of the impacted area leading to the depopulation in certain regions and crowding in others. As a further possibility, regional crowding may have been distinctly mosaic in nature, or else have occurred in 'fringe' areas along the continental periphery of the CI ash fall (see Figure 2.1). A population shift might convey the impression of a sudden appearance or 'invasion' of M/UP spectrum groups.

Second, cognitive behaviour, is a crucial component of the social fabric that leaves traces in the material record, such as mapping of resources and image making. At 40,000 BP, we are dealing with hunter-gatherer groups of enough complexity to organise life along socially constructed cognition, and with 'locales invested with association and meaning' (Gamble 1999: 425); that condition was a source of both advantages and limitations, social fragility but also social resilience.

The accentuated visibility and quick elaboration of 'art' are perhaps to be understood against this backdrop and in connection with the demands of 'crowded' social environments. We therefore object to singling out art as a proxy for human 'modernity' because it is no more than an arbitrarily overemphasised component of expressive culture, still subject to huge interpretive bias. Since art and symbolism are best explained from an information-exchange perspective and taking into account its practical value in hunter-gatherers' life (eg, Conkey 2001; Gamble 1982), they fall within the scope of CI impacts.

'The challenge', as put forth by Clark (1994: 382), 'is to identify the conditions (almost certainly demographic) that would have selected for an increasing symbolic component to human behaviour at particular places and times', making it adaptively advantageous to humans in a crowded social environment. Together with blades and bone tools, symbolism and art are not necessary and sufficient attributes for a definition of the Upper Palaeolithic, if only because they predate it. Art, like language or symboling in general, almost certainly evolved from simpler and occasional antecedents, and then subsequently *exapted* (ie, took on other functions). 'When this occurred is subject to debate, but it clearly was a "process" (and not an "event"), and almost certainly had nothing to do with genetic superiority' (Clark 2003).

In conclusion, we suggest that the CI-related feedback interfered with pre-existing conditions and processes in the western Eurasian Palaeolithic, acting on them in various possible ways – arresting, disrupting, or reorientating them as a catalytic agent, according to situations and regions. It possibly forced natural and social selection; both catalysis and selection are relevant in our provisional model. Cultural 'bottlenecks' caused by widespread ecosystemic adjustment would have accelerated and filtered developments that were already under way. Within a particular area, the CI may have constituted a single overwhelming blow, but in general sociocultural factors and ecological conditions may have interplayed with the event in such a way as to render change inevitable.

We predict that a number of occasional, incipient cultural traits, available in a rich fund of behavioural repertoires, became viable options under the CI stress. Behavioural patterns are altered creatively to solve novel problems. Certain aspects of hominin (or even anthropoid) fundamental faculties are not perhaps as trivial in the present argument as they seem to be. How low food abundance can prompt innovative behaviour such as rapid readjustment of tool use is widespread among apes and even monkeys. Living in large, stable social groups can often be shown to favour the rapid evolution of enhanced cognitive abilities, such as tracking group members' social status and marking relationships and alliances. In this chapter, we have outlined steps towards a model of accelerated, selective (ie, constraining or enhancing) change within a framework of fundamental continuity – a model of 'change within continuity' in the advanced Palaeolithic of western Eurasia. We predict that the CI critical role is best seen as precipitating conditions for change in evolving, highly multivariate human ecosystems.

ACKNOWLEDGMENTS

This chapter is a contribution of the project 'The impact of the large explosive eruptions on environment and climate: Campanian Ignimbrite, the most powerful eruption of the last 200,000 years in the Mediterranean area', supported by the Italian Ministry of Education, University and Research (grant FIRB No. RBAU01HTPA; G Orsi, coordinator). We thank John Grattan for his invitation, encouragement, and patience; an input of clear thinking by the editors of this volume greatly improved the chapter. We are grateful to Ofer Bar-Yosef, Robert Bednarik, Alberto Broglio, Paolo Gambassini, John Gowlett, Irka Hajdas, John Hoffecker, Jan Mangerud, Marco Peresani, Michael Rampino, Andrej Sinitsyn, Lawrence Straus, and Erik Trinkaus for sharing opinions and papers. We alone, however, are responsible for the use here of such information.

REFERENCES

Accorsi, CA, Aiello, E, Bartolini, C, Castelletti, L, Rodolfi, G and Ronchitelli, A (1979) 'Il giacimento Paleolitico di Serino (Avellino): stratigrafia, ambienti e paletnologia', *Atti della Società Toscana di Scienze Naturali, Memorie* 86(A), 435–87

Ahern, JCM, Karavanić, I, Paunović, M, Janković, I and Smith, FH (2004) 'New discoveries and interpretations of hominid fossils and artifacts from Vindija cave, Croatia', *Journal of Human Evolution* 46, 25–65

Ambrose, SH (1998) 'Late Pleistocene human population bottlenecks, volcanic winter, and differentiation of modern humans', *Journal of Human Evolution* 34, 623–51

Balter, M (2004) 'Dressed for success: Neandertal culture wins respect', *Science* 306, 40–41

Barberi, F, Innocenti, F, Lirer, L, Munno, R, Pescatore, TS and Santacroce, R (1978) 'The Campanian Ignimbrite: a major prehistoric eruption in the Neapolitan area (Italy)', *Bulletin of Volcanology* 4, 10–22

Bar-Yosef, O and Kuhn, SL (1999) 'The big deal about blades: laminar technologies and human evolution', *American Anthropologist* 101, 322–38

Bednarik, RG (2002) 'The human ascent: a critical review', *Anthropologie* 40, 101–05

Bednarik, RG (2003) 'Seafaring in the Pleistocene', *Cambridge Archaeological Journal* 13, 41–66

Brantingham, PJ, Kuhn, SJ and Kerry, KW (2004) *The Early Upper Paleolithic beyond Western Europe*, Berkeley: University of California Press

Cacho, I, Grimalt, JO, Pelejero, C, Canales, M, Sierro, FJ, Abel Flores, J and Shackleton, M (1999) 'Dansgaard-Oeschger and Heinrich event imprints in Alborán Sea paleotemperatures', *Paleoceanography* 14, 698–705

Cini Castagnoli, G, Albrecht, A, Beer, J, Bonino, G, Shen, Ch, Callegari, E, Taricco, C, Dittrich-Hannen, B, Kubik, P, Suter, M and Zhu, GM (1995) 'Evidence for [10]Be enhanced deposition in Mediterranean sediments 35 Kyr BP', *Geophysical Research Letters* 22, 707–10

Civetta, L, Orsi, G, Pappalardo, L, Fisher, RV, Heiken, G and Ort, M (1997) 'Geochemical zoning, mingling, eruptive dynamics and depositional processes – the Campanian Ignimbrite, Campi Flegrei caldera, Italy', *Journal of Volcanology and Geothermal Research* 75, 183–219

Clark, GA (1994) 'Comment on Byers, AM, "Symboling and the Middle-Upper Palaeolithic transition: a theoretical and methodological critique"', *Current Anthropology* 35, 382

Clark, GA (2002) 'Neandertal archaeology–implications for our origins', *American Anthropologist* 104, 50–67

Clark, GA (2003) 'Comment on Bednarik 2003', *Cambridge Archaeological Journal* 13, 56–58

Clark, GA and Willermet, C (1997) *Conceptual Issues in Modern Human Origins Research*, New York: Aldine de Gruyter

Conard, NJ (ed) (2006) *Neanderthals and Modern Humans Meet*, Tübingen Publications in Prehistory, Tübingen, Germany: Eberhard-Karls-University

Conard, NJ, Grootes, PM and Smith, FH (2004) 'Unexpectedly recent dates for human remains from Vogelherd', *Nature* 430, 198–201

Conkey, MW (2001) 'Hunting for images, gathering up meanings: art for life in hunting-gathering societies', in Panter-Brick, C, Layton, RH and Rowley-Conwy, P (eds), *Hunter-Gatherers: An interdisciplinary Perspective*, pp 267–91, Cambridge: Cambridge University Press

Fedele, FG, Giaccio, B, Isaia, R and Orsi, G (2002) 'Ecosystem impact of the Campanian Ignimbrite eruption in Late Pleistocene Europe', *Quaternary Research* 57, 420–24

Fedele, FG, Giaccio, B, Isaia, R and Orsi, G (2003) 'The Campanian Ignimbrite eruption, Heinrich Event 4, and the Palaeolithic change in Europe: a high-resolution investigation', in Robock, A and Oppenheimer, C (eds), *Volcanism and the Earth's Atmosphere*, pp 301–25, Geophysical Monograph 139, Washington, DC: American Geophysical Union

Fisher, RV, Orsi, G, Ort, M and Heiken, G (1993) 'Mobility of a large-volume pyroclastic flow – emplacement of the Campanian Ignimbrite, Italy', *Journal of Volcanology and Geothermal Research* 56, 205–20

Gambassini, P (1997) *Il Paleolitico di Castelcivita, Culture e Ambiente*, Naples: Electa Napoli

Gamble, C (1982) 'Interaction and alliance in Palaeolithic society', *Man* 17, 92–107

Gamble, C (1999) *The Palaeolithic Societies of Europe*, Cambridge: Cambridge University Press

Giaccio, B and Coppola, D (2000) 'Note preliminari sul contesto stratigrafico e paleoecologico del sito "Tana delle Iene" (Ceglie Messapica, Brindisi, SE Italia)' *Il Quaternario – Italian Journal of Quaternary Science* 13, 5–20

Giaccio, B, Di Canzio, E and Fedele, FG (2004) 'Tephrostratigraphic and palaeoenvironmental correlation between the "Middle/Upper Palaeolithic" sequence of Castelcivita Cave (Southern Italy) and the GISP2 ice-core', 32nd International Geological Congress, Florence, 20–28 August, abstracts

Giaccio, B, Hajdas, I, Peresani, M, Fedele, FG and Isaia, R (2006) 'The Campanian Ignimbrite (c. 40 ka BP) and its relevance for the timing of the Middle to Upper Palaeolithic shift: timescales and regional correlations', in Conard, NJ (ed), *Neanderthals and Modern Humans Meet*, pp 343–75, Tübingen Publications in Prehistory, Tübingen, Germany: Eberhard-Karls-Universität

Gibbons, A (2001) 'The riddle of coexistence', *Science* 291, 1725–29

Goring-Morris, AN and Belfer-Cohen, A, (eds) (2003) *More than Meets the Eye: Studies on Upper Palaeolithic Diversity in the Near East*, Oxford: Oxbow Books

Gowlett, JAJ (2001) 'Out in the cold', *Nature* 413, 33–34

Haesaerts, P, Damblon, F, Sinistyn, A and Van der Plicht, J (2004) 'Kostienki 14 (Voronezh, Central Russia): new data on stratigraphy and radiocarbon chronology', in Dewez, M, Noiret, P and Teheux, E (eds), *Proceedings of XIV UISPP Congress: The Upper Palaeolithic – General Sessions and Posters*, pp 169–80, BAR S1240, Oxford: Archaeopress

Hoffecker, JF (2002) *Desolate Landscapes: Ice Age Settlement in Eastern Europe*, New Brunswick, NJ: Rutgers University Press

Huang, C-Y, Zhao, M, Wang, C-C and Wei, G (2001) 'Cooling of the South China Sea by the Toba eruption and correlation with other climate proxies ~71,000 years ago', *Geophysical Research Letters* 28, 3915–18

INQUA Congress (2003) 'The extinction of the European Neanderthals during Isotope Stage 3', Session No 82 (Straus, LG, presiding), XVI INQUA Congress, Reno, Nevada, 23–30 July

Jöris, O and Weninger, B (1998) 'Extension of the 14-C calibration curve to ca. 40,000 cal BC by synchronizing Greenland 18O/16O ice core records and North Atlantic Foraminifera profiles: a comparison with U/Th coral data', *Radiocarbon* 40, 495–504

Jöris, O and Weninger, B (2000) 'Radiocarbon calibration and the absolute chronology of the Late Glacial', in Valentin, B, Bodu, P and Christensen, M (eds), *L'Europe Centrale et Septentrionale au Tardiglaciaire*, pp 19–54, Mémoires du Musée de Préhistoire d'Ile de France 7, Nemours: APRAIF

Kelly, R (1995) *The Foraging Spectrum: Diversity in Hunter-Gatherer Lifeways*, Washington, DC: Smithsonian Institution Press

Klein, RG and Edgar, B (2002) *The Dawn of Human Culture*, New York: Wiley

Kozłowski, JK (1990) 'A multiaspectual approach to the origins of the Upper Palaeolithic in Europe', in Mellars, P (ed), *The Emergence of Modern Humans: An Archaeological Perspective*, pp 419–38, Edinburgh: Edinburgh University Press

Kozłowski, JK, Laville, H and Ginter, B (eds) (1992) *Temnata Cave: Excavations in Karlukovo Karst Area, Bulgaria*, vol 1, Krakow, Poland: Jagellonian University Press

Kuhn, SL (1995) *Mousterian Lithic Technology: An Ecological Perspective*, Princeton, NJ: Princeton University Press

Mellars, P (2004) 'Neanderthals and the modern human colonization of Europe', *Nature* 432, 461–65

Moreno, A, Cacho, I, Canals, M, Prins, MA, Sánchez-Goñi, MF, Grimalt, JO and Weltje, GJ (2002) 'Saharan dust transport and high-latitude glacial climatic variability: the Alborán Sea record', *Quaternary Research* 58, 318–28

Mussi, M (2001) *Earliest Italy: An Overview of the Italian Paleolithic and Mesolithic*, New York: Kluwer Academic/Plenum Publishers

Narcisi, B and Vezzoli, L (1999) 'Quaternary stratigraphy of distal tephra layers in the Mediterranean – an overview', *Global and Planetary Change* 21, 31–50

Oppenheimer, C (2002) 'Limited global change due to the largest known Quaternary eruption, Toba ~74 kyr BP?' *Quaternary Science Reviews* 21, 1593–1609

Orsi, G, De Vita, S and di Vito, MA (1996) 'The restless, resurgent Campi Flegrei nested caldera (Italy): constraints on its evolution and configuration', *Journal of Volcanology and Geothermal Research* 74, 179–214

Ort, M, Orsi, G, Pappalardo, L and Fisher, RV (2003) 'Anisotropy of magnetic susceptibility studies of depositional processes in the Campanian Ignimbrite, Italy', *Bulletin of Volcanology* 65, 55–72

Palma di Cesnola, A (2001) *Le Paléolithique Supérieur en Italie*, Grenoble, France: Jérôme Millon

Paterne, M, Kallel, N, Labeyrie, L, Vautravers M, Duplessy, J-C, Rossignol-Strick, M, Cortijo, E, Arnold, M and Fontugne, M (1999) 'Hydrological relationship between the North Atlantic Ocean and the Mediterranean Sea during the past 15–75 kyr', *Paleoceanography* 14, 626–38

Perrotta, A and Scarpati, C (2003) 'Volume partition between the plinian and co-ignimbrite air fall deposits of the Campanian Ignimbrite eruption', *Mineralogy and Petrology* 79, 67–78

Pyle, DM, Ricketts, GD, Sinitsyn, AA, Praslov, N, Lisitsyn, S, Margari, V and van Andel, TH (2004) 'Y5 Tephra from the Campanian Ignimbrite eruption: a key chronostratigraphic marker for the Mediterranean and Eastern Europe' Abstract, XVI INQUA Congress, Reno, Nevada, 23–30 July 2003

Rampino, MR (2002) 'Supereruptions as a threat to civilizations on Earth-like planets', *Icarus* 156, 562–69

Rampino, MR and Ambrose, S (2000) 'Volcanic winter in the Garden of Eden: the Toba supereruption and the late Pleistocene human population crash', in McCoy, FW and Heiken, G (eds), *Volcanic Hazards and Disasters in Human Antiquity*, pp 71–82, Geological Society of America, Special Paper 345

Rampino, MR and Self, S (1992) 'Volcanic winter and accelerated glaciation following the Toba super-eruption', *Nature* 359, 50–52

Rampino, MR and Self, S (1993) 'Climate-volcanism feedback and the Toba eruption of ca. 74,000 years ago', *Quaternary Research* 40, 269–80

Renfrew, C (1972) *The Emergence of Civilisation. The Cyclades and the Aegean in the Third Millennium B.C.*, London: Methuen

Robock, A (2000) 'Volcanic eruptions and climate', *Review of Geophysics* 38, 191–219

Rolandi, G, Bellucci, F, Heizler, MT, Belkin, HE and De Vivo, B (2003) 'Tectonic controls on the genesis of ignimbrites from the Campanian Volcanic Zone, southern Italy', *Mineralogy and Petrology* 79, 3–31

Rosi, M, Vezzoli, L, Aleotti, P and De Censi, M (1996) 'Interaction between caldera collapse and eruptive dynamics during the Campanian Ignimbrite eruption, Phlegrean Fields, Italy', *Bulletin of Volcanology* 57, 541–54

Sánchez Goñi, MF, Cacho, I, Turon, J-L, Guiot, J, Sierro, FJ, Peypouquet, J-P, Grimalt, JO and Shackleton, NJ (2002) 'Synchroneity between marine and terrestrial responses to millennial scale climatic variability during the last glacial period in the Mediterranean region', *Climate Dynamics* 19, 95–105

Sánchez Goñi, MF, Turon, J-L, Eynaud, F and Gendreau, S (2000) 'European climatic response to millennial-scale changes in the atmosphere-ocean system during the Last Glacial period', *Quaternary Research* 54, 394–403

Scaillet, B, Luhr, JF and Carroll, MR (2003) 'Petrological and volcanological constraints on volcanic sulfur emissions to the atmosphere', in Robock, A and Oppenheimer, C (eds), *Volcanism and the Earth's Atmosphere*, pp 11–40, Geophysical Monograph 139, Washington, DC: American Geophysical Union

Scaillet, B and Pichavant, M (2003) 'Experimental constraints on volatile abundances in arc magmas and their implications for degassing processes', in Oppenheimer, C, Pyle, DM and Barclay, J (eds), *Volcanic Degassing*, pp 23–52, Geological Society Special Publication 213, London

Seymour, KSt, Christanis, K, Bouzinos, A, Papazisimou, S, Papatheodorou, G, Moran, E and Dénès, G (2004) 'Tephrostratigraphy and tephrochronology in the Philippi peat basin, Macedonia, Northern Hellas (Greece)', *Quaternary International* 121, 53–65

Shea, JJ (1997) 'Stone spear points from the Middle Paleolithic: an inter-regional perspective', in Knecht, H (ed), *Projectile Technologies: Archaeological and Ethnoarchaeological Perspectives*, pp 79–106, New York: Plenum Press

Sigurdsson, H (1990) 'Evidence of volcanic loading of the atmosphere and of climate response', *Palaeogeography, Palaeoclimatology, Palaeoecology* 89, 277–89

Sinitsyn, AA (2003) 'A Palaeolithic "Pompeii" at Kostenki, Russia', *Antiquity* 77, 9–14

Sinitsyn, AA, Sergin, VY and Hoffecker, JF (eds) (2002) *Kostenki v Kontekste Paleolit Evrazii*, Saint Petersburg: Russian Academy of Sciences

Thunnel, R, Federman, A, Sparks, S and Williams, D (1979) 'The origin and volcanological significance of the Y-5 ash layer in the Mediterranean', *Quaternary Research* 12, 241–53

Ton-That, T, Singer, B and Paterne, M (2001) '^{40}Ar/^{39}Ar dating of latest Pleistocene (41 ka) marine tephra in the Mediterranean Sea: implications for global climate records', *Earth and Planetary Science Letters* 184, 645–58

Torrence, R (2001) 'Hunter-gatherer technology: macro- and microscale approaches', in Panter-Brick, C, Layton, RH and Rowley-Conwy, P (eds), *Hunter-Gatherers: An interdisciplinary Perspective*, pp 73–98, Cambridge: Cambridge University Press

Trinkaus, E, Moldovan, O, Milota, Ş, Bilgấr, A, Sarcina, L, Athreia, S, Bailey, SE, Rodrigo, R, Gherase, M, Higham, T, Bronk Ramsey, C and van der Plicht, J (2003) 'An early modern human from the Peştera cu Oase, Romania', *Proceedings of the National Academy of Science* 100, 11231–36

van Andel, TH and Davies, W (eds) (2003) *Neanderthals and Modern Humans in the European Landscape during the Last Glaciation: Archaeological Results of the Stage 3 Project*, Cambridge: McDonald Institute for Archaeological Research, University of Cambridge

Vishnyatsky, LB and Nehoroshev, PE (2004) 'The beginning of the Upper Paleolithic on the Russian Plain', in Brantingham, PJ, Kuhn, SJ and Kerry, KW (eds), *The Early Upper Paleolithic beyond Western Europe*, pp 80–96, Berkeley: University of California Press

Voelker, AHL, Grootes, PM, Nadeau, M-J and Sarnthein, M (2000) 'Radiocarbon levels in the Iceland Sea from 25–53 kyr and their link to the earth's magnetic field intensity'. *Radiocarbon* 42, 437–52

Vrba, ES, Denton, GH, Partridge, TC and Burckle, LH (eds) (1995) *Paleoclimate and Evolution, with Emphasis on Human Origins*, New Haven, CT: Yale University Press

Wagner, G, Beer, J, Laj, C, Kissel, C, Masarik, J, Muscheler, R and Synal, H-A (2000) 'Chlorine-36 evidence for Mono Lake event in the Summit GRIP ice core', *Earth and Planetary Science Letters* 181, 1–6

Watts, WA, Allen, JRM and Huntley, B (1996) 'Vegetation history and palaeoclimate of the last glacial period at Lago Grande di Monticchio, southern Italy', *Quaternary Science Reviews* 15, 133–53

Watts, WA, Allen, JRM and Huntley, B (2000) 'Palaeoecology of three interstadial events during oxygen-isotope stages 3 and 4: a lacustrine record from Lago Grande di Monticchio, southern Italy', *Palaeogeography, Palaeoclimatology, Palaeoecology* 155, 83–93

Wolpoff, MH, Hawks, J, Frayer, DW and Hunley, K (2001) 'Modern human ancestry at the peripheries: a test of the replacement theory', *Science* 291, 293–97

Wulf, S, Kraml, M, Brauer, A, Keller, J and Negendank, JFW (2004) 'Tephrochronology of the 100 ka lacustrine sediment record of Lago Grande di Monticchio (southern Italy)', *Quaternary International* 122, 7–30

Wynn, T (2003) 'Comment on R Bednarik', *Rock Art Research* 20, 120–21

Zielinski, GA, Mayewski, PA, Meeker, LD, Whitlow, SI and Twickler, MS (1996a) 'A 110,000-yr record of explosive volcanism from the GISP2 (Greenland) ice core', *Quaternary Research* 45, 109–18

Zielinski, GA, Mayewski, PA, Meeker, LD, Whitlow, SI and Twickler, MS (1996b) 'Potential atmospheric impact of the Toba mega-eruption ~71,000 years ago', *Geophysical Research Letters* 23, 837–40

Chaos and Selection in Catastrophic Environments: Willaumez Peninsula, Papua New Guinea

Robin Torrence and Trudy Doelman

AN ARCHAEOLOGICAL PERSPECTIVE

Prehistoric archaeology can make a unique contribution to understanding the interaction between natural disasters and social change because the discipline is concerned with long sequences of time. Instead of looking for a one-to-one correlation between an individual event and a cultural reaction, archaeologists such as Sheets (1999, Chapter 4) and Grayson and Sheets (1979) have used a long-term, comparative perspective to investigate whether particular cultural processes operate within disaster prone environments. Theirs is a very different approach to that adopted by modern hazard managers and the public at large, where the major emphasis is placed on losses – number of deaths and casualties, size of property damage, etc. This popular concept of disasters, what we term the 'insurance perspective', has a very limited time frame. An archaeological approach, however, is concerned with whether and in what circumstances natural disasters have a significant effect on the overall shape of cultural history. Only by taking a broad view of the past can one address significant questions such as were certain kinds of cultural behaviour more likely to develop in places with frequent disasters, or does the occurrence of disasters constrain the development of societies or cause it to follow certain pathways? In this case study, we seek to illustrate the value of an archaeological perspective on natural disasters both for the future and for understanding processes of change in human societies.

The long-term cultural history of the Willaumez Peninsula of New Britain Island in Papua New Guinea (Figure 3.1) has experienced much volcanic activity throughout the c 45,000 years of human occupation, including the most recent eruptions in 2002 and 2005. Our reconstructions of past environments and cultural history are based on fieldwork and analyses by a large interdisciplinary team that explored prehistoric landscapes on Garua Island (eg, Torrence 2002a, 2002b; Torrence and Stevenson 2000; Torrence *et al* 2000) and within the isthmus region at the base of the Willaumez Peninsula (see Torrence 2000, 2001, 2002c; Torrence and Neall 2004; Torrence, Neall *et al* 2004; Torrence *et al* 1999). Following a general discussion of potential responses to volcanic activity, we turn to specific themes concerning human coping strategies for volcanic disasters: recolonisation, material culture, social exchange, conceptions of place, and patterns of land use.

CATASTROPHIC ENVIRONMENTS

It seems likely that human history within environments experiencing frequent natural disasters will differ from that in stable settings or where there is a low severity or occurrence of hazards. To model the possible outcomes, we identify a particular category of environments that we designate as 'catastrophic'. These are defined by the presence of frequent, very severe environmental perturbations. What separates these hazardous or disaster-prone settings from others that are merely 'uncertain' or 'variable' is the very high magnitude of the environmental forcing agent(s). In a catastrophic environment, the events (eg, major

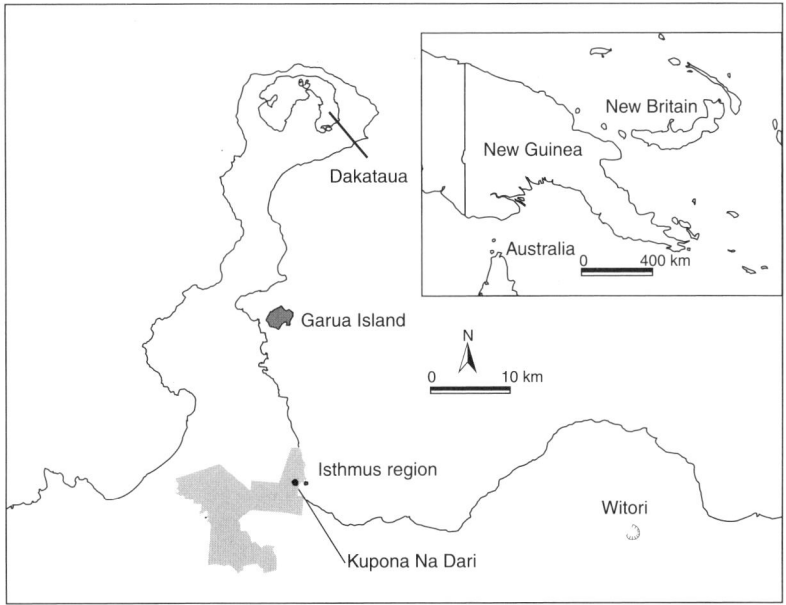

Figure 3.1 Willaumez Peninsula region of New Britain, Papua New Guinea, the volcanoes that have dominated its history and the archaeological study areas.

volcanic eruptions, large earthquakes, hurricanes or cyclones, etc), are so destructive that most cultural organisations would fail to cope on the short term, as, for example, following Hurricane Katrina when even modern disaster management failed. The severity of the events in catastrophic environments means that cultural strategies such as 'risk minimising' or 'bet hedging' may not be effective enough to ensure long-term survival in place. Consequently, different theoretical perspectives may be required. In addition to the devastating consequences for the people who experience these events, their relatively frequent occurrence may mean that the natural environment remains disturbed and is incapable of reaching maturity.

The most obvious prediction for cultural history within catastrophic environments is a chaotic, punctuated history of settlement. Given the severity of the disasters that defines these types of environments, *in situ* collapse of the local population through some combination of mortality and emigration followed later by replacement by a different group seems plausible. Prehistoric groups are unlikely to have developed adequate means for persisting in areas subject to lava and pyroclastic flows and thick depths of air-fall tephra that cause total devastation of the vegetation. Even with extensive outside assistance, regions that experienced major volcanic eruptions in recent times have been abandoned temporarily (eg, Rabaul, Papua New Guinea, 1994 and 2006) (Lentfer and Boyd 2001) and Mount Pinatubo, Philippines (1991) (Newhall and Punongbayan 1996). Will a catastrophic history only yield a chaotic pattern of crashes and recoveries, or are there opportunities for longer term trends to develop?

Insights derived from models for evolution within extreme environments are a useful starting point for thinking about long-term prehistory. Severe environmental change can result in the failure of species to cope resulting in extinctions, which would be comparable to abandonment of an afflicted area by human groups. Such events can also have positive consequences because the loss of species opens up opportunities for innovations, labelled by Hoffmann and Parsons (1997: 22, 204) as 'evolutionary novelties' (eg, Turner and Dale 1998). They argue that extinctions can be a powerful factor 'in triggering periods of evolutionary divergence' (Hoffmann and Parsons 1997: 23).

Given the intense selection created by environmental extremes, over time adaptive behaviour that enables persistence within a catastrophic environment can be expected to arise and spread. For example, biologists predict that selection under extreme conditions can lead to the success of species that 'adopt a stress-resistant life-cycle stage, or reduce the amount of effort they expand on activities such as reproduction' so that they can increase resources needed for survival (Hoffmann and Parsons 1997: 19–20).

Translating these basic biological concepts to a cultural setting, one would expect both chaotic and directional trends in long-term history. Extinction events followed by rapid evolution should be a common occurrence, given the severity of the disasters in a catastrophic environment. Over time, these would result in a chaotic pattern of change. It is also likely that groups would find better ways to cope with the effects of disasters. These behaviours would then persist or even be enhanced through intense selection operating over the long term.

Turning to our case study, it is interesting that Torrence (2002a; Torrence *et al* 2000) has noted that the prehistory of Garua Island (Figure 3.1) exhibits both chaotic and directional trends as predicted. The history of occupation is dominated by phases of abandonment and recolonisation that are associated with radical shifts in some components of the material culture assemblages. This pattern fits the chaotic model in which local extinctions and resulting culture change track the occurrence of disasters. At the same time, selection might be operating because a long, continuous thread of gradual change can be detected in lithic assemblages and patterns of land use (Torrence 1992, 2002a, 2002b) on Garua and in other parts of New Britain (Pavlides 2006). Following additional fieldwork in the isthmus region of the Willaumez Peninsula, we now turn to a more comprehensive evaluation of our models for long-term history within a catastrophic environment.

HOLOCENE VOLCANIC HISTORY

Although human populations have experienced frequent volcanic activity since initial colonisation c 45,000 BP (Torrence, Neall *et al* 2004), we concentrate on the far richer archaeological record of the Holocene. At this time, the environmental history of the Willaumez Peninsula was dominated by eruptions from two volcanoes that exploded large volumes of air-fall tephra over vast areas: Witori located to the east (W-K series) and Dakataua to the north (Dk) (Figure 3.1). The wide-scale impacts and high severity of the five eruptions are demonstrated by the high VEI ratings, estimated volume of material produced, and the large spatial scale over which tephra was spread, as summarised in Table 3.1 and described at length elsewhere (see Boyd *et al* 1999; Machida *et al* 1996; Torrence 2002a; Torrence *et al* 2000). Recent fieldwork has clarified the stratigraphic relationship between the Witori and Dakataua eruptions and led to redating of the latter event (McKee *et al* 2005). Dates for the eruptions, presented in Table 3.2, are based on a Bayesian analysis of an extensive set of radiocarbon determinations from geological and archaeological contexts (Petrie and Torrence submitted) Additional tephras from Witori and the Numundo maar volcano have been identified, but their spatial

Table 3.1 Comparison of major Holocene eruptions in the Willaumez Peninsula

Eruption	Vei[1]	Volume (km³)[2]	Tephra coverage (km²) for different depths[3]				
			>200 cm	50–200 cm	20–50 cm	<20 cm	Total
W-K4	5	6	0	1,677	2,354	8,517	12,515
DK	6	10	280	269	83	64	697
W-K3	5	6	200	1,677	1,666	1,974	5,517
W-K2	5	30	797	4,124	6,039	4,500	15,460
W-K1	6	10	0	5,078	2,189	3,609	10,876

[1] Scale of eruption as measured by the Volcanic Explosivity Index. Data from Machida *et al* (1996).
[2] Data from Machida *et al* (1996).
[3] Based on Boyd *et al* (1999), which is extrapolated from Machida *et al* (1996: 73–75).

Table 3.2 Dates for Holocene volcanic eruptions in the Willaumez Peninsula expressed in terms of the highest posterior density (HPD) region (94.5%). Based on a Bayesian analysis of radiocarbon dates by Petrie and Torrence (submitted)

Eruption	Date
W-K4	1310–1180
DK	1345–1275
W-K3	1750–1550
W-K2	3480–3190
W-K1	6150–5770

scale or thicknesses are limited and their effects on human societies have not yet been thoroughly studied (Torrence 2002c; Torrence and Neall 2004; Torrence *et al* 1999).

The depth of the air-fall tephras within the study areas (Table 3.3) provides another criterion for estimating local impacts of the eruptions. Depths in excess of 50 cm are likely to destroy all the ground-cover and strip the canopy, whereas a thickness in the 20–50 cm range will defoliate the trees, but some plants can recover (Torrence 2002a: 301; eg, Blong 1984: 316–335; Boyd *et al* 1999). Tephras from the Witori and Dakataua volcanoes have different distributions (Table 3.3; Machida *et al* 1996). The W-K1 and W-K2 air-fall tephras have been identified in both study regions, whereas a thin layer of W-K3 is only rarely detected on Garua Island and W-K4 appears to be absent there. Significant depths of the Dk air-fall tephra are restricted to the northern part of the Willaumez Peninsula including Garua Island, with only a few centimetres falling in the isthmus region.

Table 3.3 Percentage of study area covered by significant depths of *in situ* air-fall tephra from each of the major volcanic eruptions

	Willaumez Peninsula	Garua Island		
	>50 cm	>20 cm	>50 cm	>20 cm
W-K4	5	50	0	0
DK	0	0	100	100
W-K3	90	100	0	0
W-K2	100	100	100	100
W-K1	100	100	0	0

On the basis of the tephra distributions, one would predict that within the isthmus region the W-K1, W-K2, and W-K3 eruptions devastated the vegetation, the W-K4 event was considerably less serious with possibilities for quick recovery, and the Dk event would have been negligible. In contrast, on Garua Island the Dk and W-K2 eruptions must have had devastating results for the terrestrial fauna and flora, whereas the W-K3 event would not have been significant. The impact of W-K1 is more difficult to assess because the tephra is only preserved in redeposited contexts on Garua Island and has a patchy occurrence on the mainland, suggesting that it fell during the rainy season and was rapidly removed from the surface. We suggest that the W-K1 event could have had a major impact in the isthmus region, but may have been much less serious on Garua Island.

Further information about the impact of the eruptions comes from microfossil studies based on plant phytoliths and starch granules (Boyd *et al* 2005; Lentfer and Torrence 2007; Parr *et al* 2001). Although they demonstrate a general correlation between tephra depth and the degree of vegetation destruction, considerable variation across space has also been noted, as predicted by ecologists studying disasters (Turner *et al* 1998). The effects of W-1 are not clear from these studies, but the plant microfossils show that, as expected, the vegetation was seriously damaged following W-K2 in both areas, Dk on Garua Island, and W-K3 in the isthmus area.

The modern landscape in both study areas is a product of its volcanic history (Boyd and Torrence 1996; Boyd *et al* 2005; McKee *et al* 2005; Torrence 2001, 2002c; Torrence and Neall 2004; Torrence *et al* 1999). In general, the Holocene air-fall tephras have largely mantled earlier morphologies established by volcanic flows, but in the coastal regions erosion and redeposition have created new surfaces. For example, the small coastal plain on Garua Island and the larger one on the eastern side of the peninsula are the products of infilling of ponds and marshes by *in situ* and redeposited air-fall tephras. Through time, the rich resources of the estuarine ecosystems were replaced by land that would have created new opportunities for gardening (Boyd *et al* 2005). An even more drastic change took place following the W-K2 eruption when a tidal embayment on the western side of the peninsula, comprising c 200 km², was converted to swampy land, probably as a result of the massive flooding that redeposited enormous quantities of tephra derived from the Kulu river catchment. This rapid transformation of a marine bay to a freshwater mangrove swamp must have had drastic consequences for potential human land use (Torrence 2002c; Torrence and Neall 2004).

PUNCTUATED HISTORY

The volcanic history has created ideal conditions for archaeological research. Our study has capitalised on the ubiquitous stratigraphy built up by air-fall tephras from the Holocene eruptions. Using information derived from 67 test pits (1 m²) in the isthmus and 69 on Garua Island (Torrence 2002b) and Figures 3.2 and 3.3, we can trace the volcanic layers over large regions and therefore obtain relative dates for the artefacts interbedded between them (Figure 3.4).

Given that the huge scale of the Holocene volcanic eruptions in the Willaumez Peninsula led to massive destruction of natural resources following the emplacement of thick layers of tephra and the transformation of landscape through erosion and redeposition, abandonment of the region would be predicted. On the basis of a Bayesian analysis of 115 radiocarbon dates from the widely distributed test pits, Petrie and Torrence (submitted) found that after each eruption both regions were probably abandoned for varying lengths of time (Table 3.4). Using data available within our region and its hinterland (where archaeological research is severely lacking), one cannot yet reconstruct the relative levels of mortality versus emigration, but the effect was the same in all cases: local extinction of the population leading to a punctuated pattern of settlement as predicted by the chaos model.

In contrast, the history of material culture is only partly chaotic. Some changes in the occurrence of artefact types can be directly linked to disasters, but not all volcanic disasters result in extinction. In some cases, artefact types must have been preserved elsewhere, perhaps by descendants of the original population. A distinctive class of flaked obsidian artefacts, called stemmed tools (Araho *et al* 2002), survived the W-K1 eruption, but virtually disappeared after the W-K2 event. South of our region, similar tools made from chert also ceased to be manufactured after W-K2 (Pavlides 2006). As yet, we have no information about when stemmed tools were first made so we cannot

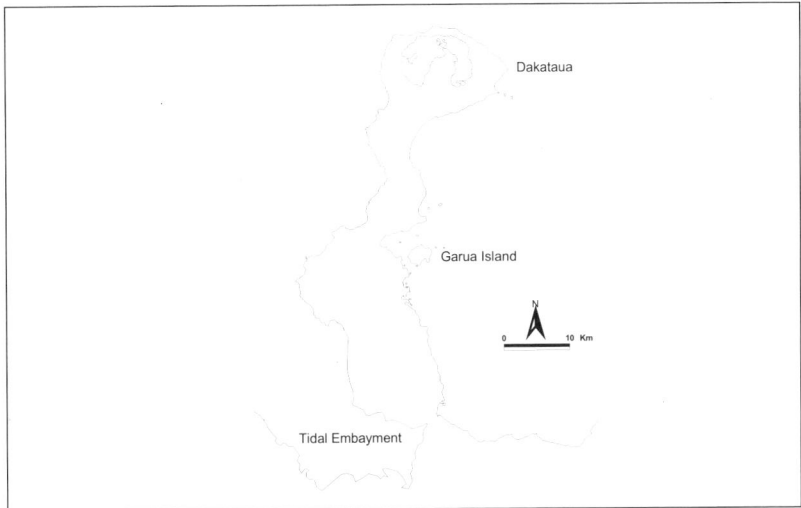

Figure 3.2   Location of archaeological test pits in the isthmus study region.

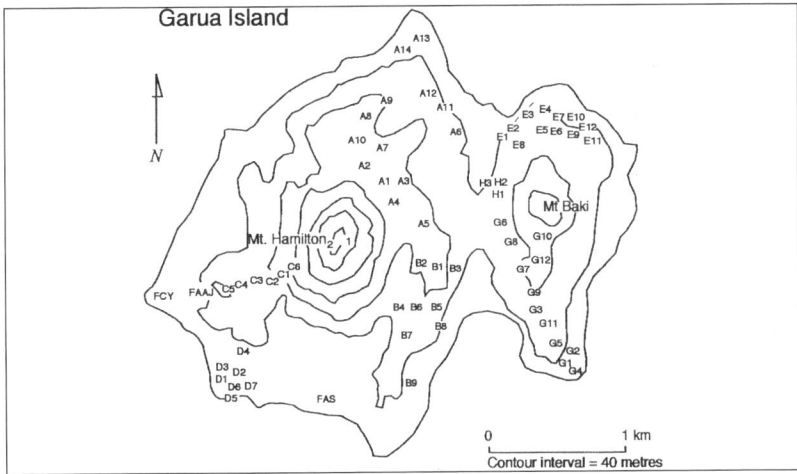

Figure 3.3   Location of archaeological test pits on Garua Island.

link their origin to the volcanic history. A few surviving stemmed tools have been found on Garua and a nearby island in sites representing earliest recolonisation (Specht and Kononenko in prep). Currently, we cannot determine whether the loss of this tool type occurred outside the region or was caused by the process of resettlement.

A potential 'evolutionary novelty' following the W-K2 event is the introduction of decorated pottery in the Lapita style. Ceramics were a short-lived and possibly unsuccessful innovation in this region. They disappeared by the time of the W-K3 eruption.

Figure 3.4 Tephrastratigraphy of Holocene volcanic events in the isthmus area of the Willaumez Peninsula is illustrated in this section from test pit XLVIII. Each of the lighter layers represents visually distinctive air-fall tephra from a particular volcanic eruption. Archaeological material is found in the darker, soil layers formed in the top part of the tephras.

Table 3.4 Length of time the Willaumez Peninsula was abandoned following volcanic eruptions expressed as the highest posterior density (HPD) region (95.4%). Derived from a Bayesian analysis of radiocarbon dates by Petrie and Torrence (submitted)

Period	Location	
	Garua	*Isthmus*
Post–W-K4		0–150
Post–DK	0–240	
Post–W-K3	0–250	0–150
Post–W-K2	0–240	0–280
Post–W-K1	0–220	1,300–2,000

RECOLONISATION

Populations that persist within catastrophic environments over the long term must be able to flee the afflicted region, survive somewhere else, and then recolonise later. Traits that facilitate refuging in safe areas and/or expansion back into the devastated regions would therefore have been quite advantageous. Since this appears to have been the case with the flexible land use and technological strategies observed during the Pleistocene period (Torrence, Neall *et al* 2004), we might expect these to have continued during the Holocene. We can also predict that changes in the nature of recolonisation that increase the speed at which populations return following local extinctions should occur (eg, Hoffmann and Parsons 1997: 186–87) because within a competitive environment, the success of a population depends on its ability to access resources. Biologists have noted that opportunistic species that colonise new areas rapidly are characteristic of environments that fluctuate wildly. To evaluate these hypotheses, we examine the timing of recolonisation in relation to the nature of the volcanic disasters and then turn to a range of factors, including social exchange, concepts of place, and patterns of land use, that might have assisted this process.

At this stage in our research, we cannot closely monitor the specific process of colonisation after each of the Holocene events, but using the results of the Bayesian analysis, we can estimate how quickly people returned to the area. Despite the breadth of the ranges for the estimations, the results, summarised in Table 3.4, reveal intriguing variation. (Note that these differ from predictions by Torrence *et al* [2000] and Torrence [2002a:297], which were based on the single, earliest date following the eruption.) Generally, the potential length of time the region was abandoned matches the potential effects of the eruptions as measured by the depth of tephra cover (eg, Tables 3.3 and 3.4), but there are notable exceptions. For example, the very long period following the W-K1 eruption in the isthmus region is surprising, given the expected greater impact from the larger, subsequent W-K2 eruption. The scarcity of organic material for radiocarbon dating in the post–W-K1 period, however, may mean we have not detected earliest recolonisation. In addition, because of the thicker Dk tephra, one would have predicted that this disaster should have had more serious effects on Garua than the W-K2 eruption (Torrence 2002a: Table 16.2, 296), but the island was probably re-occupied more quickly after Dk. Similarly, the potential gap on Garua Island after W-K3 is surprisingly long given the relatively thin depth and patchiness of the tephra on the island.

Although the history of occupation is punctuated as predicted by the chaos model, the scale of the natural disaster and human response are not perfectly correlated within the larger isthmus region. On the mainland, as opposed to the small off-shore island, the timing of recolonisation has a directional trend involving a decrease in the length of abandonment through time. Furthermore, with the exception of the isthmus post–W-K1, the period of abandonment after each eruption may have been quite brief. Rapid colonisation is a logical outcome of selection. The next step is to investigate types of behaviour that might have facilitated rapid recolonisation.

SOCIAL STRATEGIES

Social factors must have played an important role in the settlement history of the Willaumez Peninsula. Since the impacts of the Witori and Dakataua air-fall tephras were sufficiently severe that forests and cultivated plots would have been devastated, the survival of populations over the long term would have required substantial assistance from outside the region. Also, exchange networks and concepts about place could have been key factors in the process of recolonisation.

Exchange and Interaction

Networks established through exchange are crucial for groups inhabiting catastrophic environments. First, as Torrence (2002a: 303–04, 2004) argued previously, social links are important because for people to escape disasters, they must be accepted within new social territories and granted access to resources over a relatively long period of time. A second way in which social connections are important for small groups moving into unoccupied areas during recolonisation is through the maintenance of 'lifelines' back to the source population resident in a more secure environment (eg, Green 1987; Kirch 1988). A decrease through time in the length of the refuging period following a series of natural disasters may therefore imply that social ties had been strengthened and extended over time, enabling a larger portion of refuges to survive and/or to recolonise more effectively.

One way to monitor social networks is through the spatial distribution of obsidian derived from localised outcrops in three main areas in the peninsula and Mopir located near Witori itself (eg, Torrence 2004). Obsidian from all the available sources was recovered from the test pits in both study regions throughout both the Pleistocene and the Holocene. This result is particularly striking for Garua Island because non-local obsidian was imported even though good quality obsidian is available on the island, suggesting that the creation of social ties was as important, if not more so, than the exchanged goods (Torrence and Summerhayes 1997).

It is hard to connect chronological changes in the exchange systems within the Willaumez Peninsula with potential differences in their effectiveness for coping with volcanic disasters. Even if they travelled to the sources to obtain their raw material rather than acquire it through exchange, individuals would still have had to forge social relations with the groups living near the outcrops or those who 'owned' this desirable resource in order to obtain resources. For this reason, the composition of sources reflects the breadth of social relations.

Except for the period following the W-K2 event, a broad mixture of obsidian from a range of different sources occurs within the test pits. This pattern indicates that people tried to maximise the spatial distribution of social links created through the exchange of this raw material. In contrast, after the W-K2 event, only obsidian from one source was widely circulated and only small quantities of the other sources are represented in the test pits (Torrence 2004). Torrence and Summerhayes (1997) have suggested that at this time there was a single intra-regional system of exchange that integrated all communities, but that it did not persist after W-K3. After the disaster, people may have returned to an open and flexible system of exchange with each person or group acquiring its own supplies of obsidian through multiple social links. This highly structured intra-regional system was ineffective at creating social networks and/or too costly to maintain. Instead, the more open and flexible networks characteristic of previous and succeeding periods were therefore reinvented.

Concepts of Place

It seems likely that certain concepts of space associated with rapid recolonisation would come to prominence over time. The way immigrants view a region will vary depending on whether they are descendant communities of the previous inhabitants or have no ideological connection to this social place. Such differing concepts will condition the speed at which groups move in. For example, a concept of the affected areas as 'dangerous' or inhabited by powerful spirits, as might be expected following a major volcanic eruption, (see examples in Cronin and Cashman, Chapter 9 and Chester and Duncan, Chapter 10) could impede a return to the region. In contrast, positive views, such as a homeland, location of owned, and/or valued resources that might need to be (re)claimed or protected, new opportunities, a safe refuge, room to expand, etc would encourage a speedy return or spur on new colonists. Although currently we cannot reconstruct the specific ideas about place that lay behind past colonisations in this region, it is possible to hypothesise that they motivated or at least enabled the observed increased rate of recolonisation (eg, Torrence in press). A directional change in these concepts would constitute good evidence for selection operating with a catastrophic environment, although finding ways to measure these past ideologies remains a task for future research.

Memory

So far we have discussed positive changes as a result of selective processes operating within catastrophic environments. It is also worth considering characteristics that are absent. Memory commonly plays an active role in shaping cultural landscapes (eg, Campbell 2006 and cited references). Following the 1994 Rabaul eruption, people returned almost immediately to place territorial markers around their properties although their houses and gardens had been buried under metres of tephra (Lentfer and Boyd 2001: 55). A similar scenario has been documented at the Hashimuregawa site in Japan, where shell middens were re-established in the same location as those buried under tephra (Shimoyama 2002: 336). Although they could not return immediately to live in the devastated area, people still created concrete ways to retain the past. The potential for memorialising places in this way, however, could be hindered considerably by periods of abandonment greater than several generations.

With the drastic changes in physical geography that follow major volcanic events, familiar places, trails, special markers or structures, etc may be destroyed and, as in the case of the isthmus region following the W-K2 eruption, whole landscapes can be changed beyond recognition. It seems likely that within volcanically active environments the potential for remembering, reworking, and reusing the places and artefacts of the past to create new social worlds (eg, Bradley 2002) is greatly diminished. Furthermore, depending on the source of the population, memory may not have a useful role for colonising groups who need to move into new areas. It is therefore interesting that there is a notable poverty of rock art and stone arrangements in the Willaumez Peninsula, particularly given their ubiquity in neighbouring regions that were not impacted by the Holocene eruptions, such as the Bali-Witu islands (Byrne 2005; Torrence *et al* 2002). Clearly, the meaning assigned to particular places might be very different for people living in environments subject to catastrophic change compared to residents of settings where spaces are relatively permanent and markings are not repeatedly erased.

In contrast, the catastrophes themselves are such powerful experiences for human groups that they are likely to become the raw material for memory, for instance as myths and stories. In these cases, the source of the disaster (eg, a volcano) could become a powerful symbol or focus for ritual, although this might not be a positive factor in stimulating colonisation because many volcanic areas are remembered as dangerous or associated with powerful beings (eg, Blong 1982; Chester and Duncan, Chapter 10; Cronin and Cashman, Chapter 9; Dillian 2004, Chapter 12; Elson *et al* 2002; Hoffman 1999: 306; Holmberg, Chapter 13; Plunket and Uruñuela 1998a, 1998b). An extreme example of the long-term impact of a volcanic event is the Pompeii eruption, which Allison (2002) argues has been more potent in the modern world than directly following the event itself (also see Torrence and Grattan 2002: 10).

At this stage, there is no evidence in the material record for how the prehistoric disasters may have been remembered and incorporated into the lives of successive populations within the Willaumez Peninsula, but, of course, they may have lived on for some time within oral traditions. The seeming absence of such stories within the population currently resident in the region may be a product of extensive movements since European contact or may reflect the considerable time that has elapsed since the last major volcanic event over a thousand years ago. Despite the current lack of evidence for the Willaumez Peninsula, the role of memory within catastrophic environments is an aspect that deserves further research.

LAND USE

The punctuated, chaos model appears particularly appropriate for changes in land use in a region impacted frequently by devastating volcanic events, because each of these would have disrupted natural patterns of succession and returned the vegetation back to an immature state. Land use strategies based on foraging within a mature tropical forest might not be able to survive the natural disasters and would be ineffective for some time until the forest regenerated to a state in which there were adequate mature species and regular food supplies. In contrast, a flexible and highly mobile pattern of land use, such as has been hypothesised for the Pleistocene period (eg, Pavlides 2004; Torrence, Neall *et al* 2004), might be better suited and would enable recolonisation on the relatively short time scales observed in the Willaumez Peninsula (with the puzzling exception of W-K1).

In contrast, it could be argued that by destroying the forest, the volcanic eruptions would have selected for more intensive forms of land use than foraging. Although perhaps not as flexible in terms of the ability to move to places with alternate sources of prey, by creating their own sources of food-shifting cultivators can utilise a very wide range of ecosystems. As opposed to a foraging strategy that requires a mature forest to intensify production, shifting cultivation benefits from an immature ecosystem because it requires considerably less energy to prepare and maintain gardens. By reducing the forest, the eruptions would make it easier for cultivators to recolonise the region because the need for clearance would be reduced. In fact, the destruction of the forest could even attract cultivators. Following this line of reasoning, one would predict that through time people would adopt increasingly intense forms of cultivation (eg, reduced fallow periods) and depend more on cultivated rather than wild resources because they would better able to take advantage of the unoccupied lands.

Land Management

Recent analyses of phytoliths and starch taken from archaeological test pits in both study areas provide support for increasing dependence on cultivation. Lentfer and Torrence (2007) and Boyd *et al* (2005) report that the natural pattern of forest succession following the volcanic eruptions was disrupted increasingly earlier in the sequence by human interference in the form of burning and clearance. The changes they observed are progressive through time but not gradual because there is a quantum leap in the intensity of clearance and introduction of new plants following the Dk event on Garua and after W-K4 in the isthmus. At one locality on Garua Island, Lentfer and Torrence (2007) record the progressive introduction of new plants, beginning with bamboo (post W-K1), followed by an edible palm and possibly bananas (post W-K2), and ending with a major increase in the presence of bamboos and fig trees (post DK), suggesting the presence of a village surrounded by cultivated gardens. Although the sample size is small, the data indicate a clear increase through time in the intensity of land management with cultivation as the most likely mechanism.

Based on the nature of the stone tool assemblages, Torrence (1992, 2002a: 302–303, 2002b; eg, Torrence *et al* 2000) has argued that a chronological trend towards more expedient use of raw material on Garua Island, together with a higher degree of clustering in the pattern of discard, signals a reduction in the mobility of people and by inference an increase in the intensity of land management. The same trends are present in the assemblages found in the isthmus region where there is a change through time towards less retouch and higher rates of discard. Here the character of the spatial distribution of obsidian artefacts (which comprise over 95% of the assemblages) across the landscape is remarkably stable through time, despite the infilling of the large tidal embayment following the W-K2 event. The trends found on Garua Island are not so evident in the isthmus region, although the most recent period is the most clustered (ie, more of the assemblage is situated in fewer places [Figure 3.5]). The widespread distribution of material in both regions suggests that people always had a highly flexible pattern of land use.

Artefact Distributions

Another way to monitor changes in land use patterns is to examine spatial patterning in the abundance of artefacts discarded within different components of the landscape, as shown in Figure 3.6, which illustrates the proportion of the assemblage located at various distances from the coast within the isthmus region. The data are not very meaningful for periods 2, 3, and 3.2 because almost the entire region was within 4 km of a coastline at that time. However, in the following periods there is a notable difference in the predominance of material discarded in inland zones between the periods that experienced the effects of severe volanic

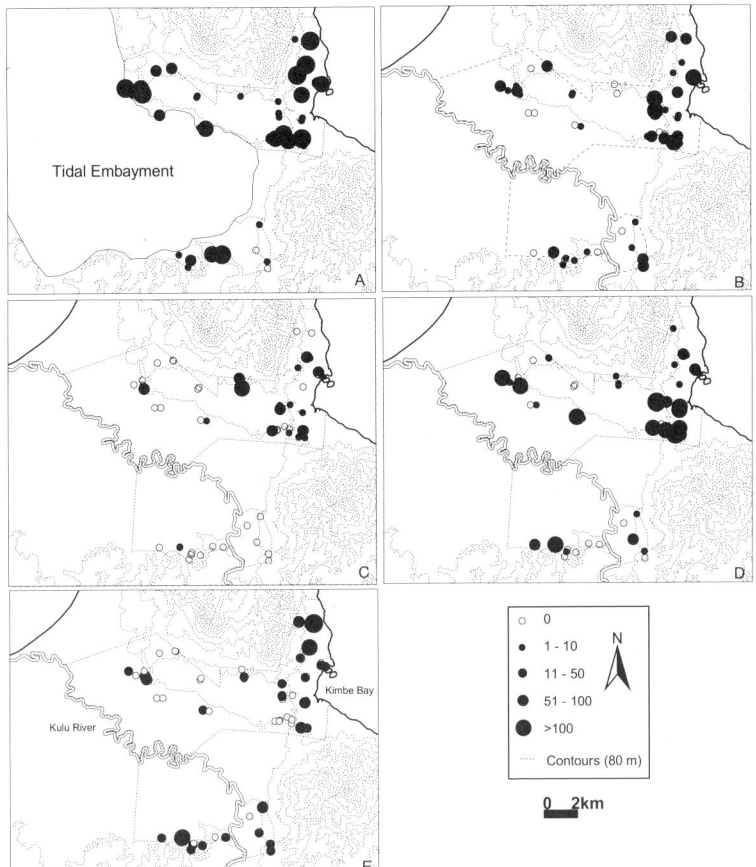

Figure 3.5 Distribution of obsidian artefacts in the test pits in the isthmus region of the Willaumez Peninsula.

activity (periods 4–6) and those when there were only very minor falls of tephra from the W-H series of the Witori volcano (periods 7, 8). During the times of high impact, land use has a significant inland component; otherwise the focus of artefact discard is near the coast. The relationship of this pattern to forms of land use is not clear at this stage of research, but could relate to the unstable nature of both the inland and coastal plains because the loose, unstable tephra eroded off the slopes, causing flooding, and formed large alluvial fans that gradually pushed the coastline outwards (Torrence 2001; Torrence and Neall 2004). Near the coast, the ground was probably very wet and swampy.

In contrast to lowlands, ridgelines might have provided a more favourable setting for gardens and settlements. This hypothesis is supported by the differential distribution of stone artefacts in relation to

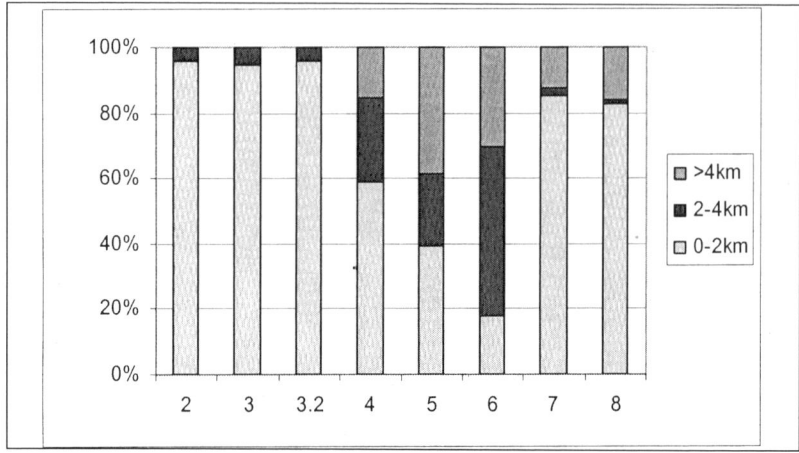

Figure 3.6 Changes in the proportion of the obsidian assemblage located at different distances from the coast. The numbers represent the following stratigraphic horizons: 2, pre–W-K1; 3, between W-K1 and W-K2; 3.2, pre–W-K2 but W-K1 is not preserved; 4, between W-K2 and W-K3; 5, between W-K3 and W-K4; 6, between W-K4 and W-H series; 7, within W-H series; 8, modern topsoil.

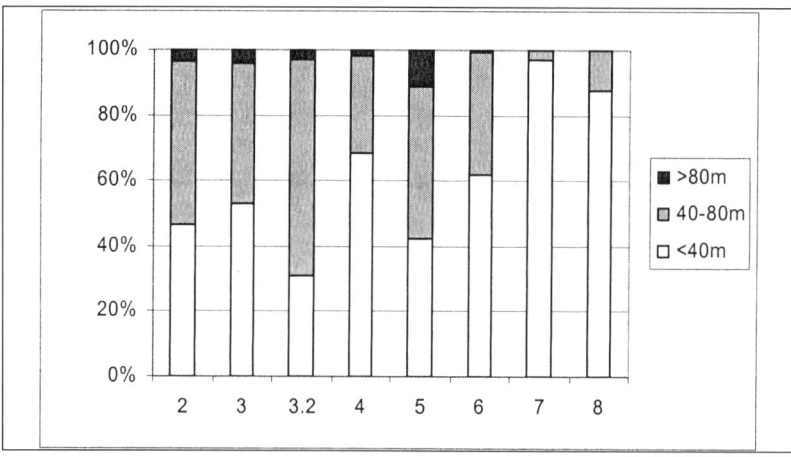

Figure 3.7 Changes in the proportion of the obsidian assemblage located at different elevations. The numbers represent the following stratigraphic horizons: 2, pre–W-K1; 3, between W-K1 and W-K2; 3.2, pre–W-K2 but W-K1 is not preserved; 4, between W-K2 and W-K3; 5, between W-K3 and W-K4; 6, between W-K4 and W-H series; 7, within W-H series; 8, modern topsoil.

elevation. As shown in Figure 3.7, through time there is a directional trend towards lower altitudes, with the exceptions of periods 3.2 and 5, which are located on higher ground than would be expected for a direct correlation between time and elevation. Period 3.2 is defined for test pits where the W-K1 tephra was not preserved and periods 2 and 3 cannot be distinguished stratigraphically. Interestingly, Petrie and Torrence (submitted) found that period 3.2 locations were re-occupied earlier after W-K1 than places where the tephra was still *in situ* (period 3). There is, then, an intriguing possibility that people preferentially used places less damaged by the air-fall tephras and consequently, where forest, garden plots, or fallow land had been better preserved. An alternative explanation is that the early colonisers did not depend heavily on cultivation for their subsistence, but were targeting preserved patches of vegetation where foraging was still possible or where the forest regenerated more rapidly. Such patches might have provided a temporary refuge prior to cultivation in areas deeply covered with tephra. Further research is needed to evaluate these different possibilities.

Period 5 is also interesting because it represents the shortest length of time following the eruption and may therefore encapsulate the earliest stage of recolonisation (eg, Table 3.2). If this assumption is correct, then the earliest activity was mainly in the upland zones, as in the case of period 3.2. Tephra on steeper slopes of the higher ground was subject to erosion and consequently was probably thinner. In these areas, the forest would have suffered less damage, as observed following the 1991 Pinatubo eruption (Vince Neall, personal communication). In contrast to periods 2.3 and 5, both of which only represent the early stages of the process, the chronological trend towards lower elevations reflects the later stages when activities, and especially cultivation, were shifted to flatter land, especially as the coastal plain became drier and more fertile with the emplacement of additional layers of air-fall and redeposited tephra.

In summary, the hypothesis that the periodic volcanic destruction of the tropical forest selected for more intensive land use involving cultivation is supported by increased burning through time, together with the introduction of new plant species and a trend towards a greater degree of artefact clustering suggesting sedentism. At the same time, people appear to have maintained flexible land use practices that enabled them to take advantage of the widest range of areas suitable for cultivation. The importance of inland, upland regions has been noted and requires further examination to separate out the likely factors involved.

PERSISTENCE IN A CATASTROPHIC ENVIRONMENT

Our case studies on Garua Island and within the isthmus region have provided an intriguing insight into the long-term effects of catastrophic environments on human history. Previously, Torrence (2002a) argued that the prehistory of the Willaumez Peninsula was created by colonisers who imported particular lifestyles following each volcanic disaster. She posited that the long-term directional trends observed were the result of processes operating outside the area affected by the eruptions and may not have been related to the volcanic history in any direct manner. Viewed as a series of short-term episodes, similar to what archaeologists usually call 'periods' or 'phases', it is certainly legitimate to envisage the prehistory of this region as the consequence of a series of separate colonisations.

In this chapter, we have adopted a different view in which the overall shape and form of the past c 10,000 years is considered as a single entity, rather than an ordered set of events. Using this perspective, we have begun to examine if and how human history might have a particular character when unfolded within environments that experience relatively frequent natural disasters with a high magnitude. Our analyses show that both chaos and selection were important in shaping the cultural history of the Willaumez Peninsula. By their very nature, catastrophic environments lead to extinction. Consequently, the population history was punctuated by periods of abandonment (extinctions) followed by colonisations that occasionally introduced new forms of material culture (evolutionary novelties), as in the case of Lapita pottery.

Over time, the frequent extinction events acted as a selective factor. The case study has identified behaviours that persisted through time because they facilitated refuging and rapid recolonisation: notably mobility, flexibility, and social exchange. Finally, a temporal trend towards higher levels of landscape management, which probably represents more intense cultivation, is also the consequence of selection because it resulted in the reduction of time when the region was abandoned.

Admittedly, our study areas represent just two small windows into a process that had a much greater spatial scale. One of the problems of studying catastrophic environments is defining the spatial limits of the affected region (Torrence 2002a: 306). Based on purely physical measures, one can map various scales of environmental change and make predictions about impacts on vegetation, marine resources, and water sources, but the cultural effects would have spread much more widely to include populations linked to the damaged region through social and economic networks, areas to which people fled for refuge, and the source areas for the subsequent colonising populations. These need not be adjacent to the impacted area. The problem is how to identify the relevant places.

Clearly, there would have been complex interactions between the long-term history of the directly impacted regions, its periphery, and areas with a social connection. In a study of the Yombon region to the south of the Willaumez Peninsula, where there were smaller but still significant falls of tephra, Pavlides (2006; Torrence *et al* 2000) has found much the same pattern of continuity despite a punctuated settlement history. Future research should investigate the relationship between the character of long-term prehistory and these volcanic events out- side the Willaumez Peninsula and particularly beyond the area of sig- nificant tephra falls.

Looking farther afield, one cannot help but wonder about the rela- tionship between the arrival of Lapita pottery in the Bismarck Archipelago and the largest of the volcanic disasters – the W-K2 erup- tion – which occurred only shortly beforehand. The people bearing this distinctively decorated pottery were certainly colonisers. Throughout their range they are primarily found in areas that were previously uninhabited. Whether the pottery represents a local devel- opment that blended external features with well-established trad- itions in the Bismarck Archipelago (eg, Allen and White 1989; Green 1991) or migrants from outside (see summary in Kirch 1997), the ideol- ogy represented in the iconography of the pottery was compatible with and probably assisted rapid colonisation. It is even possible that the social and ideological systems that created the pottery are in some way directly related to the W-K2 natural disaster, possibly through people finding and adapting new forms of social expression as a con- sequence of seeking social networks and perhaps assistance from out- side the affected area.

Moving later in time, Lilley (2004a, 2004b) has proposed that changes within the cultural history of the Siassi Islands, located just off the western end of New Britain, represent a follow-on effect from the W-K3 and W-K4 eruptions. In particular, there may have been an influx of people to the islands beginning around 1,700 BP, following a period of abandonment and correlated with the onset of an active trading system involving pottery from the mainland of New Guinea to the west and obsidian from the Willaumez Peninsula. It seems prob- able that Lilley has identified the arrival of either refugees from the eruptions or perhaps the same colonists who were responsible for the repopulation of New Britain following the disasters. If the latter is the case, then we may need to look outside New Britain itself for the source of the various colonising populations. The evidence for the expansion of exchange following the repopulation of the Siassi Islands is also quite interesting given the constant role of obsidian exchange in the Willaumez Peninsula.

EXTENDING THE MODELS

Our data show that cultural groups found effective ways for persist- ing within the catastrophic environment of the Willaumez Peninsula, but additional archaeological research should evaluate and expand our observations. In future work, we hope to enhance the analyses and also address important attributes whose archaeological signatures have yet to be tackled (eg, memory, concepts of place, and social organisation). Further considerations of evolutionary processes in settings that have experienced extreme environmental change could also be valuable (eg, Hoffmann and Parsons 1997; Turner and Dale 1998; Turner *et al* 1998).

It would also be interesting to see how far the results of our case study can be extended to other catastrophic environments. For example, Sheets's (1999, Chapter 4) conclusion that simple, egalitarian social systems have survived volcanic events in Central America better than hierarchical organisations may be explained by their similarities to early colonisers. The greater flexibility of egalitarian societies may also help account for their persistence throughout prehistory in the Arenal region of Costa Rica, which experienced frequent volcanic eruptions (Sheets and McKee 1994; Sheets *et al* 1991). Furthermore, recent archaeological research in Mexico (Gonzalez and Huddart, Chapter 5; Gonzalez *et al* 2000; Siebe *et al* 1996), Japan (Shimoyama 2002: 328–35), Italy (Allison 2002; Mastrolorenzo, Mastrolorenzo *et al* 2002; Mastrolorenzo, Petrone *et al* 2006) and the Philippines (Gaillard *et al*, Chapter 11) has revealed intriguing variations through time in the ways societies have coped with volcanic disasters. To what degree these also exhibit unique combinations of chaos and selection as, in the case of the Willaumez Peninsula, is an important research question. It is hoped that in the future, archaeologists working in volcanic environments such as these will take greater advantage of the potential of their long records for testing ideas about the potential impacts of natural disasters on human history.

ACKNOWLEDGMENTS

We are grateful for funding from the Australian Research Council, Australian and Pacific Science Foundation, Pacific Biological Foundation, Australian Museum, AINSE, Earthwatch Institute, and New Britain Palm Oil Ltd. We also received assistance from the National Research Institute (PNG), National Museum and Art Gallery (PNG), West New Britain Provincial Cultural Centre, University of Papua New Guinea, Mahonia Na Dari Research Station, Kimbe Bay Shipping Agencies, and Walindi Resort. Grateful thanks to all project members for their enthusiasm and hard work and especially long-term collaborators, Bill Boyd, Hugh Davies, Carol Lentfer, Chris McKee, Vince Neall, Jeff Par, Cameron Petrie, Ed Rhodes, Jim Specht, and Peter White and to Huw Barton for useful comments. The research would not have been possible without the continuing support of residents in West New Britain.

REFERENCES

Allen, J and White, J (1989) 'The Lapita homeland: some new data and an interpretation', *Journal of the Polynesian Society* 98, 129–46

Allison, P (2002) 'Recurring tremors: the continuing impact of the AD 79 eruption of Mt. Vesuvius', in Torrence, R and Grattan, J (eds), *Natural Disasters and Cultural Change*, pp 107–25, London: Routledge

Araho, N, Torrence, R and White, J (2002) 'Valuable and useful: Mid-Holocene stemmed obsidian artefacts from West New Britain, Papua New Guinea', *Proceedings of the Prehistoric Society* 68, 61–81

Blong, R (1982) *The Time of Darkness: Local Legends and Volcanic Reality in Papua New Guinea*, Seattle: University of Washington Press

Blong, R (1984) *Volcanic Hazards: A Sourcebook on the Effects of Eruptions*, Sydney: Academic Press

Boyd, W, Lentfer, C and Luker, G (1999) 'Environmental impacts of maor catastrophic Holocene volcanic eruptions in New Britain, PNG: a preliminary model for palaeoenvironmental change', in Kesby, J, Stanley, J, McLean, F and Olive, L (eds), *Geodiversity: Readings in Australian Geography at the Close of the 20th Century*, pp 361–72, Canberra: Australian Defence Force Academy

Boyd, W, Lentfer, C and Parr, J (2005) 'Interactions between human activity, volcanic eruptions and vegetation during the Holocene at Garua and Numundo, West New Britain, PNG', *Quaternary Research* 64, 384–98

Boyd, W and Torrence, R (1996) 'Periodic erosion and human land use on Garua Island, PNG: a progress report', *Tempus* 6, 256–74

Bradley, R (2002) *The Past in Prehistoric Societies*, London: Routledge

Byrne, S (2005) 'Recent survey and excavation of the monumental complexes on Uneapa Island, West New Britain, Papua New Guinea', *Papers from the Institute of Archaeology* 16, 95–101

Campbell, M (2006) 'Memory and monumentality in the Rarotongan landscape', *Antiquity* 80, 102–17

Dillian, C (2004) 'Sourcing belief: using obsidian sourcing to understand prehistoric ideology in northeastern California, U.S.A.', *Mediterranean Archaeology and Archaeometry* 4, 33–52

Elson, M, Ort, M Hesse J and Duffield, W (2002) 'Lava, corn, and ritual in the northern Southwest', *American Antiquity* 67, 119–35

Gonzalez, S, Pastrana, A, Siebe, C and Duller, G (2000) 'Timing of the prehistoric eruption of Xitle volcano and the abandonment of Cuicuilco Pyramid, Southern Basin of Mexico', *Special Memoir, Geological Society of London* 171, 205–24

Grayson, D and Sheets, P (1979) 'Volcanic disasters and the archaeological record', in Sheets, P and Grayson, D (eds), *Volcanic Activity and Human Ecology*, pp 623–32, New York: Academic Press

Green, R (1987) 'Obsidian results from the Lapita sites of the Reef/Santa Cruz Islands', in Ambrose, W and Mumery, J (eds), *Archaeometry: Further Australasian Studies*, pp 239–49, Canberra: Australian National University

Green, R (1991) 'The Lapita cultural complex: current evidence and proposed models', in Bellwood, P (ed), *Indo-Pacific Prehistory 1990: Proceedings of the 14th Congress of the Indo-Pacific Prehistory Association*, pp 295–305, Canberra: Indo-Pacific Prehistory Association

Hoffmann, A and Parsons, P (1997) *Extreme Environmental Change and Evolution*, Cambridge: Cambridge University Press

Hoffman, S (1999) 'After Atlas shrugs: cultural change or persistence after a disaster', in Oliver-Smith, A and Hoffman, S (eds), *The Angry Earth: Disaster in Anthropological Perspective*, pp 302–26, London: Routledge

Kirch, P (1988) 'Long-distance exchange and island colonisation: the Lapita case', *Norwegian Archaeological Review* 21, 103–17

Kirch, P (1997) *The Lapita People*, Oxford: Blackwell

Lentfer, C and Boyd, W (2001) *Maunten Paia: Volcanoes, People and Environment: The 1994 Rabaul Eruption*, Lismore, Australia: Southern Cross University Press

Lentfer, C and Torrence, R (2007) 'Holocene volcanic activity, vegetation succession and ancient human land use: unraveling the interactions on Garua Island, Papua New Guinea', *Review of Palaeobotany and Palynology* 143: 83–105

Lilley, I (2004a) 'Diaspora and identity in archaeology: moving beyond the Black Atlantic', in Meskell, L and Preucel, R (eds), *A Companion to Social Archaeology*, pp 287–312, Oxford: Blackwell

Lilley, I (2004b) 'Trade and culture history across the Vitiaz Strait, Papua New Guinea', in Attenbrow, V and Fullagar, R (eds), *A Pacific Odyssey: Archaeology and Anthropology in the Western Pacific. Papers in Honour of Jim Specht*, pp 89–96, Records of the Australian Museum, Supplement 29, Sydney: Australian Museum

Machida, H, Blong, R, Specht, J, Torrence, R, Moriwaki, H, Hayakawa, Y, Talai, B, Lolok, D and Pain, C (1996) 'Holocene explosive eruptions of Witori and Dakataua caldera volcanoes in West New Britain, Papua New Guinea', *Quaternary International* 34–36:65–78

Mastrolorenzo, G, Palladino, D, Vecchio, G and Taddeucci, J (2002) 'The 472 AD Pollena eruption of the Somma-Vesuvius (Italy) and its environmental impact at the end of the Roman Empire', *Journal of Volcanology and Geothermal Research* 113, 19–36

Mastrolorenzo, G, Petrone, P, Pappalardo, L and Sheridan, M (2006) 'The Avellino 3780-yr-B.P. catastrophe as a worst-case scenario for a future eruption at Vesuvius', *Proceedings of the National Academy of Sciences* 103, 4366–70

McKee, C, Patia, H, Kuduon, J and Torrence, R (2005) *Volcanic Hazard Assessment of the Krummel-Garbuna-Welcker Volcanic Complex, Southern Willaumez Peninsula, WNB, Papua New Guinea*, Geological Survey of Papua New Guinea Report 2005/4

Newhall, C and Punongbayan, S (eds) (1996) *Fire and Mud: Eruptions and Lahars of Mount Pinatubo, Philippines*, Seattle and Quezon City: University of Washington Press and Philippine Institute of Volcanology and Seismology

Parr, J, Lentfer, C and Boyd, W (2001) 'Spatial analysis of phytolith assemblages at an archaeological site in West New Britain, Papua New Guinea', in Clark, G, Anderson, A, Vunidilo, T (eds), *The Archaeology of Lapita Dispersal in Oceania*, pp 125–34, Canberra: Pandanus Books

Pavlides, C (2004) 'From Misisil Cave to Eliva hamlet: rediscovering the Pleistocene in interior West New Britain', in Attenbrow, A and Fullagar, R (eds), *A Pacific Odyssey: Archaeology and Anthropology in the Western Pacific. Papers in Honour of Jim Specht*, pp 97–108, Records of the Australian Museum, Supplement 29, Sydney: Australian Museum

Pavlides, C (2006) 'Life before Lapita: new developments in Melanesia's long-term history', in Lilley, I (ed), *Archaeology of Oceania: Australia and the Pacific Islands*, pp 205–27, Blackwell: Oxford

Petrie, C and Torrence, R (submitted) 'Measuring the effects of volcanic disasters using a Bayesian approach to radiocarbon dating'

Plunket, P and Uruñuela, G (1998a) 'The impact of Popocatépetl volcano on Preclassic settlement in central Mexico', *Quaternaire* 9, 53–59

Plunket, P and Uruñuela, G (1998b) 'Preclassic household patterns preserved under volcanic ash at Tetimpa, Puebla, Mexico', *Latin American Antiquity* 9, 287–309

Sheets, P (1999) 'The effects of explosive volcanism on ancient egalitarian, ranked, and stratified societies in Middle America', in Oliver-Smith, A and Hoffman, S (eds), *The Angry Earth: Disaster in Anthropological Perspective*, pp 36–58, New York: Routledge

Sheets, P, Hoopes, J, Melson, W, McKee, B, Sever, T, Mueller, M, Cheanult, M and Bradley, J (1991) 'Prehistory and volcanism in the Arenal area, Costa Rica', *Journal of Field Archaeology* 18, 445–65

Sheets, P and McKee, B (eds) (1994) *Archaeology, Volcanism, and Remote Sensing in the Arenal Region, Costa Rica*, Austin: University of Texas Press

Shimoyama, S (2002) 'Volcanic disasters and archaeological sites in Southern Kyushu, Japan', in Torrence, R and Grattan, J (eds), *Natural Disasters and Cultural Change*, pp 326–42, London: Routledge

Siebe, C, Abrams, M, Macías, J and Obenholzner, J (1996) 'Repeated volcanic disasters in prehispanic time at Popocatépetl, central Mexico: past key to the future?' *Geology* 24, 399–402

Specht, J and Kononenko, N (in prep) 'Late Holocene stemmed tools in central New Britain: persistence or change?'

Torrence, R (1992) 'What is Lapita about obsidian: a view from the Talasea sources', in Galipaud, J-C (ed), *Poterie Lapita et Peuplement*, pp 111–26, Noumea, New Caledonia: ORSTOM

Torrence, R (2000) 'Archaeological fieldwork in West New Britain, PNG, May–June 2000', report submitted to the PNG National Museum and various organisations in PNG

Torrence, R (2001) 'Archaeological fieldwork in West New Britain, PNG, June–July 2001'. Report submitted to the PNG National Museum and various organisations in PNG

Torrence, R (2002a) 'What makes a disaster? A long-term view of volcanic eruptions and human responses in Papua New Guinea', in Torrence, R and Grattan, J (eds), *Natural Disasters and Cultural Change*, pp 292–310, London: Routledge

Torrence, R (2002b) 'Cultural landscapes on Garua Island, PNG', *Antiquity* 76, 766–76

Torrence, R (2002c) 'Archaeological fieldwork in West New Britian, PNG, June–August 2002', report submitted to the PNG National Museum and various organisations in PNG

Torrence, R (2004) 'Now you see it, now you don't: changing obsidian source use in the Willaumez Peninsula, Papua New Guinea', in Cherry, J, Scarre, C and Shennan, S (eds), *Explaining Social Change: Studies in Honour of Colin Renfrew*, pp 115–25, McDonald Institute Monographs, Cambridge: McDonald Institute for Archaeological Research

Torrence, R (in press) 'Punctuated landscapes: creating cultural places in volcanically active environments', in David, B and Thomas, J (eds), *Handbook of Landscape Archaeology*, Walnut Creek, CA: Left Coast Press

Torrence, R and Grattan, J (2002) 'Trends in the archaeology of disasters', in Torrence, R and Grattan, J (eds) *Natural Disasters and Cultural Change*, pp 1–18, London: Routledge

Torrence, R and Neall, V (2004) 'Archaeology fieldwork in West New Britain Province, Papua New Guinea, June–July 2004', report submitted to the PNG National Museum and various organisations in PNG

Torrence, R, Neall, V, Doelman, T, Rhodes, E, McKee, C, Davies, H, Bonetti, R, Guglielmetti, A, Manzoni, M Oddone, M, Parr, J and Wallace, C (2004) 'Pleistocene colonisation of the Bismarck Archipelago: new evidence from West New Britain', *Archaeology in Oceania* 39, 101–30

Torrence, R, Pavlides, C, Jackson, P and Webb, J (2000) 'Volcanic disasters and cultural discontinuities in the Holocene of West New Britain, Papua New Guinea', *Special Memoir, Geological Society of London* 171, 225–44

Torrence, R, Specht, J and Boyd, B (1999) 'Archaeological fieldwork on Numundo and Garu Plantations, West New Britain, PNG', report submitted to the PNG National Museum and various organisations in PNG

Torrence, R, Specht, J and Vatete, B (2002) 'Report of an archaeological survey of the Bali-Witu Islands, West New Britain, PNG', report submitted to the PNG National Museum and various organisations in PNG

Torrence, R and Stevenson, C (2000) 'Beyond the beach: changing Lapita landscapes on Garua Island, PNG', in Murray, T and Anderson, A (eds), *Australian Archaeology. Collected Papers in Honour of Jim Allen*, pp 324–45, Canberra: Archaeology and Natural History, Research School of Pacific and Asian Studies, Australian National University

Torrence, R and Summerhayes, G (1997) 'Sociality and the short distance trader: intra-regional obsidian exchange in the Willaumez region, Papua New Guinea', *Archaeology in Oceania* 32, 74–84

Turner, M, Baker, W, Peterson, C and Peet, R (1998) 'Factors influencing succession: lessons from large, infrequent natural disturbances', *Ecosystems* 1, 511–23

Turner, M and Dale, V (1998) 'Comparing large, infrequent disturbances: what have we learned?' *Ecosystems* 1, 493–96

CHAPTER 4

People and Volcanoes in the Zapotitan Valley, El Salvador

Payson Sheets

EXPLORING COPING STRATEGIES

Variation in a wide range of physical, social, and cultural aspects should be considered when exploring the impact and repercussions of explosive volcanic eruptions on Precolumbian societies because they can affect the resistance and the vulnerability of a society attempting to cope with sudden massive stress. Natural phenomena such as the magnitude of the eruption, speed of onset, geochemical and physical characteristics of the ejecta, the flora and fauna, climate, and soils in the area of impact are relevant. Pertinent cultural and social phenomena include demography, societal complexity, adaptation, the built environment, and the political landscape (Sheets 1980).

To explore patterning and variation in the interaction between natural-social components of volcanic disasters, I have documented 25 cases of volcanic disasters in the region stretching from Mexico through Central America (Sheets 1999) (Table 4.1). The database is intriguing, but still very lean, particularly when dealing with relationships amongst many significant variables and long time spans. My review emphasises the interplay among the factors of the volcanic impact, effects on flora, fauna, soils, and human societies, as well as the societal recoveries or lack of recoveries as they are documented in the archaeological, historic, and geological records.

Torrence and Grattan (2002) have noted the difficulties archaeologists face in evaluating ancient natural disasters. Too often, scholars have emphasised the dramatic apocalyptic aspects, and at worst treated all volcanic eruptions as natural disasters. Torrence and Grattan wisely advocate a healthy dose of skepticism and a careful case-by-case examination of the long-term effects of disasters caused by extreme geophysical events. An important component is scale. Eruptions vary immensely in scale, from a highly localised event to huge eruptions,

Table 4.1 Names, dates, countries, types of eruptions, magnitudes, and comments on effects and recoveries

Name	Date	Country	Type	Magnitude	Comments
Coatepeque	77–55k y.a.	El Salvador	Plinian	Large	Prior to human occupation
Ilopango TBJ	c 4th century	El Salvador	Plinian, phreatomagmatic	Large	Long-lasting cultural effects
Popocatépetl	1st century	Mexico	Plinian	Large	Re-ordered social dynamics in Highlands
Loma Caldera	7th century	El Salvador	Phreatomagmatic	Small	Intense effects in tiny area
Boquerón	c 9th century	El Salvador	Phreatomagmatic	Medium	Moderate social effects for short time
Playón	1658–59	El Salvador	Magmatic, cinder cone	Small–medium	Intense effects in 6 km^2
San Marcelino	1722	El Salvador	Magmatic	Small	Devastated agricultural land 15 km^2
Izalco	1779–1996	El Salvador	Composite	Each small	Intense effects in a few km^2
San Salvador	1917	El Salvador	Magmatic	Small	Devastated 16 km^2 of agricultural land
Arenal	2000 BC–1968 AD	Costa Rica	Plinian	Small–medium	Abandonments with successful reoccupations
Barú	c 7th century	Panama	Plinian	Small	Major effects on local chiefdoms

usefully arranged on the Volcanic Explosivity Index (VEI) of Newhall and Self (1982). Similarly, in terms of the scale of human society affected, an eruption can be disastrous for the isolated human household living too close to the volcanic vent, but be negligible at the societal level in either the short or especially the long run.

But how can archaeologists explore the record for variation concerning impacts of different eruptions and make at least initial attempts to discern patterns? I suggest that a useful way to examine coping strategies to volcanic eruptions is to utilise the human behavioural adjustments that people commonly use when faced with extreme geophysical events, as presented in Table 4.2 (modified from Burton *et al* 1978).

Burton and his colleagues developed a set of adjustments and thresholds based on their extensive documentation of many cases from a wide range of natural disasters. The minimal adjustment is loss absorption, defined as making incidental adjustments with no need for a conscious, overt programme of changes. Evidence of this is not likely to be preserved in the archaeological record. However, when the scale or intensity of impact increases and the first threshold of awareness is crossed, the second level of adjustment is loss acceptance. Loss acceptance involves conscious awareness of the problem, and the losses of the victims are often borne by a larger group of people. If the stress is greater and the second threshold, direct action, is crossed, then loss reduction is the result. The material correlates of loss reduction would have a better chance of being preserved in the archaeological record. If the stress is yet greater, then the third threshold, intolerance, is crossed with the result being radical action, which can involve *in situ* significant to fundamental adaptive changes, or more drastic adjustments such as migration, and these have the greatest chance of being detectable by archaeologists. An example of radical action is migration to a location far from the disaster, as in the Volcán Barú discussed below.

Burton *et al* (1978) focus on the negative societal impacts of natural disasters on people and societies that survived. I will expand their

Table 4.2 Range of impacts and human/cultural responses

Beneficial Effects of Natural Disasters (dusting of tephra)
Loss Absorption
Awareness threshold
Loss Acceptance
Direct action threshold
Loss Reduction
Intolerance threshold
Radical Action
Devastation, no human survival

framework one step in both directions. We also need to study the geographic and demographic scales of cases when people did not survive and beneficial effects of volcanic eruptions (eg, when thin tephra layers contribute aeration and nutriments to soils and asphyxiate insect pests).

Disasters can also have creative effects on cultures, because people adjust to their changed circumstances by making modifications in their settlements, adaptations, oral histories, and religious beliefs and practices. This is somewhat analogous to some new thinking in biology in which scholars explore how natural disturbances can help maintain biodiversity and productivity (Lindenmayer *et al* 2004) and re-create structural complexity as well as landscape heterogeneity within biologic communities.

Volcanically induced disasters are not necessarily egalitarian, so even when the stresses were the same, the effects were differential. A good example is modern El Salvador, where the elite often live at higher elevations, with commoners in lower areas at greater risk from density flow phenomena such as pyroclastic flows, mudflows, and lava flows. In addition, today as in the Precolumbian past, the elite regularly had more substantial housing than commoners.

VOLCANOES AND PEOPLE IN THE ZAPOTITAN VALLEY

The Zapotitan valley of El Salvador (Figures 4.1 and 4.2) provides an excellent basis for a comparative examination because it has been affected by numerous explosive and effusive (eg, lava flows) volcanic eruptions during the past two millennia, although the natural and cultural records of the interplays among volcanic eruptions and societies are far from complete. These are significant islands of knowledge, but unfortunately they are surrounded by seas of ignorance that need sustained research.

Coatepeque

The Coatepeque eruptions emanated from the Santa Ana volcanic complex during the late Quaternary period. Recent volcanological research has dated these eruptions c 77,000 to c 55,000 years ago (Pullinger 1998; Rose *et al* 1999), but it is unlikely that people were affected by them.

Ilopango TBJ

The Coatepeque and Ilopango eruptions were physically similar, massive plinian eruptions of acidic tephra that were certainly devastating to the tropical flora and fauna of central and western El Salvador. They are

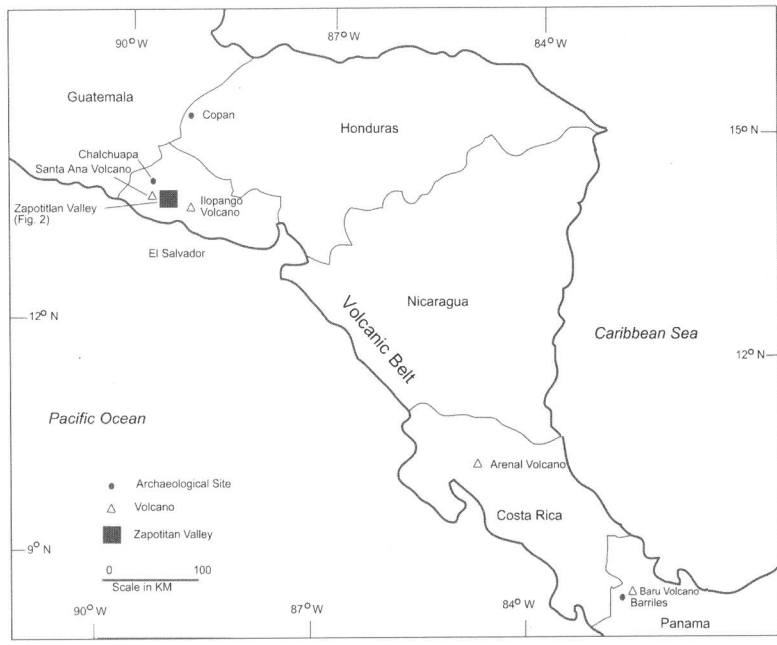

Figure 4.1 Central America showing the location of the Zapotitan valley.

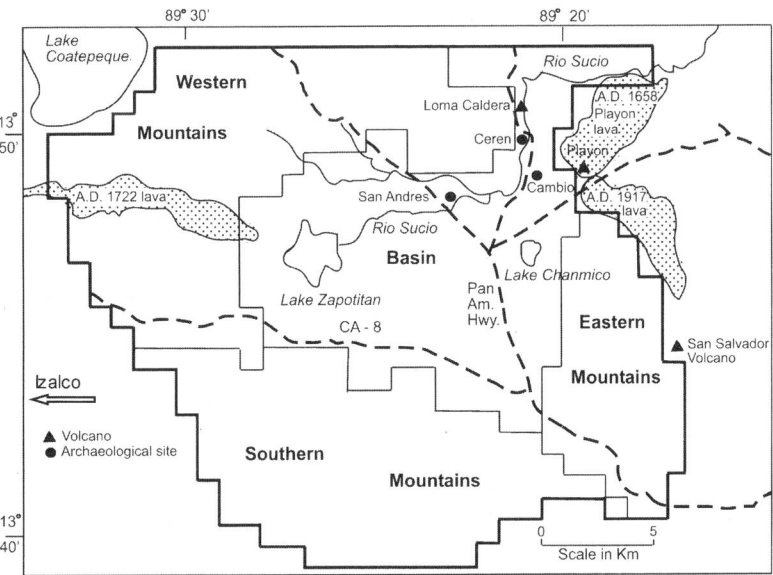

Figure 4.2 Map of the Zapotitan valley showing the four physiographic provinces (Basin, Western Mountains, Eastern Mountains, and Southern Mountains). The dashed lines are the principal highways. The thick irregular line is the limit of the 550 km^2 archaeological survey (Black 1983).

also similar in that each source had multiple eruptions. Rose *et al* (1999: 107, 109) describe four eruptions from Ilopango, called the TB4, TB3, TB2, and the TBJ (ranged in order from earliest to the most recent). The TB4 probably occurred prior to human habitation of the area, but the TB3 and TB2 eruptions may have affected people.

Porter (1955) excavated some three thousand potsherds, 47 obsidian artefacts, two metates, and charcoal in a paleosol buried by a thick deposit of white volcanic ash at Barranco Tovar in San Salvador. She noted the ash layer, which resembles the Ilopango TBJ tephra with an associated pyroclastic flows, was 10–20 m thick around San Salvador. She obtained the date of 1040 ± 360 BC in the early 1950s when radiocarbon dating was in its infancy. There are two obvious possibilities: (1) the site is buried by the TBJ and the date is considerably too early or

Figure 4.3 Ciudad Credisa site, a site buried by 6.7 m of Ilopango TBJ tephra in the San Salvador area. The site is 13 km from the source. Sherds were found at the base of this cut, in the paleosol (dark layer at bottom). Payson Sheets is cleaning the cut for photography and drawing.

(2) the date may be reasonable and thus the site is earlier and buried by tephra from the TB2 or 3 eruption.

Almost a century ago, the Salvadoran scholar Jorge Larde (1926) discovered various exposures in the San Salvador region where white volcanic ash had covered over ceramic and lithic artefacts. He reasoned an ancient eruption had buried archaeological sites. He showed the deposits and artefacts in what he called the 'C horizon' to Sam Lothrop, who described the ash and artefact exposures in considerable detail. Lothrop (1927) focused on the Cerro Zapote site, in south-eastern San Salvador. (The Ciudad Credisa site shown in Figure 4.3 is 3 km farther east, in the south-east corner of San Salvador.) He recognised the early character of the ceramics, dating to what is now called the Preclassic period. Lothrop also identified the artefacts stratigraphically superior to the volcanic ash layer as Maya in character, in what he called the 'A horizon,' thus indicating the cultural affiliation of the group that reoccupied the area. Later research has confirmed Lothrop's suspicion of source, as it is now known that the archaeological sites were buried by tephra from Ilopango, called the TBJ (*tierra blanca joven*, ie, young white earth) tephra (summarised in Sheets 1983a). Subsequent research demonstrated that most tephra deposits mentioned in these studies were from the same massive eruption: the TBJ tephra from Ilopango (Hart and Steen-McIntyre 1983).

The Ilopango TBJ eruption was dated to 260 ± 114 AD by multiple radiocarbon analyses (calibrated 1-sigma composite) conducted in the 1970s (Sheets 1983b), but these may be too early, based on recent efforts to obtain more precise results by careful sampling and AMS (accelerator mass spectrometry) dating. They indicate that the eruption may have occurred at 421 AD (429) 526 calibrated one sigma, or 408 AD (429) 536 calibrated two sigma (Dull *et al* 2001).

The TBJ eruption was one of the greatest Holocene eruptions in Central America and its ecological and societal impacts were felt throughout El Salvador and adjoining Guatemala and Honduras. The plinian and phreatomagmatic eruption resulted in the elimination of flora and fauna over tens of thousands of km^2. Although population density prior to the eruption is unknown, a conservative estimate is at least 30 people per km^2 (based on Black 1983). The eruption ejected a total volume of approximately 18 km^3 of dense rock equivalent (DRE) into the air, creating a zone of total devastation in an area of some 1,000 km^2 surrounding Ilopango where no people, animals, or plants would have survived (Dull *et al* 2001). The area was devastated by huge pyroclastic flows that uprooted and carbonised trees as much as 25 km away from the vent. The death toll in this area is estimated at 30,000 people (Dull *et al* 2001).

Beyond the area of total devastation is an area at least 10,000 km², with an estimated population of 300,000, where the ash blanket was more than 50 cm thick, termed the 'zone of depopulation' by Dull *et al* (2001). Based on data from Ceren (see below), all roofs would have been collapsed. People could have survived by taking protective measures, such as breathing through cotton cloth to filter out the fine tephra, but many must have perished from asphyxiation. Certainly, no survivors could have continued to inhabit an area where fresh water was contaminated and agricultural fields were deeply buried by the sterile white ash. Segerstrom (1950) found that volcanic ash depths greater than 10–25 cm from the recent eruption of Paricutin exceeded the coping abilities of traditional Mesoamerican maize agriculturalists. Survivors would have had to migrate to arable land outside the zone, thus crossing the intolerance threshold and adopting the radical action response mode (Burton *et al* 1978).

Dull (2004) investigated the paleoecological records of lake sediments in western El Salvador. He found evidence for intensive agriculture in the centuries before the eruption followed by ecological devastation. He saw no chance of human survival in the 100–200 km² area near the source (ie, the San Salvador area and eastern portion of the Zapotitan valley).

> It appears that the TBJ eruption was not merely a temporary setback or inconvenience for the people of western El Salvador. The bulk of the evidence (eg, settlement patterns, ceramic affinities, paleoecological records) suggests that the TBJ eruption forced a wholesale social and economic reconfiguration of western El Salvador and southeastern Guatemala. (Dull 2004: 242)

The Sierra de Apaneca had been occupied and farmed for three millennia, from 2,500 BC until the TBJ eruption, but the lake core indicates an absence of either occupation or agriculture in the succeeding centuries of the Precolumbian era. As proposed by Dull, the population pressures of the Preclassic period that led to occupation in highland mountainous areas probably did not occur in the late Classic and Postclassic periods. Another significant factor is the slower weathering rates of tephra and the slower vegetative recolonisation rates at higher elevations and thus slower vegetative and soil recoveries in the higher mountains.

The favoured loci for settlement in the Preclassic through the Postclassic periods were low-lying valleys, alluvial floodplains, and basins (Black 1983; Sheets 1984) because of easy access to freshwater, higher water table for natural vegetation and cultigens, and the best arable soils for agriculture. These were the zones most deleteriously impacted by the pyroclastic flows and lahars during the Ilopango TBJ eruption.

Generations of people had benefitted from living in the favoured areas, but their descendents paid the price when Ilopango erupted.

Areas beyond the zone of disruption would have dealt with the thinner ash levels by loss acceptance. Agricultural yields would have diminished under the tephra-induced stress, but maize and other culti- gens could have survived and people could still have dug through the ash blanket to plant crops. With weathering and incorporation of the tephra, soils could have improved within a few years of the eruption. Still farther from the eruption, in areas like the western Maya highlands, northern Honduras, and the southern Maya Lowlands, the thin dusting of TBJ tephra could have improved agricultural productivity in the first year of emplacement by providing a mulch layer and increasing soil porosity, adding nutrients, and perhaps suffocating some insect pests. As tephra weathered in the years following the eruption and became incorporated into soils by human and bioturbation, the zone of benefi- cial effects of the eruption would have moved steadily towards the source.

The Ilopango TBJ eruption was about the same scale as the 1st cen- tury AD eruption of Popocatépetl in central highland Mexico, also rated as VEI 6 (Plunket and Uruñuela 2006). Since large areas of the volcano's flanks had to be abandoned, Plunket and Uruñuela (2006) propose that the social dynamics of two of the largest settlements in central highland Mexico at Cholula and Teotihuacan benefitted from the arrival of refugees fleeing the eruption because they may have been coerced into the massive construction programmes.

Recovery from TBJ

As a general rule, recoveries from tephra-induced stresses occurred earlier at greater distances from Ilopango, because areas with only a few mm or cm of tephra would have recovered rapidly and benefitted from the dusting. Some areas such as the Apaneca mountains (Dull 2004) never recovered during Precolumbian times, probably because of elevation, topography, and perhaps lower population pressures after the eruption. The general source of the people who resettled the region is clear. As Lothrop (1927) suggested for central El Salvador, the re- occupation of the Zapotitan valley emanated from the north, judging from the close similarities in architecture and ceramics of the 6th century immigrants with those from the general Copan area of Honduras. The close cultural and economic relationships of western Salvador and highland Guatemala that existed prior to the eruption were perman- ently severed by the volcanic disaster and were never re-established. Demographic recovery from the Ilopango eruption occurred within about a century, but cultural recovery in the sense of resilience and re-establishment of the same pre-eruption cultural tradition never occurred (see Dull 2004). In the Preclassic, and barely into the Classic, indigenous cultural florescence of the 'Miraflores' cultural tradition in central and western El Salvador was permanently truncated (Dull *et al* 2001). Although the western component of the 'Miraflores' tradition continued in the Guatemalan highlands, the eastern component was devastated and never recovered in El Salvador.

The Ceren Site

The Ceren site provides detailed information about the timing and the nature of recovery from the Ilopango eruption. Located in the northern end of the Zapotitan valley (Sheets 1992), the site called 'Joya de Ceren' in El Salvador, was a thriving agricultural village on the left bank of the Río Sucio (Sheets 1992) (Figure 4.4). It was founded during the 6th century AD by immigrants to the region devastated by the Ilopango event and then occupied for perhaps a century before its burial by tephra from the Loma Caldera volcanic vent. The depth of the Ilopango tephra was measured as 1.2 m at the Cambio site, only 2 km south of Ceren (Hart and Steen-McIntyre 1983: 23), although erosion and compaction would have decreased its original thickness. This depth would place the Ceren area and most of the Zapotitan valley within the zone of complete devastation from the Ilopango eruption. Overnight, the eruption would have changed a tropical gallery forest environment interspersed with

Figure 4.4 The Ceren site. The well-weathered fertile Preclassic soil is the dark stratum below the white TBJ tephra from the Ilopango eruption. Structure one of Household one was built out of that Preclassic soil. The Loma Caldera tephra is about 5 m deep, entombing the structure and its environs.

swidden and intense agriculture and dense human populations into a white sterile desert. The acidic (sialic) nature of the Ilopango TBJ tephra would have slowed soil formation and vegetation recolonisation. The recovery of floral, faunal, and soils recovery would have preceded the return of maize-based agriculture. Perhaps a century of time following each eruption was needed for the natural processes of weathering, soil formation, vegetation recovery, and faunal reoccupations, a time estimate consistent with what has been found at Ceren.

Because Ceren is the earliest known community established in the Zapotitan valley after the Ilopango eruption, it provides insight into the nature of social recovery from Ilopango. Ceren was a village of commoners, with perhaps 100–150 inhabitants and no elite or attached occupational specialists (Sheets 2002). To date, four households have been excavated and only one of them in its entirety, so the sample is very small. In addition to the households, a civic centre of the village, consisting of a formal plaza surrounded at least on its south and west sides by solid earthen-walled public buildings, has been identified. A religious core of two special structures has also been found at the easternmost edge of the site, at the highest elevation within the site and on the bank overlooking the Río Sucio. The two buildings have a different orientation from the abovementioned 30° east of north for the household and public buildings. In addition, with their white painted walls with red hematite decoration, their unique construction of earthen columns, and the multiple floor levels from outside to the most elevated back room, they contrasted with all other structures at the site.

The ethnicity of the Ceren immigrants, and thus their home territory, is probably best explored by examining two categories of material culture: architecture and artefacts. Each Ceren household built at least three functionally differentiated buildings: a domicile, a storehouse, and a kitchen (Sheets 2002). This differentiation of household structure and use of built space is a Maya characteristic that existed in the Copan valley at the time of Ceren (Webster *et al* 1997). The Chorti Indians of the same area near Copan continued to construct functionally differentiated household structures up into the last century (Wisdom 1940). The artefacts with the highest ethnic sensitivity are the polychrome food and drink serving vessels that make up about 22% of each household's fired clay vessel inventory (Beaudry-Corbett 2002). The majority are of the Copador type, evidently made in one or more localities in the Copan area and imported into the Zapotitan valley.

The close association with these features with the Maya of Copan puts an ethnic, cultural, and perhaps a linguistic distance between the pre-TBJ eruption peoples and those who recolonised the Zapotitan valley and surrounding areas. The vibrancy of the pre Ilopango 'Miraflores cultural sphere' was never re-established in El Salvador. Demographic recovery of prime habitable areas in the valley and most of central and western El Salvador was complete within three centuries after the eruption, whereas other areas never regained the ancient population levels during Precolumbian times.

Loma Caldera

The Loma Caldera vent, only 600 m north of the Ceren village, opened under the Río Sucio and began erupting early one evening in August. Inference from the nature of the artefacts and their relations to particular activities enables reconstruction of the timing. Time of year is derived from vegetation maturity (Sheets 1992). Vegetation, including annuals such as maize and seasonally sensitive trees such as guayaba, identify the middle of the rainy season, most likely the month of August. Artefacts indicate a meal had just been served before the eruption, but the dishes were not yet washed, as evidenced by the finger swipes of food left in hemispherical serving bowls. In tropical climates, dishes generally are washed soon after a meal is completed to avoid attracting noxious insects, so I believe the eruption occurred very soon after the meal was served. Because dried and stored maize had been placed in soaking vessels to soften overnight, agricultural tools were back from the fields, and pots were removed from the three-stone hearths, it evidently was the evening meal. Following this line of reasoning would mean the Loma Caldera eruption began about 6–7 pm, if Cerenians ate dinner about when traditional Salvadoran *campesinos* do now. Ironically, the dating by year is more challenging than dating to month and time of day. Calibrated radiocarbon dates, when averaged, yielded a composite 2-sigma calibrated range from 610 to 671 AD (McKee 2002).

In volcanological terms, the eruption from the nearby Loma Caldera vent was miniscule in comparison to the Ilopango TBJ eruption, as it devastated only a few km^2 (Miller 1992). People beyond the small devastated zone, between 10 and 20 km^2 in area, would have crossed the intolerance threshold and abandoned their homes to take up farming elsewhere. The estimated 165–440 people per km^2 in this very fertile basin province (Black 1983) would indicate some 1,500–9,000 displaced people. When compared to the Ilopango eruption, natural recovery from the Loma Caldera eruption was faster, because the tephra was more mafic and therefore weathered more rapidly. Also, the source areas for recolonising flora and fauna were no farther than a few km from any point on the tephra blanket. Human demographic and cultural recoveries probably only required a few decades.

The Cambio site, 2 km south of Ceren, provides a more complete stratigraphic record of the physical and cultural processes following the Loma Caldera eruption (Chandler 1983). Cambio (Figure 4.5) received a thin layer of the Loma Caldera tephra, measured by Miller (2002) at 10 cm. Weathering, compaction, and human disturbance likely had thinned it in the decades after its emplacement. Based on our inspection of the stratigraphy, human reoccupation of Cambio occurred soon after the Loma Caldera eruption, following the formation of a juvenile 'A' horizon soil. This process probably took a few decades at most, and perhaps only a few years.

Beaudry-Corbett (2002) found that the 'before and after' ceramic assemblages were identical, except that a new kind of polychrome pottery called 'Arambala' is present in small amounts after the Loma Caldera eruption at Cambio but was absent at Ceren. Other than that small addition to the ceramic inventory, the material culture is indistinguishable before Loma Caldera erupted as seen at Ceren.

Boqueron

During the Late Classic period, the San Salvador volcano ('Boqueron') blasted a fine wet tephra through a large crater lake in a phreato-magmatic eruption that coated the Zapotitan valley (Hart 1983). The

Figure 4.5 Stratigraphy at the Cambio site. Tom Peebles provides scale. Note partial vessel at his feet, from Preclassic occupation. The Ilopango TBJ tephra overlies the well-developed Preclassic paleosol, and has a juvenile paleosol. The Loma Caldera ('LC') tephra overlies the TBJ and has a moderately developed paleosol. The San Andres talpetate tuff ('SA') and a juvenile paleosol underly a building foundation dating to about 1,000 years ago. The Playon tephra dates to 1658–1659 AD.

tephra dried into a tough compact unit. The tephra is 6 m thick atop Boqueron (which means 'big mouth'). Hart (1983) named the tephra the 'San Andres Talpetate Tuff' after the site where it had been noted by archaeologists. The tephra was thick enough on the eastern side of the valley to eliminate flora, fauna, and people, but it thinned rapidly on the western side. The depth of the tuff at over 30 cm thick at the San Andres site (Hart 1983), probably explains why archaeologists in the 1990s found evidence of a hiatus in occupation following the tephra emplacement (Chris Begley, personal communication).

Sofield (2004) reports that the eruption began with a pyroclastic flow that extended some 3 km north-west of the crater followed by the main eruption that covered about 293 km² with 10 cm or more of indurated tephra, a more accurate estimate than Hart (1983). Sofield calculated the total eruptive volume at almost 0.5 km³. Contrary to statements in Sofield (2004), the eruption has not yet been dated.

The tephra was over 20 cm thick at Cambio (Chandler 1983), and would have had almost as long a deleterious effect there as it did at San Andres. Because this hardened tephra is so compact, it would have been more damaging to traditional agriculture than would a loose, dry air-fall tephra. I suggest that c 10 cm of this hard tephra was beyond the maximum with which traditional maize-based agriculture could cope (see Segerstrom 1950). This eruption, therefore, caused agricultural difficulties that exceeded the intolerance threshold all over San Salvador volcano including the eastern and much of the central part of the Zapotitan valley, which therefore must have been abandoned for a few decades. Vegetation recolonisation, soil formation, and faunal recoveries probably came gradually and steadily, mainly from the north-west, where the tephra blanket was thinner.

We can generate a rough estimate of the number of people forced into radical action when the intolerance threshold was crossed and out-migration became necessary. Black's (1983: 82) Zapotitan valley survey estimated the overall regional population densities in the Late Classic to be between 70 and 180 people per km². Using Sofield's (2004) calculation for the area affected establishes a range of 20,510–52,740 people who would have had to migrate. In comparison, Boqueron caused a much greater disaster than the Loma Caldera eruption, but was an order of magnitude smaller than the Ilopango TBJ event.

Unfortunately, dating the eruption, studying its immediate effects, and documenting recovery processes is hindered because virtually all of the archaeological data recovered at San Andres remain unpublished. The best dating is provided by the stratigraphy at the Cambio site, where the tuff is sandwiched between two Late Classic occupation layers that postdate Ceren and the Loma Caldera tephra (Chandler 1983). On this basis, the eruption most likely dates to the 9th century AD or shortly thereafter (contra Schofield's date of 800 BP).

Playon

On 3 November 1658, a millennium after the Loma Caldera eruption, the area was rocked by major earthquakes that killed many people and destroyed settlements including Quetzaltepeque and San Salvador. The Playon eruption (Figure 4.6) began in earnest the next day as volcanic ash spewed from the vent and an andesitic lava began flowing downhill to the north (Meyer-Abich 1956). According to colonial documents, both lava and tephra caused great destruction to agriculture, the local indigo industry, and livestock (Gallardo 1997). The lava came within 2 km of the town of Quetzaltepeque and largely encircled and partially buried the town of Nexapa, clearly exceeding their intolerance threshold and forcing people to relocate their settlement to its present location at Nejapa. The lava covered at least 8 km^2 (Hart 1983). At its northernmost extent, the lava blocked the Río Sucio (formerly known as the Río Nixapa, or 'river of ash' in Nahuat) and formed a sizeable lake in the centre of the Zapotitan valley.

Browning (1971: 100–04) describes the struggles of the displaced Pipil Indian community of Nexapa, which occupied the bottom of the colonial social hierarchy, when their common lands were largely buried under the lava. They moved eastwards, to what is now Nejapa and

Figure 4.6 Aerial view from helicopter above San Salvador volcano, looking north. The black lava flow from the 1917 eruption of 'Boqueron' is in the foreground, flowing downhill to the north-west (left) and towards the north-east. The 'P' marks Playon volcano, and the '1658' is placed mid-distance on the lava flow that started in that year, and flowed north to block the Río Sucio and form a lake for a few months. 'C' marks the Ceren site, with the Loma Caldera vent immediately to the right.

petitioned the colonial government for land owned by The Crown to be changed for use to sustain them. Their struggles for land continued for almost 80 years, until 1736 when they were finally given legal title to the land on which the town stood and a narrow strip of common land up the slope of San Salvador volcano. Even at this early historic time land was not freely available and struggles for ownership were particularly difficult for disenfranchised disaster victims.

San Marcelino

Pullinger (1998: 78) described the 1722 lava flow that emerged from the eastern side of the Santa Ana volcanic complex. The flow headed eastwards for 11 km from two vents situated at 1,325 m, destroying the traditional village of San Juan Tecpan and approximately 15 km^2 of prime agricultural land used for subsistence crops and cacao. The burial of residences and agricultural land exceeded the intolerance threshold, and people were forced to evacuate. I have not been able to find more information about the repercussions of this eruption and I suspect that colonial documents contain information about impact, refugees, and legal struggles over replacement lands, perhaps analogous to Nejapa residents after the Playon eruption. At present, the soils and vegetation on top of this lava flow are patchy and sparse, similar to those on top of the Playon lava. Centuries of weathering and plant succession are still needed before this area regains its productivity.

Izalco Eruptions

From the 18th century into the 20th, continuous eruptions from Izalco, El Salvador, were visible well into the Pacific Ocean, earning it the title 'Lighthouse of the Pacific'. Meyer-Abich (1956) conducted the earliest serious geological study of Izalco, and Pullinger (1998: 79–81) recently surveyed current knowledge. In 1779, the volcano started building itself from a vent at 1,300 m on the southern flank of Santa Ana volcano. Eruptions ceased in 1966 when the crater rim reached 1,952 m.

From the two centuries of eruptions, lava flows from Izalco were in the 5–8 km long range. Ash falls affected people and vegetation within a few km of the volcano on and just beyond the southeastern edge of the Zapotitan valley, exceeding many communities' intolerance thresholds. A pyroclastic flow buried the traditional village of Matazano in 1926, killing 56 people. The net decline in the agricultural production of the highly productive, gentle, lower southern slope of the Santa Ana complex because of the volcano had, and continues to have, considerable implications for a portion of this crowded nation.

San Salvador

The large crater of San Salvador volcano contained a crater lake in the early 20th century. However, in June 1917, the lake boiled away in the same week that an earthquake did considerable damage to cities on all sides of the volcano, and lava began to flow from the volcano's north side (Meyer-Abich 1956). Sofield (2004: 152) estimates the earthquake at about 5.6 on the Richter scale and he notes the lava emanated from three vents. The lava flows destroyed highly productive coffee farms. The black basaltic lava covered about 16 km², which is still devoid of vegetation other than some lichens and occasional grasses. The principal use of the lava flow today is to filter rainwater that supplies San Salvador. A grim use of the area developed in the civil war from 1979 through 1992 as paramilitary squads used the lava flow as their favoured place to dispose of bodies.

Additional Events

The record of volcanic eruptions affecting the environment and residents of the Zapotitan valley during the last two thousand years seems impressive but it is very incomplete. Some sizeable lava flows are not included here for lack of documentation, such as that near Cerro Chino, which covers over 10 km². Miller (2002) noted a few eruptions in the past two to three thousand years that affected the northern Zapotitan valley, but these are not yet well studied. Sofield (2004) has documented at least 42 volcanic events at San Salvador volcano in the past 40,000 years and at least 13 monogenetic magmatic eruptions during the past 1,800 years, but very few have been investigated archaeologically. This is a humbling window on how much more archaeological research is required to adequately document the dynamics of human-volcanic interactions of just one volcano in Mesoamerica.

CULTURAL COMPLEXITY AND WARFARE

Human societies vary considerably in their vulnerability to disasters because of differences in hazard perception or their ability to cope with a sudden massive stress, make adjustments, and attempt to recover. A study of ancient Mexican–Central American cases of explosive volcanism found that 'simple' egalitarian villages were considerably more resilient to explosive volcanism than their more complex chiefdom and state neighbours (Sheets 1999). The striking resilience of egalitarian ancient Costa Rican societies to some 10 explosive eruptions of Arenal volcano (Melson 1994) is attributed to low population densities, availability of refuge areas, low dependence on agriculture, reliance on a wide variety of wild food sources in the rich tropical rain-forest for a majority of the diet, lack of hostility among groups, and minimal investment in the 'built environment' of facilities constructed for living, storage, workshops, or agriculture. Elaborate burial rites involving multiple communities could have facilitated exchanges and alliances that were useful when refuge areas were needed. In contrast, complex state-level societies in ancient Mesoamerica evidently were more vulnerable to sudden massive stress because of their greater reliance on the 'built environment' as well as high population dens-ities, scarcity of refuge areas, reliance on intensive agriculture, and domesticated dietary staples.

A second case highlights the importance of understanding cultural factors in the impacts of disasters (Sheets 1999). Volcán Barú in Panama (Figure 4.1) erupted, probably in the 7th century AD, and deposited a relatively thin tephra layer, perhaps some 10–15 cm on settlements in the Barriles area (Linares and Ranere 1980), where the society consisted of a series of culturally similar but politically independent polities, prob-ably centralised chiefdoms (contra Hoopes 1996) packed into the flood-plain of the Río Chiriqui Viejo.

The tephra depth from the Volcán Barú eruption is comparable to tephra depths deposited on egalitarian villages towards the western end of Lake Arenal, Costa Rica, where egalitarian communities recovered rapidly, probably within a few decades, with no detectable changes in their architecture, artefacts, adaptation, or any other measure that could be found. In contrast, the stress from the thin tephra blanket from Volcán Barú forced a drastic cultural adjustment. The Barriles chief-doms, located in a highland tropical moist seasonal environment in the upper reaches of the river, abandoned their territories, never to return. They migrated over the continental divide to an environment with much greater precipitation and had to change their architecture and agriculture to adapt to the changed conditions (Linares and Ranere 1980: 244–45).

These examples illustrate that physical factors such as tephra thick-ness, chemistry, natural environment, flora, or fauna do not by them-selves explain the dramatic differences in human reactions to a volcanic disaster because cultural factors are relevant. The Barriles societies relied on more intensive agriculture than did Arenal societies and they built more elaborate architecture, especially in the chief's villages. In my view, the much higher population density, intensive agriculture, and greater built environment evidently put the Barriles societies at greater risk to sudden stresses than at Arenal, but one other factor should be considered.

Although the Barriles groups shared a common culture, the polities were competitive and often at war with each other. Evidence for hos-tility is recorded in sculptures with severed heads (Linares and Ranere

1980: 50) and special axes made for decapitations. The hostile political landscape obviated tephra-stressed people taking refuge in nearby areas, in contrast to impacted Arenal societies, and they were forced to emigrate when faced with a natural disaster.

CONCLUSIONS, CONSIDERATIONS, AND IMPLICATIONS

To compare eruptions and human responses, I propose an overarching model called 'scalar vulnerabilities'. When the magnitudes of eruptions, societal complexity, and intensities of agriculture are considered, some tentative patterns begin to emerge. As the scale of eruptions, hostility, elaborateness of redistributive economies, or the built environment increase, so do the vulnerabilities of the societies. For the extreme eruptions, VEI 6 or greater, no pre-industrial human societies survive unscathed (eg, TBJ, Popocatépetl). At considerably lesser orders of magnitude, radical action in terms of out-migration may be necessitated, but both egalitarian and complex societies can recover (Arenal, Late Classic Boqueron). Relatively small eruptions can have deleterious effects on the settlements directly affected, but they do not have long-lasting societal repercussions (Loma Caldera, historic San Salvador). In general, in the cases considered here, egalitarian societies with low levels of built environments, minimal reliance on intensive agriculture and domesticated staples, and low population densities exhibited the greatest resilience to sudden massive volcanic stresses.

Along the scale of societal complexity, at least in the small sample described here, the more complex societies were the most vulnerable to volcanically induced stresses. Apart from social complexity, the scale of political hostility was shown in the Bariles case to dramatically affect a society's vulnerability. A small perturbation among societies with chronic warfare can require more extreme coping strategies than would be predicted by the other physical science and societal complexity scales.

The predictability or unpredictability of changes in a society's environment can have different implications. An important cross-cultural study found that unpredictable sudden stresses were much more difficult for non-Western societies to cope with than were long-term stresses such as sustained droughts (Ember and Ember 1992). Unpredictable stresses or disasters led to warfare to obtain resources from others more often than chronic stresses and more predictable disasters. Viewed in this light, the Barriles chiefdoms had engineered a high level of predictable stress in the form of chronic hostility and this left them vulnerable to the relatively small but unpredictable stress cause by the eruption of Volcán Barú.

In my review of volcanic disasters, I have found that various components of volcanic eruptions affected human populations in quite different ways. Lava flows covering a few km^2 eliminated agriculture for many centuries and caused resettlements. Tephra deposits affected a few km^2 to thousands of km^2, but only the really great tephra deposits such as the Ilopango TBJ fundamentally altered the cultural trajectory of ancient societies. The medium- to smaller size tephra aprons caused short-term difficulties, but weathering, plant succession, and human recovery were relatively rapid and no long-term cultural effects resulted. The biggest eruptions created such vast stresses in all domains of ecology, economics, politics, and society that recoveries of the original social orders simply did not occur. The demographic and cultural reoccupations following the mega-disasters were by different societies.

Tropical botanists are considering the 'intermediate disturbance hypothesis', which proposes that intermediate scales of disturbance promote maximum species diversity (Molino and Sabatier 2001). An analogue with cultures could be proposed: small- to medium-scale disasters can have creative aspects in addition to their much-heralded (and over-reported) destructive aspects. Some disasters are so immense that they overwhelm all cultures, but medium-scale disasters could promote cultural diversity, redundancy of social units, and societal resilience.

Volcanic/human disasters are not egalitarian because they do not affect all components of societies uniformly. It is my personal observation that in complex societies, freedom of choice to avoid perceived hazards is not uniform. Members of the upper class, with greater wealth and power, have more leeway as to where to locate their residences and other facilities than do the poor or politically disenfranchised.

The population of what is now El Salvador has changed dramatically (Sheets 1992: 125). In Precolumbian times, the population occupying this area ranged c 250,000–750,000, with regional population densities ranging 10–50 per km^2. After the decimation from imported Old World diseases in the 16th and 17th centuries, the population recovered back to its Precolumbian peak of 750,000 in 1900. The population of El Salvador has burgeoned at the end of the 20th century to more than 6,000,000 people, with a density of about 300 per km^2 and is continuing to surge upwards. Consequently, more people are forced to live and work in harm's way, where volcanic and other disasters will take progressively greater tolls of death and destruction. Currently, many of the urban and rural poor are crowded into especially hazardous locations, such as lowlying *barrancas* that are particularly vulnerable to floods, lahars, and pyroclastic flows.

Current disaster research is clearly multidisciplinary, but very few projects are interdisciplinary. Because different disciplines vary considerably in their assumptions and theories about human-environmental interaction and individual disciplines change over the decades, comparative studies are often difficult. Not only do different disciplines view hazards and disasters in disparate ways, so, too, do different societies (see Chapters 9–13). Future research needs a better integration of the physical and social sciences to better understand the complex interactions encapsulated in the case studies that I have presented here.

ACKNOWLEDGMENTS

I want to express my appreciation to my colleagues who critiqued and increased the social science aspects of an early version of this chapter: E James Dixon, Cathy Cameron, Linda Cordell, Steve Lekson, Art Joyce, and Frank Eddy. Dr Dixon's additional critique is greatly appreciated. Carlos Pullinger was very helpful in volcanological and dating details of the Coatepeque eruptions. Brian McKee gave the manuscript a very careful copyediting, for which I am most grateful. Comments by Dan Miller and John Hoopes were helpful. I appreciate Chris Begley's observations on his ongoing research at the San Andres site. I am very appreciative of the kind invitation by the editors to contribute and their identification of weak, redundant, and irrelevant sections that improved the chapter.

REFERENCES

Beaudry-Corbett, M (2002) 'Ceramics and their use at Ceren', in Sheets, P (ed), *Before the Volcano Erupted: The Ancient Ceren Village in Central America*, pp 117–38, Austin: University of Texas Press

Black, K (1983) 'The Zapotitan valley archaeological survey', in Sheets, P (ed), *Archeology and Volcanism in Central America: The Zapotitan Valley of El Salvador*, pp 62–97, Austin: University of Texas Press

Browning, D (1971) *El Salvador: Landscape and Society*, Oxford: Clarendon Press

Burton, I, Kates, R and White, G (1978) *The Environment as Hazard*, New York: Oxford University Press

Chandler, S (1983) 'Excavations at the Cambio site', in Sheets, P (ed), *Archeology and Volcanism in Central America: The Zapotitan Valley of El Salvador*, pp 98–118, Austin: University of Texas Press

Dull, R (2004) 'Lessons from the mud, lessons from the Maya: paleoecological records of the Tierra Blanca Joven eruption' in Rose, W, Bommer, J, Lopez, D, Carr, M and Major, J (eds), *Natural Hazards in El Salvador*, pp 237–44, Geological Society of America Special Paper 375, Boulder, CO: Geological Society of America.

Dull, R, Southon, J, and Sheets, P (2001) 'Volcanism, ecology, and culture: a reassessment of the Volcan Ilopango tbj eruption in the southern Maya realm', *Latin American Antiquity* 12, 25–44

Ember, C and Ember, M (1992) 'Resource unpredictability, mistrust, and war' *Journal of Conflict Resolution* 36, 242–62

Gallardo, R (1997) *El Obraje de Añil de San Andres*, Mexico City: Litográfica Turmex

Hart, W (1983) 'Classic to Postclassic tephra layers exposed in archaeological sites, eastern Zapotitan valley', in Sheets, P (ed), *Archaeology and Volcanism in Central America: The Zapotitan Valley of El Salvador*, pp 44–51, Austin: University of Texas Press

Hart, W and Steen-McIntyre, V (1983) 'Tierra Blanca Joven tephra from the AD 260 eruption of Ilopango caldera', in Sheets, P (ed), *Archeology and Volcanism in Central America: The Zapotitan Valley of El Salvador*, pp 13–34, Austin: University of Texas Press

Hoopes, J (1996) 'Settlement, subsistence, and the origins of social complexity in Greater Chiriquí: a reappraisal of the Aguas Buenas tradition', in Lange, F (ed), *Paths to Central American Prehistory*, pp 15–48, Niwot: University Press of Colorado

Larde, J (1926) 'Arqueologia Cuzcatleca' *Revista de Ethnologia, Arqueologia, y Lingüística* (San Salvador, El Salvador) 1, 3–4

Linares, O and Ranere, A (eds) (1980) *Adaptive Radiations in Prehistoric Panama*, Cambridge, MA: Peabody Museum, Harvard University

Lindenmayer, D, Foster, D, Franklin, J, Huner, M, Noss, R, Schmiegelow, F and Perry, D (2004) 'Salvage harvesting policies after natural disturbance', *Science* 303, 1303

Lothrop, S (1927) 'Pottery types and their sequence in El Salvador', *Indian Notes and Monographs* 1 (4), 165–220

McKee, B (2002) 'Household 2 at Ceren: the remains of an agrarian and craft-oriented corporate group', in Sheets P (ed), *Before the Volcano Erupted: The Ancient Ceren Village in Central America*, pp 57–71, Austin: University of Texas Press

Melson, W (1994) 'The eruption of 1968 and tephra stratigraphy of Arenal Volcano', in Sheets, P and McKee, B (eds), *Archaeology, Volcanism, and Remote Sensing in the Arenal Region, Costa Rica*, pp 24–47, Austin: University of Texas Press

Meyer-Abich, H (1956) 'Los volcanes activos de Guatemala y El Salvador', *Anales del Servicio Geologico Nacional de El Salvador* 3, 1–102

Miller, D (1992) 'Summary of 1992 geological investigations at Joya de Ceren', in Sheets, P and Kievit, K (eds), *1992 Investigations at the Ceren Site, El Salvador: A Preliminary Report*, pp 5–9, Boulder: Department of Anthropology, University of Colorado

Miller, D (2002) 'Volcanology, stratigraphy, and effects on structures', in Sheets, P (ed), *Before the Volcano Erupted: The Ancient Ceren Village in Central America*, pp 11–23, Austin: University of Texas Press

Molino, J and Sabatier, D (2001) 'Tree diversity in tropical rain forests: a validation of the intermediate disturbance hypothesis', *Science* 294, 1702–04

Newhall, C and Self, S (1982) 'The volcanic explosivity index (VEI): an estimate of explosive magnitude for historical volcanism', *Journal of Geophysical Research* 87, 1231–38

Plunket, P and Uruñuela, G (2006) 'Social and cultural consequences of a late Holocene eruption of Popocatépetl in central Mexico', *Quaternary International* 151, 19–28

Porter, M (1955) 'Material preclásico de San Salvador', *Communicaciones del Instituto Tropical de Investigaciones Científicas* 3/4, 105–12

Pullinger, C (1998) 'Evolution of the Santa Ana volcanic complex, El Salvador', unpublished MA thesis, Department of Geology, Houghton, Michigan Technological University

Rose, W, Conway, F, Pullinger, C, Deino, A and McIntosh, W (1999) 'A more precise age framework for late Quaternary silicic eruptions in northern Central America', *Bulletin of Volcanology* 61, 106–20

Segerstrom, K (1950) 'Erosion studies at Paricutin', *US Geological Survey Bulletin 965A*, Washington DC: US Geological Survey

Sheets, P (1980) 'Archaeological studies of disaster: their range and value', Working Paper 38, Boulder: Hazards Center: Institute of Behavioral Science, University of Colorado

Sheets P (1983a) 'Summary and conclusions', in Sheets, P (ed), *Archeology and Volcanism in Central America: The Zapotitan Valley of El Salvador*, pp 275–94, Austin: University of Texas Press

Sheets, P (1983b) 'Introduction', in Sheets, P (ed), *Archeology and Volcanism in Central America: The Zapotitan Valley of El Salvador*, pp 1–13, Austin: University of Texas Press

Sheets, P (1984) 'The prehistory of El Salvador: an interpretive summary', in Lange, F and Stone, D (eds), *The Archaeology of Lower Central America*, pp 85–112, Albuquerque: University of New Mexico Press

Sheets, P (1992) *The Ceren Site: A Prehistoric Village Buried by Volcanic Ash in Central America*, Fort Worth, TX: Harcourt Brace

Sheets, P (1999) 'The effects of explosive volcanism on ancient egalitarian, ranked, and stratified societies in Middle America', in Oliver-Smith, A and Hoffman, S (eds), *The Angry Earth: Disaster in Anthropological Perspective*, pp 36–58, New York: Routledge

Sheets P (ed) (2002) *Before the Volcano Erupted: The Ancient Ceren Village in Central America*, Austin: University of Texas Press

Sofield, D (2004) 'Eruptive history and volcanic hazards of Volcan San Salvador', in Rose, WJ, Bommer, J, Lopez, D, Carr, M and Major, J (eds), *Natural Hazards in El Salvador*, Geological Society of America Special Paper 375, pp 147–58, Boulder, Colorado: Geological Society of America

Torrence, R and Grattan, J (2002) 'The archaeology of disasters: past and future trends', in Torrence, R and Grattan, J (eds), *Natural Disasters and Cultural Change*, pp 1–18, Routledge: London

Webster, D, Gonlin, N and Sheets, P (1997) 'Copan and Ceren: two perspectives on ancient Mesoamerican households', *Ancient Mesoamerica* 8, 43–61

Wisdom, C (1940) *The Chorti Indians of Guatemala*, Chicago: University of Chicago Press

CHAPTER 5

Paleoindians and Megafaunal Extinction in the Basin of Mexico: The Role of the 10.5 K Upper Toluca Pumice Eruption

Silvia Gonzalez and David Huddart

INTRODUCTION

With its fertile soils and plentiful water resources, the Basin of Mexio has been an attractive place for animals and humans alike since the end of the Pleistocene. In the Late Pleistocene and the Early Holocene, this region experienced major environmental change associated with volcanic processes. These included major Plinian eruptions from the Tlaloc/ Telapon, Popocatépetl, and the Nevado de Toluca volcanoes, which produced pyroclastic flows, lahars, and ash falls. Superimposed on these volcanic changes were the effects of climatic change, including phases of colder, drier, and wetter climates than today. Together, the volcanic and climatic changes affected the development of geomor-phological processes and landforms, vegetation, and soil processes (Huddart and Gonzalez 2006).

The area around the lake is particularly rich in Late Pleistocene sedi-ments that contain extinct megafauna including mammoths, sabre-toothed cats, bisons, camels, horses, and glyptodons. There are at least 30 reported localities with mammoth bones (*Mammuthus columbi*), with about half of the sites showing evidence of human presence (eg, cut marks, worked bone, or Lerma-type lithic points) (Armenta-Camacho 1959; Arroyo-Cabrales *et al* 2006; Aveleyra de Anda 1955; Aveleyra de Anda and Maldonado-Koerdell 1952a, 1952b; Gonzalez, Morett Alatorre *et al* 2006; Lorenzo and Mirambell 1986). In these prehistoric sites, the majority of the deposits consist of lake sediments, volcanic ashes, and derived volcanic mudflows (lahars).

Our geoarchaeological studies at several important Paleoindian sites (eg, Tepexpan, Tlapacoya, Tocuila, and Santa Isabel Iztapan I and II) (Figure 5.1) indicate that during the Late Pleistocene several major

Plinian type volcanic eruptions produced three important tephra marker horizons across the basin: (1) the Great Basaltic Ash (GBA) at 29,000 BP; (2) the Pumice with Andesite (PWA) at 14,450 BP from the Popocatépetl volcano, and (3) the Upper Toluca Pumice (UTP) at 10,500 BP from the Nevado de Toluca volcano. (We use uncalibrated radiocarbon dates in this chapter.) The earliest evidence for human presence in the Basin of Mexico are lithic tools and hearths at the Tlapacoya site dated to 21,700·± 500 and 24,040 ± 320 BP (García-Bárcena 1986; Mirambell 1973). However, the oldest directly radiocarbon dated humans in the area are Peñon Woman III found in a small island in the middle of Lake Texcoco (Figure 5.1) with a C^{14} date of 10,755 ± 55 and Tlapacoya Man dated to 10,200 ± 65 BP (Gonzalez *et al* 2003; Gonzalez, Jiménez López *et al* 2006).

In this chapter, we discuss the impact of the Plinian eruption that produced the UTP, or Tripartite Pumice, one of the main tephra markers produced by the Nevado de Toluca volcano on Paleoindian and megafaunal communities in the Basin of Mexico at around 10,500 BP.

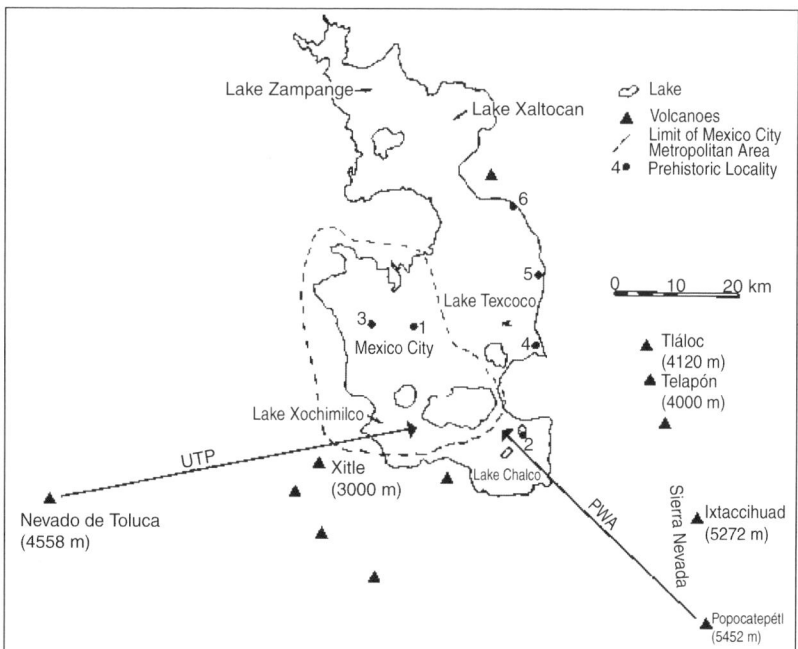

Figure 5.1 Location of Paleoindian main sites around the Basin of Mexico discussed in this chapter: (a) Peñon Woman III, (b) Tlapacoya Man, (c) Metro Man Skull, (d) Chimalhuacan Man, (e) Tocuila Mammoths, (f) Tepexpan Man. UTP = Upper Toluca Pumice, PWA = Pumice with Andesite. The arrows indicate the main sources and directions followed by the ash falls.

Paleoindian remains are found in association with this particular tephra layer (Metro Balderas Man, Chimalhuacan Man, and Tlapacoya Man). After the ash fall, a series of lahar events incorporated many megafaunal bones in catastrophic events that produced widespread 'bone beds'. After this series, the Pleistocene megafauna are gone from the Basin of Mexico, indicating that this particular volcanic event was a very important factor in their extinction in central Mexico. Evidence for humans also disappears in this region for nearly four thousand years. To understand the relationship between volcanic activity, megafauna, and humans, we present new data from several Paleoindian sites directly associated with the presence of the UTP: (1) human remains from the Metro Balderas, Chimalhuacan, and Tlapacoya sites and (2) the most spectacular example of mammoth remains embedded in lahars from the UTP at Tocuila, Texcoco, Mexico State (Figure 5.1).

THE UPPER TOLUCA ERUPTION

The Nevado de Toluca volcano, 4,680 m above sea level, is the fourth-highest peak in Mexico. It is 22 km from the city of Toluca and 80 km from Mexico City. The deposits of Nevado de Toluca have been studied in detail since the 1970s (Bloomfield and Valastro 1974, 1977; Bloomfield *et al* 1977; Cantagrel *et al* 1981). These authors agree in assigning a Late Pleistocene age for the volcano, with three large volcanic eruptions recognised: a Vulcanian eruption that occurred c 28,000 BP and two Plinian-type eruptions that resulted in the emplacement of the Lower Toluca Pumice at 24,000 BP and the UTP (c 10,500 BP). An initial tephrochronological framework for the Toluca Basin has been established by Newton and Metcalfe (1999), which includes the major element glass geochemistry of the UTP. This allows direct correlation with our work on this tephra in the Basin of Mexico.

Stratigraphic studies of the volcano have found a complex volcanic story of construction and destruction of central dacitic domes and sector collapses that started before 50,000 BP, with the latest eruptive activity occuring at 3,300 BP (Arce 1999; Arce *et al* 2003; Arce *et al* 2005; Capra and Macías 2000; Macías and Arce 1997; Macías *et al* 1997). The volcano is currently dormant, but its past volcanic history suggests that a Plinian eruption scenario might be possible in the near future. Since such an event would put 30 million people at risk in the modern metropolises of Toluca and Mexico City, it is important to try to understand in detail the effects on the paleoenvironment and ecosystems from past Plinian eruptions associated with Nevado de Toluca.

The UTP was produced by the largest Plinian eruption of the Nevado de Toluca volcano. The ash fall, found up to 110 km to the north-east of the volcano covered a minimum area of 2,000 km² (Figure 5.1). This, together with the total volume (dense rock equivalent) of erupted magma of 14 km³ (Arce, Macías *et al* 2003), rank this volcanic event as one of the largest recorded in central Mexico during the Late Pleistocene. Recent detailed stratigraphic studies by Arce *et al* (2003) found charcoal embedded in the UTP with an age of 10,455 ± 95 BP. We know that humans were already present in the Basin of Mexico at the time of the UTP eruption, as the oldest directly radiocarbon dated Paleoindian so far is the semi-complete skeleton of Peñon Woman III, with an age of 10,755 BP (Gonzalez, Jiménez López *et al* 2006; Gonzalez *et al* 2003). Two Paleoindian remains found in association with the UTP deposits – Metro Balderas Man and Chimalhuacan Man – cannot be dated because of their high degree of mineralisation, but they may be the first reported human victims associated with volcanic activity in the Basin of Mexico.

HUMAN REMAINS

Metro Balderas Man

The complete skull of Metro Balderas Man (Figure 5.2a) was found in 1970 at a depth of 3.10 m during construction for the Metro Balderas station in the centre of Mexico City. It was reported to be embedded in the UTP, or Tripartite Ash (Mooser 1967). Volcanic ash samples taken from the inside of the skull were studied at the Electron Microprobe Unit of the University of Edinburgh, using their standard protocol. The analysis gave values of SiO_2 from 63.1 to 71.8% that are similar to the UTP 1 (see Newton and Metcalfe 1999). The association with the tephra has been the only way of indirectly dating the skull as collagen has not been preserved.

Chimalhuacan Man

An almost complete male skeleton, with an age at death estimated between 30 and 35 years, was found by chance in 1984 in Colonia

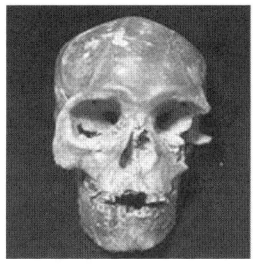 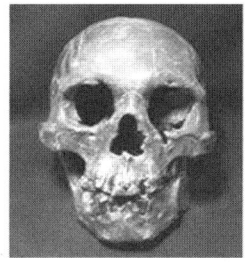 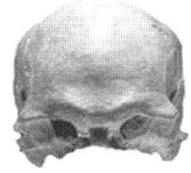

Figure 5.2 Paleoindian skulls found associated with the Upper Toluca Pumice: (a) Metro Balderas Man skull; (b) Chimalhuacan Man skull; (c) Tlapacoya Man calvarium.

Embarcadero, Chimalhuacan, State of Mexico, in association with bone tools and obsidian flakes, but there are no published records of the stratigraphy associated with the find. There is no collagen preservation in this specimen, which shows the characteristic black colour associated with strong mineralisation of the bones. However, indirect dating was possible by analysing the sediments found inside the skull (Figure 5.2b), which were a mixture of lake sediments, diatoms, and volcanic ash from the UTP. The SiO_2 value ranged from 62.5 to 77.7%, largely in agreement with the values associated with the Upper Toluca Pumice found *in situ* in other Paleoindian sites, like Tlapacoya and the Tocuila mammoth site. However, it is likely that, as at these other sites, there is reworking of older tephras with isolated grains from the PWA and rhyolitic tephras from the Tlaloc/Telapon tephras, which are much older (Huddart and Gonzalez 2006). Therefore, we estimate the date of Chimalhuacan Man to be around 10,500 years and not 33,000 years as has been previously suggested using obsidian hydration methods (Pompa y Padilla and Carreto 2001).

New Stratigraphical Evidence from Tlapacoya

Although much work has been published on the Tlapacoya Paleoindian site (Lorenzo and Mirambell 1986; Mirambell 1973, 1986), there have been doubts about the archaeological evidence and dating (Waters 1985). New trenches were excavated in 1998 by Gonzalez and Huddart to try to resolve some of these issues. The stratigraphy presented in Figure 5.3 comprises a succession of clays, silts, diatomites, volcanic ashes and reworked ash, organic-rich sediments, and lahars. No further data were found to support the view for human activity at 24,000 BP. There was no evidence for a lithic assemblage, and it is considered that the 2,500 andesite flakes from the original excavations must be derived from the local bedrock at Tlapacoya Hill, which fractures into naturally sharp flakes. Evidence of reworking of the PWA tephra from the slopes above at several locations emphasises the importance of slope processes in the nearshore area at Tlapacoya. Very important, however, is the presence of the UTP tephra *in situ*, dated to 10,500 BP (Arce *et al* 2003). A 'stratified' human calvarium was reported from this layer in 1971, originally dated by the excavators to 9,920 ± 250 BP (García-Bárcena 1986), which agrees well with the new AMS C^{14} date of 10,200 ± 65 BP for the unstratified human calvarium (Figure 5.2c) (Gonzalez *et al* 2003; Gonzalez, Jiménez Lopez 2006). Based on our new evidence, humans were definitely present at the Tlapacoya site around 10,500–10,000 BP at the time of the UTP eruption and subsequent lahars.

TOCUILA MAMMOTH SITE

Tocuila is one of the most important Late Pleistocene and early Holocene paleontological sites in the Basin of Mexico (Morett *et al* 1998a) because of the thick paleontological bone layer (~1.75 m). It is located to the north-east of Mexico City in a western suburb of Texcoco

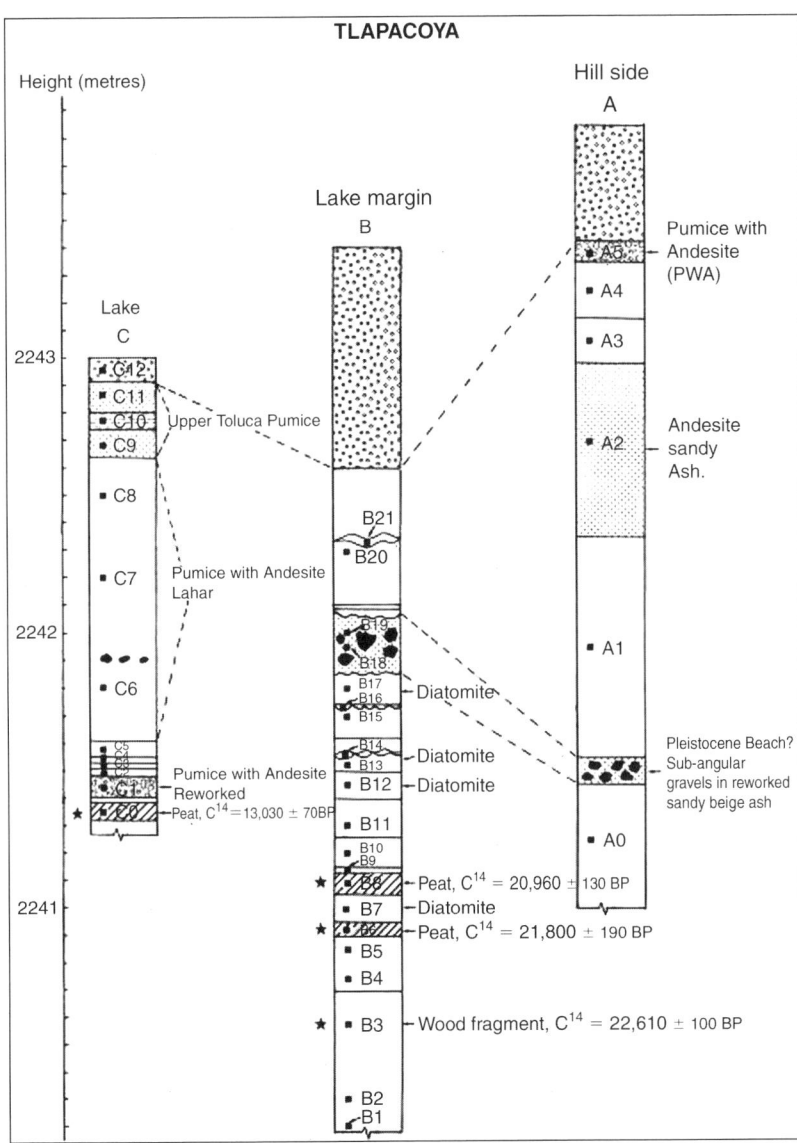

Figure 5.3   General stratigraphy at the Tlapacoya Paleoindian site showing main ash layers.

City at San Miguel Tocuila in a nearshore of the Late Pleistocene, a higher stage of Lake Texcoco (Figure 5.1). The site was discovered by chance by Luis Morett Alatorre and his team from the University of Chapingo, in collaboration with Oscar Polaco and Joaquín Arroyo from the Instituto Nacional de Antropología e Historia (INAH) whilst digging for a water tank; it was excavated in 1996. Remains of at least seven mammoths (*Mammuthus columbi*) were identified approximately 3 m below the surface in an area of 30 m^2 (Morett *et al* 1998b) (Figure 5.4a). Further mammoth remains also in UTP lahar deposits were found in 2003 (Figure 5.4b). Sedimentological and tephrochronological studies by Siebe *et al* (1997, 1999) and us have placed the archaeological and palaeontological findings in a paleoenvironmental framework and provide the basis for linking the material to the UTP eruption.

Approximately one thousand bones have been excavated at the Tocuila site, mostly plains mammoth, including three almost complete skulls, two incomplete skulls, and four mandibles (Morett *et al* 1998b) (Figure 5.4). These represent at least seven individuals, which range in age from young to adult. There are also some bones of horse, bison, camel, and rabbit, whereas in the top lacustrine unit there are fish, ducks, and flamingo (Arroyo *et al* 2003; Corona-M and Arroyo-Cabrales 1997).

The radiocarbon dates from the directly dated mammoth bones (Gonzalez, Morett Alatorre *et al* 2006) and from charcoal fragments incorporated within the lahar (Arroyo-Cabrales *et al* 2003; Siebe *et al* 1999)

Figure 5.4 Examples of mammoth remains at the Tocuila site: (a) main mammoths trench showing at least seven mammoths (*Mammuthus columbi*) embedded in a lahar deposit, photo taken in 2001; (b) new mammoth tusk discovered by chance in a drainage ditch in 2003 about 250 m north of the main mammoths trench.

help constrain the age of the deposit. The directly dated mammoth skull (OxA-7746) is dated to $11,100 \pm 80$, whereas a mammoth rib (OxA-10307) has a date of $11,255 \pm 75$. The full list in Table 5.1 shows that the dates are not in a stratigraphic sequence. The value of these dates is debateable, and when one considers the origin of the deposit within a lahar, it is not surprising to find this variability. An average age of c 11,188 BP has nevertheless been given for this deposit (Arroyo-Cabrales et al 2003; Morett et al 1998b), but we do not consider this method appropriate.

Of particular interest is the evidence of human presence at Tocuila indicated by dynamic impact fracturing features on mammoth long bone segments and fracture debris (Arroyo et al 2006; Arroyo-Cabrales et al 2001; Johnson et al 2001). The small assemblage consists of five worked bone specimens but includes a bone core with a prepared platform and scars from the removal of a number of large cortical flakes and a cortical flake that conjoins with the central flake scar on the bone core (Johnson 2001). There is no evidence of cut marks on any of the bone, and Haynes (2002) has suggested alternative explanations for such features. However, they show similar features to those typical of the North American Late Pleistocene tradition (Johnson 2001) and resemble experimentally generated cases (Stanford et al 1981). The Tocuila–worked mammoth bone assemblage, although limited, has been interpreted as human bone quarrying to produce cores for transport elsewhere and can be either concurrent with butchering or as an independent activity that requires fresh mammoth bone. It

Table 5.1 Radiocarbon dates from the Tocuila Mammoth Trench. Both dated mammoth bone samples were located towards the base of the lahar

	Depth from top (cm)	C^{14} date	Type	Sample no.
Toc.1–59	158	$10,430 \pm 75$	AMS	AA-23763
Toc.1–213 charcoal	170–173	$11,277 \pm 139$	Standard	INAH-1658
Toc.1–122 charcoal	170–205	$11,274 \pm 116$	Standard	INAH-1659
Toc.1–241	204	$10,220 \pm 75$	AMS	A-23161
Toc.1–261 and Toc.1–333 seeds and charcoal in	205–230	$11,541 \pm 196$	Standard	INAH-1660
Tocuila lahar	220	$10,650 \pm 75$	AMS	AA-25775
Toc.1–277 seeds	230–270	$11,296 \pm 186$	Standard	INAH-1661
Toc.1–355	251	$10,850 \pm 150$	AMS	AA-23764
Toc.1–355	271	$11,100 \pm 130$	AMS	AA-23765
Toc.1–424 and 1–436	270–300	$10,553 \pm 186$	Standard	INAH-1662
Toc.1–555	305–306	$12,615 \pm 95$	AMS	A-23162
Toc. –793	M. columbi skull	$11,100 \pm 80$	AMS	OxA-7746
Toc. S/N	M. columbi rib	$11,255 \pm 75$	AMS	OxA-10307

seems likely that the mammoth bones were scavenged immediately after death. There are no lithics associated with the deposit.

Research has focused in two areas where 11 trenches, ranging in thickness from 1.63 to 3.59 m, have been logged and sampled for grain size, tephra, and macro- and micro-fossils. These two areas are the main Mammoths Trench at the Tocuila Mammoth Museum and more recently discovered mammoths in trenches to the south. The oldest layer at the site is the Great Basaltic ash (GBA) tephra marker horizon, which in the main trench is overlain by a river channel completely infilled by lahar deposits in which the megafaunal bones are embedded, in a layer 1.75 metre thick (Figure 5.5). The bones tend to concentrate at the bottom of the lahar sediments forming a concentrated 'paleontological bone bed'.

Siebe *et al* (1999) interpreted the lahar sequence infilling the channel as derived from the PWA volcanic eruption dated to 14,450 BP, because chemical analysis of some pumice fragments found in the deposit correspond with the composition of a volcanic ash which originated from a Plinian eruption from Popocatépetl. However, radiocarbon dating of the sediments associated with the 'bone bed' has showed that the lahar deposit was formed no later than 10,650 ± 75 BP. Siebe *et al* (1999) explained the large discrepancy of nearly three thousand years between the tephra deposition, the dated mammoth bones in the sequence, and the lahar event being caused by the lack of water availability during the Late Pleistocene when permafrost and glacial ice were prevalent and the lack of water prohibited secondary lahar formation until much later, but this seems unlikely.

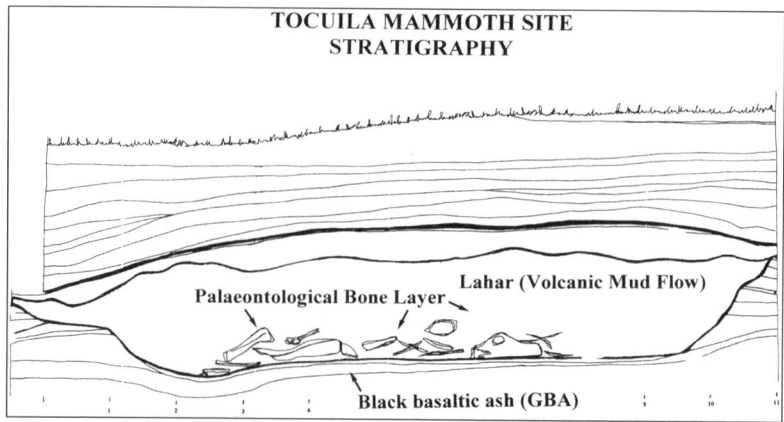

Figure 5.5(a) River channel infilled with lahar and megafaunal bones at Tocuila, in the main mammoths trench at Tocuila.

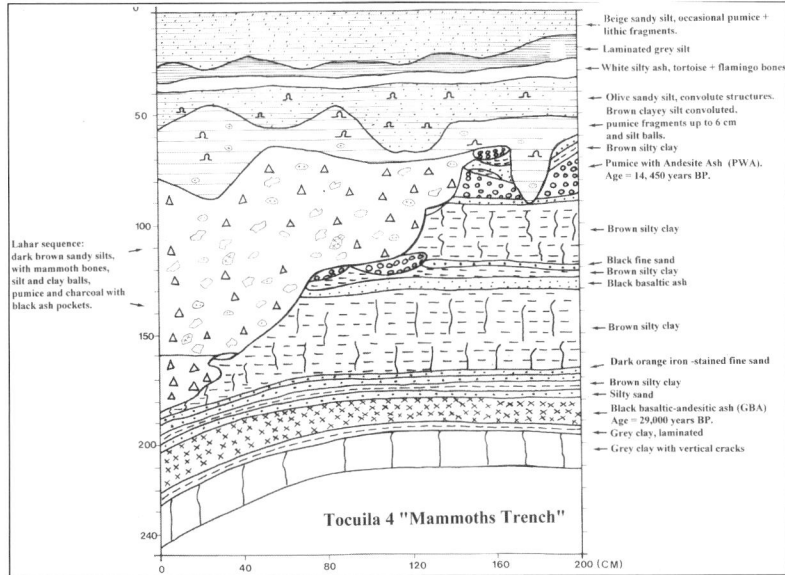

Figure 5.5(b)   Log 4 showing the margin of the river channel and the associated deposits.

A reanalysis of their data combined with additional work in new trenches where a series of known tephras have been preserved *in situ*, including the UTP ash, has led us to a different interpretation. The tephrochronology from the Tocuila sediments is complicated because of reworking but there is evidence for the presence *in situ* of key tephra markers found in the Basin of Mexico: GBA, PWA, UTP, and a younger, white rhyolitic ash (*pómez de grano fino*) (Figure 5.6). There is evidence of reworking for some of the pumice layers from channel margins and although the Tocuila lahar incorporated PWA tephra grains into its matrix, it was predominantly derived from the remobilised UTP ash from the eastern lake margins and slopes from the Tlaloc and Telapon volcanoes.

A number of studies in other regions demonstrate the patterns produced when several depositional units have been deposited over a short time (eg, Cronin *et al* 1999; Hodgson and Manville 1999). Our detailed studies of the stratigraphy in the new trenches and re-interpretation of previous reports suggest that the Tocuila lahar deposit was deposited quickly. As the sediments below and above the lahar unit are lacustrine, it is probable that the lahar event eroded a channel in the nearshore lake zone and was deposited in only a few hours. The uppermost units with convolute lamination in silts indicate a soupy mixture of sediment coming out of suspension in the lake after the main lahar unit had been

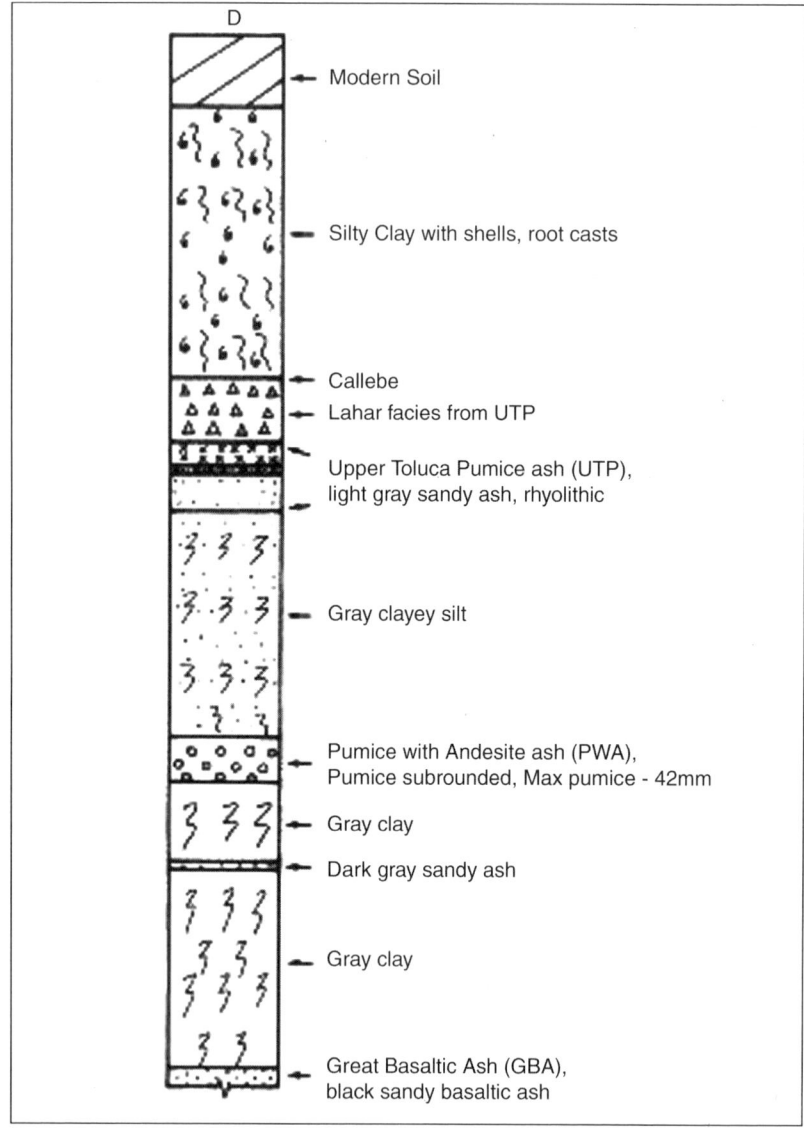

Figure 5.6 Stratigraphy in log D, to the south of the mammoths trench at Tocuila.

deposited. This sediment was deposited from hyperconcentrated flows from the body or tail of a composite sediment-laden flow (Sohn *et al* 1999). The C^{14} dates in the lahar sediments (see next section) likewise indicate erosion of pre-existing sediment and cannot be used to argue a case for a long time period of deposition in one lahar unit as suggested in Siebe *et al* (1999). Although it is common for lahars to be remobilised over decades after ash deposition (see Lirer *et al* 2001), time periods of this length proposed by Siebe *et al* (1999) are rare. A more inclusive interpretation of the evidence points to rapid deposition for the Tocuila lahar.

It seems unlikely that the lahars actually killed the mammoths found in the Tocuila area. Despite the fact that the skeletons are largely disarticulated, they are not worn by transport. It is suggested that the skeletons were lying close to the lake margins and were reworked by the distal lahars and concentrated in the lahar channels in the lake nearshore. The suggested time gap between death of the mammoths and the time of the reworking by the UTP lahar is important because it means that the mammoth skeletons were lying around the lake shores for several hundred years. The fact that bone quarrying requires fresh bone suggests that humans may have killed at least some of the mammoths before they were incorporated into the lahar, as was the case at nearby mammoth sites of Santa Isabel Iztapan I and II (Martínez del Río 1952), where there was a direct association of stone artefacts and obsidian blades with two mammoth skeletons, or at least with humans scavenging the skeletons close to time of death.

DISASTERS AND EXTINCTIONS

It is clear that the major Plinian event from the Navado de Toluca volcano, dated c 10,500 BP, had a major effect on the landscape of the Basin of Mexico and caused a widespread ecological crisis in the basin. It blanketed the topography with a layer up to 50 cm of rhyolitic ash over large areas of the basin, followed by lahars that flowed into the central lake. The whole ecosystem – including the vegetation, the soils, the hydrology, and the sub-aquatic system – would have been changed dramatically over a short time. In modern times, large plinian eruptions (eg, Pinatubo and Mount St Helens) have led to total disequilibrium in the ecosystem and to an ecological crisis. It has been recognised that in the Basin of Mexico these eruptions caused impacts for the interpretations of the palaeoclimatic record, although it has been suggested that because most of the changes seem to be synchronous events in the Texcoco-Chalco sub-basins, they were caused by climate events and not volcanic activity (Lozano-García *et al* 1993).

Based on radiocarbon dating of the Peñon Woman III skeleton at 10, 755 ± 55 BP, humans were already in the area at the time of the UTP eruption. We do not know for certain whether people were killed by this eruption, but at least in two cases – at the Metro Balderas and the Chimalhuacan sites – human skulls were found in association with UTP ash, and it is tantalising to imagine that these individuals may have been victims of the volcanic event. Because their landscape had been drastically altered, the ability of these Late Pleistocene humans to obtain food and fresh water would have been significantly affected. The small population must have either died out or migrated out of the affected areas.

Despite the great amount of excavation associated with the development and major expansion of Mexico City in the last 60 years, there is no evidence for prehistoric archaeological sites dating after the UTP eruption, and lahar events took place until human occupation at 4,500 BP at the San Vicente Chicoloapan site in the eastern basin (Gonzalez *et al* 2003) and the Tlatilco village site in the western basin at around 3,200 BP. The question that must be posed is whether this is a genuine gap in the archaeological record, which would suggest that humans were not present in the Basin of Mexico. Currently, we cannot provide a conclusive answer.

Later events suggest that the UTP had a major impact on the Paleoindian population. For example, Gonzalez *et al* (2000) and Siebe (2000) have shown that the eruption of the Xitle volcano, in the south of the basin, which resulted in tephra and lava flows, caused the abandonment of the Cuicuilco pyramid at about 1,670 BP and this, together with the Popacatépetl eruption of around 2,200–2,000 BP, caused important human migrations in the Amecameca-Chalco region. This explains the major population shift to the north-east of the basin and the rise in importance of Teotihuacan. Sanders *et al* (1979) suggested that between 100 BC and 100 AD about 80–90% of the population of the entire Basin of Mexico had nucleated in the north-east and that the population in the southern part of the basin had been reduced substantially. These two volcanic eruptions must have had a role in this process. A similar example can be taken from a Plinian eruption of Popocatépetl between 50 and 100 AD when the Tetimpa region to the east of that volcano was devastated and abruptly abandoned (Panfil *et al* 1999; Siebe *et al* 1996). It took about four hundred years before recolonisation took place, and probably between an estimated two to three thousand people migrated and contributed to the emerging urban centre of Cholula, in the Puebla Basin (Plunket and Uruñuela 2000).

The second casualty of the volcanic disaster was the megafauna. After this eruption and subsequent lahars, there is no evidence for the presence of the characteristic megafaunal assemblage that was present around the lake shorelines throughout the Late Pleistocene. It seems highly likely that this eruption was either a direct cause of faunal extinction, or, even more likely, that it contributed to this process indirectly through negative impacts on the ecosystem and hence food availability. The widespread destruction of vegetation over large areas probably affected grazing animals like mammoths, camels, bison, and horses. The equilibrium of the lakes must have been totally disrupted with the rapid deposition of acid volcanic ash, the remobilisation in the form of secondary lahars into the lakes, and the resulting rapid changes in pH. Commonly, after a major ash fall, ostracod and diatomite layers indicate that the palaeoenvironment had changed dramatically and conditions were conducive to the rapid production of these organisms. But the increase in supply of silica to the lakes through their catchments is probably the cause of the diatom changes, so the impact of the tephras is only thought to have lasted several decades (Telford *et al* 2004).

It is possible that the UTP eruption was the final 'push' in the extinction process of the mammoth and other megafauna species (such as camels, horses, glyptodonts) on the Mexican grasslands around the lake basin. Animal populations, which may have been in decline already because of the climatic changes associated with the transition from the Late Pleistocene to the Holocene, may have been pushed into extinction by the effects of the UTP, although the presence of humans may have had some effects, too. It is difficult to unravel these two likely effects. The evidence from the Tocuila site suggests that perhaps the mammoths had been utilised by humans prior to the UTP ash fall event and were then reworked by the secondary lahar.

FUTURE IMPLICATIONS

During the last century, archaeologists and geologists working on many Paleoindian and preceramic sites in the Basin of Mexico did not understand, or sometimes realise, the importance of volcanic processes in creating the sediments and sequences in which they were working and their impacts on the early populations. This has led to a partial understanding and often misinterpretation of these sequences. We have tried to show that the best way to properly understand such information is to integrate both the archaeological and volcanological data and apply geoarchaeology. In this way, the important sites can be better understood.

The record of past volcanic eruptions in the Basin of Mexico during the last 30,000 years indicates that another Plinian eruption is a likely scenario in the future, but this time such an event will put at risk a human population of around 24 million inhabitants in Mexico City. The Nevado de Toluca volcano is currently quiescent, and although its most recent eruptive activity occurred at around 3,500 BP, it may be capable of producing another Plinian eruption in the future (Arce *et al* 2003). Popocatépetl has also been active since 1997 and is capable of another large-scale Plinian type eruption that could affect over a quarter of the Mexican population who live within its catchment. Although extinction is unlikely, the effects of such an event on human populations and ecosystems would be even more catastrophic than the UTP event that we have documented.

ACKNOWLEDGMENTS

We acknowledge the initial help by Louis Morett and Joaquin Arroyo-Cabrales in introducing us to the Tocuila mammoth site and the helpful discussions of

José Luis Macías and José Luis Arce in relation to the volcanic processes at this location. Grants from Liverpool John Moores University and NERC EFCHED NER/T/S/2002/0467, 'Human dispersals and environmental controls during the Late Pleistocene/Early Holocene in Mexico', are gratefully acknowledged.

REFERENCES

Arce, JL (1999) 'Reinterpretation de la erupción pliniana que dio origen a la Pómez Toluca', unpublished Master's thesis, Universidad Nacional Autónama de México, Mexico City

Arce, JL, Cervantes, JL, Macías JL and Mora, JC (2005) 'The 12.1 ka Middle Toluca Pumice: A dacitic Plinian-subplinian eruption of Nevado de Toluca in central Mexico', *Journal of Volcanological and Geothermal Research* 147, 125–43

Arce, JL, Macías, JL and Vázquez-Selem, L (2003) 'The 10.5 ka Plinian eruption of Nevado de Toluca volcano, Mexico: stratigraphy and hazard implications', *Bulletin of the Geological Society of America* 115, 230–48

Armenta-Camacho, J (1959) 'Hallazgo de un artefacto asociado con mamut en el Valle de Puebla', *Instituto Nacional de Antropología y Historia de México. Dirección de Prehistoria Publicaciones* 7, 1–30

Arroyo-Cabrales, J, Gonzalez, S, Morett, AL, Polaco, O, Sherwood, G and Turner, A (2003) 'The Late Pleistocene paleoenvironment of the Basin of Mexico – evidence from the Tocuila Mammoth site', *Deinsea* 9, 267–72

Arroyo-Cabrales, J, Johnson, E and Morett, L (2001) 'Mammoth bone technology at Tocuila in the Basin of Mexico', in Cavarretta, G, Gioia, P, Mussi and Palombo, MR (eds), *Proceedings of the 1st International Conference, The World of Elephants*, pp 419–23, Rome: Consiglio Nazionale delle Ricerche

Arroyo-Cabrales, J, Polaco, OJ and Johnson, E (2006) 'A preliminary view of the existence of mammoth and early peoples in Mexico', *Quaternary International* 142, 79–86

Aveleyra de Anda, L (1955) 'El segundo mamut fósil de Santa Isabel Iztapan, México y artefactos asociados', *Cuardernos de Trabajo del Departamento de Prehistoria* 32, 1–151

Aveleyra de Anda, L and Maldonado-Koerdel, M (1952a) 'Asociación de artefactos con mamut en el pleistoceno superior de la Cuenca de México', *Revista Mexicana de Estudios Antropológicos*, 13, 3–29

Aveleyra de Anda, L and Maldonado-Koerdel, M (1952b) 'Association of artifacts with mammoth in the Valley of Mexico', *American Antiquity* 18, 332–40

Bloomfield, K, Sánchez-Rubio, G and Wilson, L (1977) 'Plinian eruptions of Nevado de Toluca volcano, central Mexico', *Geologische Rundschau* 66, 120–46

Bloomfield, K and Valastro, S Jr (1974) 'Late Pleistocene eruptive history of Nevado de Toluca volcano, central Mexico', *Bulletin of the Geological Society of America* 85, 901–06

Bloomfield, K and Valastro, S Jr (1977) 'Late Quaternary tephrachronology of Nevado de Toluca volcano, central Mexico', *Overseas Geology and Mineral Resources* 46, 1–15

Cantagrel, JM, Robin, C and Vincent, P (1981) 'Les grandes étapes d'évolution d'un volcan andesitique composite. Example du Nevado de Toluca', *Bulletin of Volcanology* 44, 177–88

Capra, L and Macías, JL (2000) 'Pleistocene cohesive debris flows at Nevado de Toluca volcano, central Mexico', *Journal of Volcanology and Geothermal Research* 102, 149–68

Corona-M, E and Arroyo-Cabrales, J (1997) 'New record for the flamingo (*Phoenicopterus cf. P.ruber Linnaeus*) from Pleistocene-Holocene transition sediments in Mexico', *Current Research in the Pleistocene* 14, 137–38

Cronin, SJ, Neall, VE, Lecointre, JA and Palmer, AS (1999) 'Dynamic interactions between lahar and stream flow: a case study from Ruapehu volcano, New Zealand', *Bulletin of the Geological Society of America* 111, 28–38

García-Bárcena, J (1986) 'El hombre y los proboscideos de América', *Instituto Nacional de Antropología y Historia de México, Colección Científica* 188, 41–79

Gonzalez, S, Huddart, D, Morett-Alatorre, L, Arroyo-Cabrales, J and Polaco, OJ (2001) 'Volcanism and early humans in the Basin of Mexico during the Late Pleistocene/ Early Holocene', in Cavarretta, G, Gioia, P, Mussi, M and Palombo, MR (eds), *1st International Congress, The World of Elephants*, pp 704–06, Rome: Consiglio Nazionale delle Ricerche

Gonzalez, S, Jiménez López, JC, Hedges, R, Huddart, D, Ohman, J, Turner, A, and Pompa y Padilla, JA (2003) 'Earliest humans in the Americas: new evidence from Mexico', *Journal of Human Evolution* 44, 379–87

Gonzalez, S, Jiménez López, JC, Hedges, R, Pompa y Padilla, JA, and Huddart, D (2006) 'Early humans in Mexico: new chronological data', in Jiménez López, JC, Pompa y Padilla, JA, Gonzalez, S and Ortiz F (eds), *Proceedings of the 1st International Symposium on Early Humans in America*, pp 67–76, Mexico City: INAH

Gonzalez, S, Morett Alatorre, L, Huddart, D and Arroyo-Cabrales, J (2006) 'Mammoths from the Basin of Mexico: stratigraphy and radiocarbon dating', in Jiménez López, JC, Pompa y Padilla JA, Gonzalez, S and Ortiz F (eds), *Proceedings of the 1st International Symposiumon Early Humans in America*, pp 263–74, Mexico City: INAH

Gonzalez, S, Pastrana, A, Siebe, C and Duller, G (2000) 'Timing of the prehistoric eruption of Xitle volcano and the abandonment of Cuicuilco Pyramid, southern Basin of Mexico', in McGuire, WJ, Griffiths, DR, Hancock, PL and Stewart, IS (eds), *The Archaeology of Geological Catastrophes*, Geological Society of London, Special Publication 171, 205–24

Haynes, G (2002) *The Early Settlement of North America. The Clovis Era*, Cambridge: Cambridge University Press

Hodgson, KA and Manville, VR (1999) 'Sedimentology and flow behaviour of a rain-triggered lahar, Mangatoetenui Stream, Ruapehu volcano, New Zealand', *Bulletin of the Geological Society of America* 111, 743–54

Huddart, D and Gonzalez, S (2006) 'A review of environmental change in the Basin of Mexico (40,000–10,000 BP): implications for early humans', in Jiménez López, JC, Pompa y Padilla JA, Gonzalez S and Ortiz F (eds) *Proceedings of the 1st International Symposium on Early Humans in America*, pp 77–105, Mexico City: INAH

Johnson, E (2001) 'Mammoth bone quarrying on the late Wisconsinan North American grasslands', in Cavarretta, G, Gioia, P, Mussi, M and Palombo, MR (eds), *Proceedings of the 1st International Congreso, The World of Elephants*, pp 499–43, Rome: Consiglio Nazionale delle Ricerche

Johnson, E, Morret, AL, and Arroyo-Cabrales, J (2001) 'Late Pleistocene bone technology at Tocuila, Basin of Mexico', *Current Research in the Pleistocene* 18, 13–15

Lirer, L, Vinci, A, Alberico, I, Gifuni, T, Belluci, F, Petrosino, P and Tinterri, R (2001) 'Occurrence of inter-eruption debris flow and hyperconcentrated flood-flow deposits on Vesuvio volcano, Italy', *Sedimentary Geology* 139, 151–67

Lorenzo, JL and Mirambell, L (1986) 'Mamutes excavados en la Cuenca de México', *Instituto Nacional de Antropología y Historia de México, Cuadernos de Trabajo del Departamento de Prehistoria* 32, 1–151

Lozano-García, MS, Ortega-Guerrero, B, Caballero-Miranda, M and Urrutia-Fucugauchi, J (1993) 'Late Pleistocene and Holocene Palaeoenvironments of Chalco Lake, Central Mexico', *Quaternary Research* 40, 332–42

Macías, JL and Arce, JL (1997) 'The Upper Toluca Pumice: a major plinian event occurred ca 10,5000 yrs ago at Nevado del Toluca, central Mexico', *EOS, Transactions of the American Geophysical Union* 78, F823

Macías, JL, Arce, JL, García, PA, Siebe, C, Espíndola, JM, Komorowski, JC and Scott, K (1997) 'Late Pleistocene-Holocene cataclysmic eruptions at Nevado de Toluca and Jocatitlan volcanoes, central Mexico', in Link, KP and Kowallis, BJ (eds), *Proterozoic to Recent Stratigraphy, Tectonics and Vulcanology, Utah, Nevada, Southern Idaho and*

Central Mexico, pp 493–52, Salt Lake City, UT: Brigham Young University Geology Studies

Martínez del Río, P (1952) 'El Mamut de Santa Isabel Iztapan', *Cuardernos Americanos 9*, 149–70

Mirambell, L (1973) 'Excavaciones en sitios pleistocénico en Tlapacoya, Estado de México', *Boletín del INAH* 29, 37–41

Mirambell, L (1986) 'Los excavaciones', in Lorenzo, JL and Mirambell, L (eds), *Tlapacoya: 35,000 años de historia del Lago de Chalco*, pp 13, 56, Prehistoric Series, Mexico City: INAH

Mooser, F (1967) 'Tefracronología de la Cuenca de México para los últimos treinta mil años', *Boletín del Instituto Nacional de Antropología y Historia de México* 30, 12–15

Morett, L, Arroyo-Cabrales, J and Polaco, OJ (1998a) 'El sitio paleontológico de Tocuila', *Arqueología Mexicana* 5, 57

Morett, L, Arroyo-Cabrales, J and Polaco, OJ (1998b) 'Tocuila mammoth site', *Current Research in the Pleistocene* 15, 118–20

Newton, AJ and Metcalfe, SE (1999) 'Tephrochronology of the Toluca Basin, central Mexico', *Quaternary Science Reviews* 18, 1039–59

Panfil, MS, Gardner, TW and Hirth, RG (1999) 'Late Holocene stratigraphy of Tetimpa archaeological sites, northeast flank of Popocatépetl volcano, central Mexico', *Bulletin of the Geological Society of America* 111, 204–18

Plunket, P and Uruñuela, G (2000) 'The archaeology of a Plinian eruption of the Popocatépetl volcano', in McGuire, WJ, Griffiths, DR, Hancock, PL and Stewart, IS (eds), *The Archaeology of Geological Catastrophes*, pp 195–203, Geological Society of London, Special Publication

Pompa y Padilla, JA and Carreto, ES (2001) 'Los más antiguos americanos', *Arqueología Mexicana* 52, 36–41

Sanders, WT, Parsons, JR and Santley, RS (1979) *The Basin of Mexico. Ecological Processes in the Evolution of a Civilization*, New York: Academic Press

Siebe, C (2000) 'Age and archaeological implications of Xitle volcano, southwestern Basin of Mexico', *Journal of Volcanology and Geothermal Research* 104, 45–64

Siebe, C, Abrams, M, Macías, JL and Obenholzner, J (1996) 'Repeated volcanic disasters in Prehispanic time at Popocatépetl, central Mexico, past key to the future?', *Geology* 24, 399–402

Siebe, C, SCAF, P and Urrutia-Fucugauchi, J (1999) 'Mammoth bones embedded in a late Pleistocene lahar from Popocatépetl volcano, near Tocuila, central Mexico', *Bulletin of the Geological Society of America* 111, 1550–62

Siebe, C, SCAF, P, Urrutia-Fucugauchi, J, Morett-Alatorre, L, Arroyo-Cabrales, J and Obenholzner, J (1997) 'Mammoth bones embedded in a late Pleistocene lahar deposit from Popocatépetl volcano, near Tocuila, Valley of Mexico', *Geological Society of America Abstracts with Programs* 29, A-164

Sohn, YK, Rhee, CW and Kim, BC (1999) 'Debris flow and hyperconcentrated flood-flow deposits of an alluvial fan, northwestern part of the Cretaceous Yongdong Basin, Central Corea', *Journal of Geology* 197, 111–32

Stanford, S, Bonnichsen, R and Morlan, RE (1981) 'The Ginsberg experiment: modern and prehistoric evidence of a bone-flaking technology', *Science* 212, 438–40

Telford, RJ, Barker, P, Metcalfe, S and Newton, A (2004) 'Lacustrine responses to tephra deposition: examples from Mexico', *Quaternary Science Reviews* 23, 2337–53

Waters, M (1985) 'Early man in the new world: an evaluation of the radiocarbon dated pre-Clovia sites in the Americas', in Mead, JI and Meltzer, DJ (eds), *Environments and Extinctions: Man in the Late Glacial North America*, pp 125–43, Orono: Center for the Study of Early Man, University of Maine

CHAPTER 6

Living with the Volcano: The 11th Century AD Eruption of Sunset Crater

Mark D Elson, Michael H Ort, Kirk C Anderson,
and James M Heidke

INTRODUCTION

In the summer of 1930, archaeologists from the Museum of Northern Arizona uncovered a pit structure sealed beneath a thick layer of black cinders at a prehistoric site just north of Flagstaff, Arizona (Colton 1932a, 1960). The excavation of this structure provided the first definitive evidence that Sunset Crater volcano erupted during the prehistoric occupation of northern Arizona (Figure 6.1). Although no datable tree ring or radiocarbon samples were recovered, associated ceramics placed the occupation within the past 1,000 years (Colton 1946: 125). It would not be an exaggeration to say that this finding forever changed the nature of northern Arizona archaeology (Downum 1988: 110).

Starting with this initial work and continuing into the modern period, the Sunset Crater eruption has become a critical factor in any interpretation of prehistoric settlement in the Flagstaff region. For example, to explain a dramatic increase in site density at the end of the 11th century AD, Colton proposed that thin layers of cinders deposited by the eruption acted as a water-retaining mulch, allowing previously unproductive areas to be farmed. According to Colton, the opening of new lands for settlement resulted in large-scale migration of pueblo groups into the Flagstaff area, where interaction between migrants and indigenous populations significantly changed the nature of the local settlement (Colton 1932a, 1946, 1949). More recent researchers have adopted Colton's theory, with minor modifications, to suggest that the cinder mulch was an important factor in the initial settlement and subsequent population growth evident in Wupatki National Monument, an area that contains some of the largest and most recent stone-built (pueblo) sites in the general Flagstaff region (Anderson

Figure 6.1 Location of Sunset Crater volcano in the greater Southwest United States.

1990; Downum 1988; Downum and Sullivan 1990; Sullivan 1984; Stone and Downum 1999; Sullivan and Downum 1991). Although other models suggest that it was changing climatic conditions, not the cinder mulch, which allowed new lands to be farmed (Hevly *et al* 1979; Pilles 1979, 1996), all researchers agree that the eruption of Sunset Crater had an enormous impact on the prehistoric inhabitants of north-central Arizona and probably the greater northern Southwest United States.

This chapter presents new results of research concerning the effects of the Sunset Crater eruption on local populations. We discuss four responses by the prehistoric inhabitants to this volcanic event: (1) population movement; (2) use of new agricultural methods; (3) changes in ceramic production and exchange networks; and (4) initiation of volcano-related ritual behaviour.

SUNSET CRATER VOLCANO

Sunset Crater, located about 25 km north-east of Flagstaff, is a 300 m high cinder-cone volcano (Figure 6.2). Unlike stratovolcanoes (eg, Mount St Helens and Pinatubo), cinder cones generally do not produce large explosions. Instead, they are typically formed by a single event that begins with an eruption from a small crack in the earth. They grow into

Figure 6.2 Sunset Crater cinder cone and the Bonito lava flow, looking south-east (photograph by Peter Pilles).

cone-shaped features through lava-fountaining, which throws spatter, cinders, and ash hundreds to thousands of metres into the air. Lava emanates from vents near the base of the cone.

The Sunset Crater eruption was likely foreshadowed by ominous signs: evidence from modern cinder cone volcanoes indicates that earthquakes occur for several weeks or months prior to the eruption, often increasing in frequency and magnitude through time. For example, at a cinder cone volcano located near the town of Parícutin in Michoacán, Mexico, and very similar to Sunset Crater, the first noticeable earthquake occurred 45 days prior to the eruption in 1943. A week later, daily earthquakes were occurring and these increased to 25 to 30 a day the week before the eruption. Finally, the day before the eruption, 300 earth tremors were felt (Luhr and Simkin 1993: 3; Yokoyama and De la Cruz-Reyna 1990).

During an eruption, the cloud of ash and steam creates its own weather system and thunder claps and lightning around the volcano are common occurrences. At Parícutin, which was surrounded by several small villages of Purépeche (Tarascan) Indian farmers, nobody was killed by the lava or cinder fall itself, but three people and a number of cattle and horses were killed by lightning strikes (Luhr and Simkin 1993: 7; Nolan 1979: 328). Additionally, the heavy ash and cinder fall from Sunset Crater, along with the smoke from accompanying forest fires, must have darkened the daytime sky, whilst at night the horizon would have glowed a fiery red. The roar of volcano eruptions, which is sometimes accompanied by whistling, hissing, and knocking sounds from 'the centre of the earth,' can at times be heard hundred kilometres away, and has been described as sounding like thunder, heavy surf, or artillery fire (Gadow 1930 in Luhr and Simkin 1993: 366). An eye-witness to the 1538 AD birth of Monte Nuovo volcano in Italy, described 'a horrid mouth, from which was vomited, furiously, smoke, fire, stones, and mud composed of ashes; making at the time of its opening a noise like loud thunder' (Bullard 1984: 289).

To people who had previously never heard a sound louder than a clap of thunder, the prolonged roar of the eruption would have been a highly frightening occurrence. Cinder-cone volcanoes can grow at an amazingly fast rate: at Parícutin, the cinder cone grew to a height of 167 m in only six days (Rees 1979: 251). By the end of the first year, Parícutin had reached 336 m, or 80% of its final height (Luhr and Simkin 1993: 8). For the prehistoric inhabitants of the Flagstaff area, the eruption of Sunset Crater would have been not only an awe-inspiring sight, but one also likely to have been believed to have a supernatural origin.

Sunset Crater lava flowed from two separate primary vents, covering an area of around 8 km² to depths ranging between 2 and 30 m (Figures 6.2 and 6.3). Given the high density of prehistoric settlement in

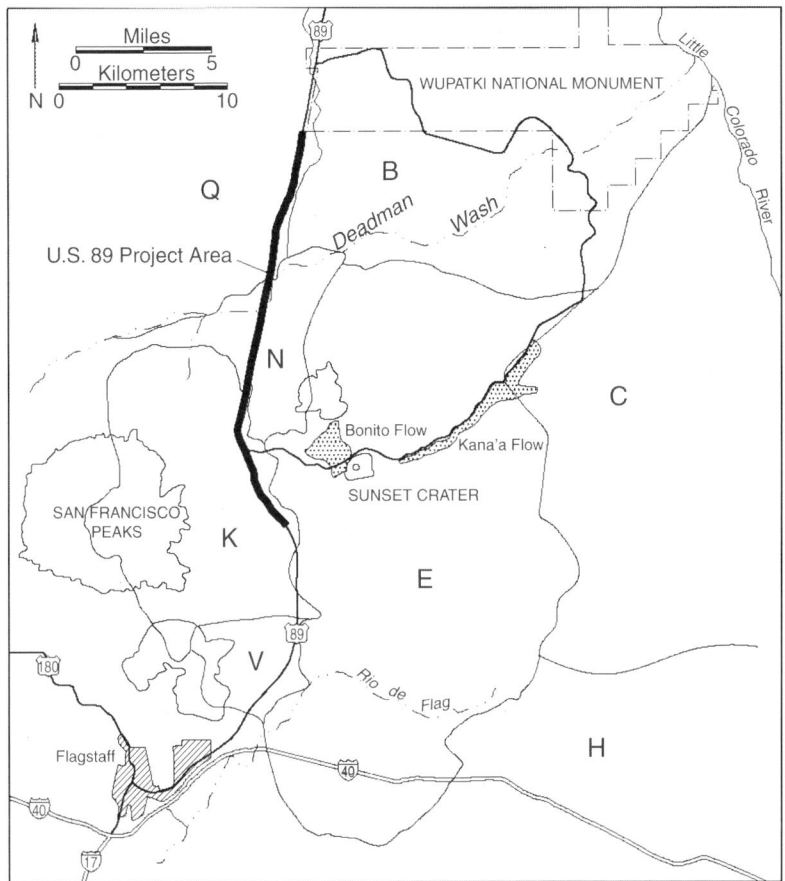

Figure 6.3 Location of the U.S. 89 project area and ceramic temper sources in relation to Sunset Crater.

areas just outside the flow and the confinement of much of the flows to the typically farmed basins, it is likely that numerous habitation and agricultural field sites were buried (Downum and Gumerman 1998: 8; Pilles 1978, 1979). In addition, the eruption spewed ash and cinders over an area of about 2,300 km². Ash and cinder fall was heaviest in areas closest to the volcano, but deposits as thick as 40 cm were also found in the fill of a pithouse at a site around 5.5 km south-west of the volcano (Swartz and Elson 2006) and deposits over 12 m in depth have been recorded closer to the cone (Amos 1986). Significant accumulations occurred at distances as great as 20 km, such as at Wupatki National Monument, where Sunset Crater cinder deposits between 5 and 10 cm have been recorded (Colton 1932a; Hooten *et al* 2001: 15).

The visibility of the eruption was calculated using digital elevational modeling based on the estimated minimum and maximum heights of the ash plume (3–5 km) and lava fire fountain (260–660 m). These data indicate that on a clear day the ash plume could have been seen from distances as great as 400 km, including high points near Palm Springs (California), Las Vegas (Nevada), Durango (Colorado), west-central New Mexico, and along the Arizona-Mexico border south of Tucson (see Figure 6.1). The fire fountain, which would have been particularly spectacular at night, would have been visible across a much smaller area, but still could have been seen from points some 250 km away in southern Utah, and east, west, and central Arizona, including the top of the Bradshaw Mountains about 75 km north of Phoenix. Areas of significant prehistoric population, such as Chaco Canyon, the Phoenix Basin, and the Four-Corners area, would have been within sight of the ash plume. Although it is not known how the eruption affected people living outside the immediate Flagstaff area, most groups living in the greater Southwest in the mid-to-late 11th century AD would have been aware that something very unusual was occurring.

The cinder and ash fall, along with the forest fires, must have disrupted prehistoric life to a significant degree. At Parícutin, an estimated 4,500 cattle and 550 horses died from breathing the fine ash emitted early in the eruption (Rees 1979: 259). At Sunset Crater, it is likely that the ash fall detrimentally affected human (and animal) health, particularly of the very young and elderly, and smoke from lava-ignited forest fires would have added to the air pollution problems. Sunset Crater lies in a relatively arid, pine-forested region, receiving less than 500 mm of precipitation per year; it is possible that forest fires spread beyond the area directly impacted by the lava, causing at least temporary displacement of both human and animal populations. Damage from cinder fall could also have occurred in areas far removed from the volcano. Data from modern eruptions indicate that 10 cm of ash, particularly when wet, is heavy enough to collapse modern roofs (Blong 1984), suggesting that structures at sites as distant as 15–20 km may have sustained significant damage.

The eruption has been dated by dendrochronology, paleomagnetic secular variation, and archaeological association. Although a date of 1064 AD for the initial eruption has become entrenched in both the archaeological and geological literatures, recent research suggests that the dating is not conclusive and can be questioned on a number of grounds (Boston 1995: 22; Elson *et al* 2002; Sheppard *et al* 2004). The 1064 AD date comes from Smiley (1958), who tree-ring dated several beams used in constructing the 100-room Wupatki Pueblo, situated about 20 km north-east of Sunset Crater. Eight of Smiley's (1958: 190) samples showed suppressed and 'complacent' rings after 1065 AD, which he interpreted as being caused by the eruption and therefore dating the volcano to the winter of 1064–65 AD. Indeed, a comparison of dated archaeological sites with and without Sunset Crater ash appears to constrain the initial eruption to sometime in the mid-to-late 11th century AD (see reviews in Boston 1995: 22–54 and Downum 1988: 222–49). However, a recent re-examination of Smiley's data indicates that several of his tree-ring samples are duplicates from the same tree, and Smiley's samples actually consisted of only one ponderosa pine and two Douglas-fir trees (Elson *et al* 2002; Robinson *et al* 1975). Therefore, this data set is not as strong or as well replicated as has been assumed. More importantly, the provenance of the analysed trees is not known; at only 1,495 masl, Wupatki is too hot and dry to support the growth of pine and fir, both today and in the prehistoric past (Sullivan and Downum 1991: 272). The closest these species are found today is on O'Leary Peak just north of Sunset Crater, around 20 km south of Wupatki, and in the San Francisco Peaks, around 30 km south-west of Wupatki (Figure 6.3). Although O'Leary Peak did receive significant ash accumulation, the San Francisco Peaks had less than 1 cm. Therefore, the trees could have come from anywhere in the higher elevations, possibly from areas not heavily affected by ash and cinder fall from the eruption. Furthermore, many factors other than volcanic eruptions can affect tree-ring growth, including localised drought, proximity to a forest fire, earthquakes, changes in the water tabled, extreme cold, and even insect infestation (Boston 1995; Elson *et al* 2002; Sheppard *et al* 2004; Sheppard *et al* 2005). Perhaps most significantly, amongst the hundreds of relevant tree-ring samples analysed since Smiley's work, none have contained this unique signature (Boston 1995: 37; Jeffrey Dean, personal communication).

The duration of the Sunset Crater eruption is also critical for understanding the resulting prehistoric response and adaptation, but again is controversial. In general, cinder-cone eruptions do not normally last more than a few years: 50% of known historic period eruptions lasted less than 30 days and 95% less than one year (Vespermann and Schmincke 2000: 688). Parícutin, which erupted for nine years between 1943 and 1952, is one of the longest cinder cone eruptions on record.

Early paleomagnetic investigations at Sunset Crater allowed an eruptive period of 100 and possibly as long as 200 years (Champion 1980; Shoemaker and Champion 1977). Like the 1064 AD date, a long eruption span has become entrenched in the literature, and there are now several reconstructions beginning in 1064 AD and ending in 1200 or 1250 AD (Holm and Moore 1987; Pilles 1979). Recent paleomagnetic work, however, indicates an eruptive period of no more than 50 years and possibly less (Ort *et al* 2002). In addition, although stratigraphic investigations at Sunset Crater have documented at least eight different eruptive episodes (Amos 1986), the lack of erosion or dune/ripple formation between episodes suggests cinder emplacement over a very short time frame, with no time for wind or water to rework the deposits. Data for a short eruption also accord better with the archaeological evidence because primary cinder deposits from some time in the late 11th century are relatively common, but similar evidence from the late 12th or early 13th centuries is ambiguous and difficult to interpret.

Therefore, although the exact date and length of the eruption are currently unknown, it is safe to say that Sunset Crater erupted sometime between 1050 and 1125 AD and probably lasted a few years or less. The accuracy of the dating is sufficient to conclude, however, that at this time the area around Sunset Crater was densely inhabited by small groups of prehistoric farmers who must have been significantly impacted by the eruption.

NEW INSIGHTS INTO NORTHERN ARIZONA ARCHAEOLOGY

Our research is based on recent archaeological investigations at around 40 prehistoric sites located along a 27 km stretch of US 89 highway just north of Flagstaff, Arizona (Figure 6.3). This project, undertaken prior to highway improvement, resulted in the excavation of over 70 structures, hundreds of smaller features, and the collection of around 100,000 artefacts, making it by far the largest excavation project ever undertaken in the Flagstaff area. The project area is within 6 km of Sunset Crater, and all project sites are within the zone where the depth of cinder and ash was greater than 5–10 cm.

A wide range of site types were present in the project area (eg, permanent habitations containing up to 20 masonry rooms and pithouses, small one–two room seasonally occupied fieldhouses, and 0.4–2 ha agricultural field systems with water-control features, such as linear rock alignments used as check dams and terraces) (Elson 2006a, 2006b). Palynological and macrobotanical data indicate subsistence was based primarily on corn agriculture, along with some cultivation of beans, squash, and cotton (Diehl 2006; Smith 2006). A number of environmental zones are present, as the project area runs from juniper-sage grassland at around 1,675 masl in the north to ponderosa pine forest at around 2,225 masl in the south. Dates ranging from 400 to 1200 AD were obtained through associated ceramic types combined with tree-ring and radiocarbon determinations. The great majority of sites were occupied between 1050 and 1150 AD, a period that includes the approximate time of the Sunset Crater eruption but also extends several decades later. Volcanic ash found on several excavated floor and extramural surfaces indicates that the eruption clearly impacted the prehistoric inhabitants.

One very significant aspect of the project area is that the 27 km transect passes through what has long been defined as a cultural frontier or boundary between two different prehistoric groups. This boundary was originally placed by Colton (1939, 1946: Figure 3, 1968) at Deadman

Wash and more recently refined to be about 7 km south of this point at the Coconino Divide, which is the highest point along the US 89 highway area within the pass between the San Francisco Peaks to the west and Sunset Crater/O'Leary Peak to the east (Downum and Gumerman 1998; also see Figure 6.3). The two groups are commonly called the Sinagua, who lived to the south and east and produced a brown-ware pottery type called Alameda Brown Ware, and the Cohonina, who lived to the north and west and produced a gray-ware pottery type called San Francisco Mountain Gray Ware. Colton (1946: 14–17) also noted the presence of a third group in the general area, the Kayenta Anasazi (or Ancestral Pueblo), who produced a different gray-ware pottery called Tusayan Gray Ware. However, because Kayenta sites are generally confined to the Wupatki area and points north-east of Flagstaff, we will not discuss them further in this chapter.

Cultural differences between the Sinagua and Cohonina noted by Colton (1946) and other researchers are based almost exclusively on differences in ceramic wares. Although there are traits in other material classes and architecture that may also distinguish Sinagua from Cohonina (eg, see Cameron 1999; Colton 1946; McGregor 1941, 1951; Samples 1990), much of this is based on either limited information or unpublished and unquantified data. When these other traits are examined closely with data from excavated contexts, the pattern is neither as strong nor as conclusive as initially surmised and, in some cases, falls apart completely. Of the 22 US 89 highway sites that were intensively investigated, half lie in what would be considered Sinagua territory and half in Cohonina (Figure 6.4). Our research addresses the interrelationships between the effects of the Sunset Crater eruption and the nature of the groups associated with the different ceramic styles.

Research Background

Data from excavated sites in our project area that yielded more than 100 sherds show a very clear line separating sites with a higher frequency of Alameda Brown Ware to the south, which traditionally would be called Sinagua, from those with a higher frequency of San Francisco Mountain Gray Ware, or Cohonina, to the north (Figure 6.4). As recent researchers have observed, the break between these occurs at the Coconino Divide.

It is important to note, however, that *all sites contained both wares* and that, although Figure 6.4 shows a very clear line, the differences in ware frequency were not great. For example, sites just north of the divide have, on average, between 40 and 60% San Francisco Mountain Gray Ware and 20–40% Alameda Brown Ware, changing to 60–80% San Francisco Mountain Gray Ware, and 1 20% Alameda

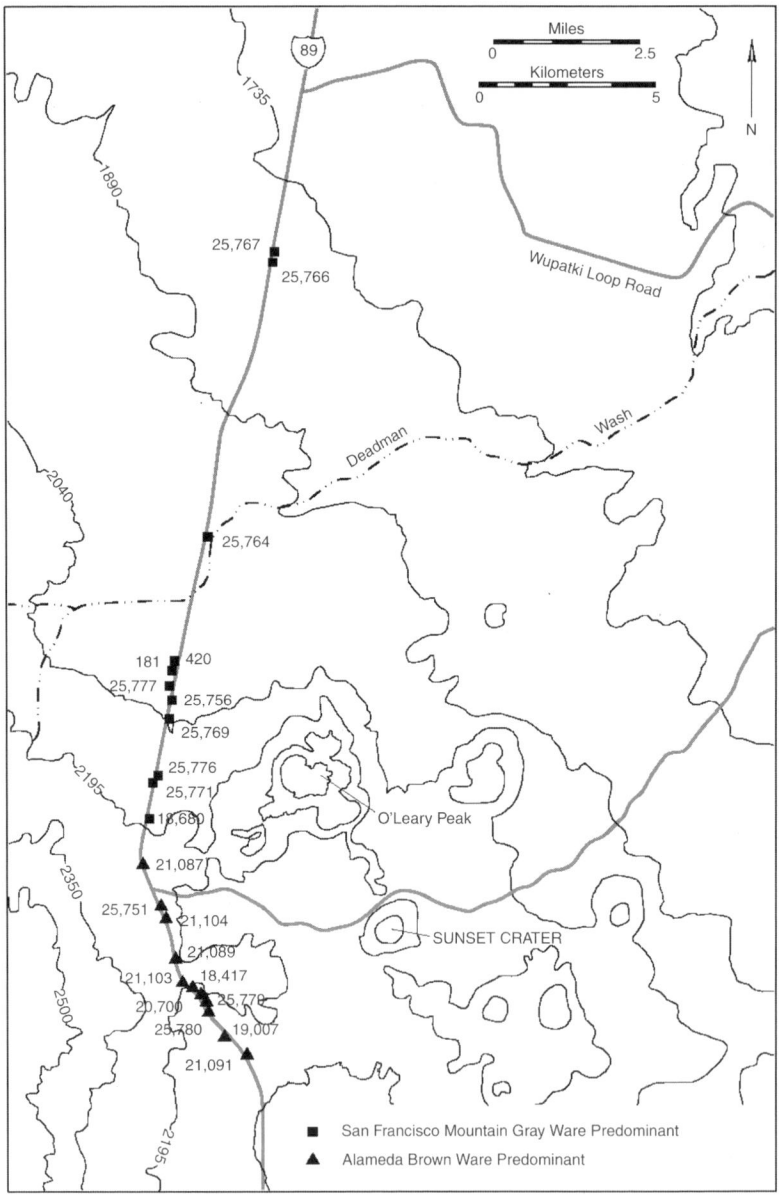

Figure 6.4 Dominant ceramic ware at excavated prehistoric sites in the U.S. 89 project.

Brown Ware in the Deadman Wash area, 7 km to the north. Similar frequencies, but in reverse, occur at sites south of the divide.

Another way to examine the impact of the Sunset Crater eruption is to examine changes in the number of excavated rooms in the US 89 project area through time, as shown in Figure 6.5.a There are no surprises here; researchers beginning with Colton have recognised that there is a significant increase in both rooms and sites in the period just following the eruption of Sunset Crater (Colton 1936, 1949; see also Downum and Sullivan 1990).

Although Colton (1936: 341) attributed this population increase to the beneficial mulching effect of the cinders in areas where they were thinly deposited, others (Hevly *et al* 1979; Pilles 1979: 479–80, 1996: 65) have proposed that it was because of favourable climatic change and the advent of seasonally occupied fieldhouse architecture, thereby increasing site or room numbers but not greatly increasing population. Our research suggests that both of these factors probably contributed to the pattern documented in the archaeological record. Although there are a few fieldhouse sites in the US 89 project area prior to the eruption, they are much more common following it (Elson 2006a, 2006b). It is also true that there is a moister climatic regime between 1050 and 1090 AD. During this period, precipitation was average or above average nearly 80% of the time, with over 60% of years having above average rainfall. In the normally arid and often unpredictable Southwest, this rainfall increase represents particularly favourable conditions for dryland farming (Salzer and Dean 2006; Salzer and Kipfmueller 2005).

Furthermore, agricultural experiments by Colton (1965), Maule (1963), and Waring (2006), strongly support the importance of the cinder mulch for local agriculture. At an experimental plot planted by Waring

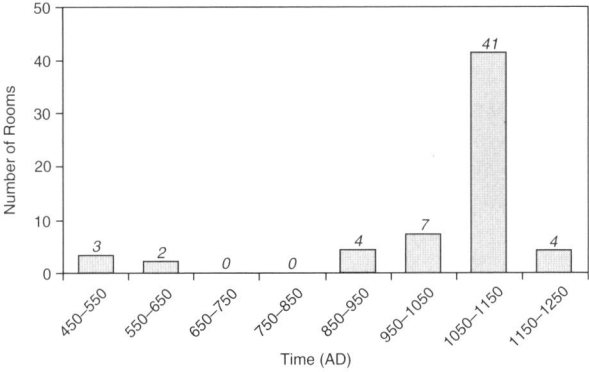

Figure 6.5(a) Change in number of rooms through time at excavated sites in the US 89 project area.

near Wupatki National Monument, which is at the very northern end of the US 89 project area at approximately 1,735 masl, no corn germinated without a cinder mulch cover, similar to results obtained by Maule in the early 1960s at Wupatki Pueblo itself. These data all strongly suggest that a cinder cover of 3–10 cm is highly beneficial to corn growth. Conversely, the data also indicate that a >15 cm thick cinder cover is detrimental to agriculture, and that corn will not grow in cinders deeper than 20 or 25 cm. This is at least partly because the silica-rich ash does not contain enough available nutrients, particularly nitrogen, to allow for successful plant propagation (Elson *et al* 2002: 123). That cinders can be both beneficial and harmful to agriculture is also supported by observations from recent eruptions, such as that of Parícutin volcano (Luhr and Simkin 1993).

Flagstaff area archaeologists have long known that the higher elevations were occupied earlier (Colton 1946; Pilles 1979). This is supported by data from the US 89 project, where the average site elevation dropped dramatically after 1050 AD (Figure 6.6). The 950–1050 AD period in Figure 6.5b is heavily skewed by a single low-elevation fieldhouse site at 1,735 m; without this site, the average elevation would be above 2,135 m, like all time periods prior to 1050 AD and the eruption of Sunset Crater.

Survey records of sites outside the US 89 project area within the Coconino National Forest and Wupatki National Monument indicate that the 1,890 masl elevation is the lower limit for most pre-eruptive settlement, at least in the area north of the San Francisco Peaks. This elevation limit is supported by data from modern climate records: for corn to successfully propagate, it needs at least 250 mm of annual precipitation, but most importantly, 150 mm of this must occur during the growing season (Muenchrath and Salvador 1995: 311). Although the 1,890 masl elevation receives approximately 370 mm of annual rain, only 150 mm of this falls during the growing season, making this the minimum elevation where corn agriculture is currently possible (Elson *et al* 2004: 24–27). These data indicate that prior to the eruption, areas below 1,890 masl, which did not receive enough rain for reliable dry farming, were probably not farmable, except on a sporadic or opportunistic basis during unusually wet periods.

Cinder Mulch Agriculture

Our data strongly support arguments made by previous researchers that the thin cinder blanket deposited by the eruption of Sunset Crater was critical to successful agriculture in this extremely arid region (see Anderson 1990; Colton 1946, 1960; Sullivan and Downum 1991). This is supported by modern agricultural research that indicates

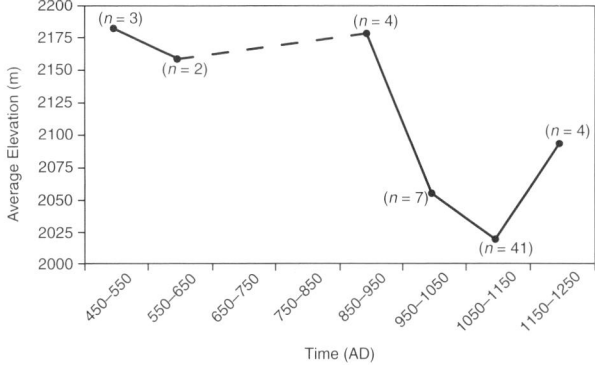

Figure 6.5 (b) Average elevation of rooms through time at excavated sites in the US 89 project area.

Figure 6.6 Small prehistoric agricultural field system located within cinder deposits in the Sunset Crater area (photograph by Kirk Anderson).

that lithic mulches reduce evaporation and increase infiltration, thereby increasing soil moisture and water availability for plant uptake (Lightfoot and Eddy 1995). Lithic mulches are common in arid and semi-arid landscapes where water for crop growth is limited. Water may be in the form of rain, fog, or dew; for example, moisture from fog was captured by stone-lined pits in the Atacama Desert of Chile and Peru, whereas dew was captured by lithic-mulch plots in the Negev of southern Israel (Lightfoot 1994). The earliest known use of lithic mulches was by the Nabatean culture in the Negev starting about 200 BC, but perhaps as early as 2000 BC (Lightfoot 1994).

In the prehistoric Southwest United States, the Classic period (post–1150 AD) Hohokam in southern Arizona used rock-pile lithic mulch fields to cultivate agave, whilst the Anasazi grew maize in pebble-mulch fields in New Mexico during the 14th and 15th centuries AD (Fish *et al* 1985; Maxwell and Anschuetz 1992). The most common prehistoric agricultural system in the general Flagstaff area was composed of small, 1–5 ha field areas containing short, linear, rock alignments, commonly within Sunset Crater cinder deposits, as shown in Figure 6.6 (Berlin *et al* 1977; Edwards 2004; Elson 2006a, 2006b, 2006c). The largest known field area consists of a 200 ha ridge and mound system within Sunset Crater cinders, approximately 20 km east of the volcano (Berlin *et al* 1990). An extensive system of small cinder fields associated with numerous types of rock alignments is also present in Wupatki National Monument, just north of the US 89 project area (Travis 1990). The only other known use of volcanic deposits as a lithic mulch is in the Canary Islands, where it was used to grow numerous vegetable and fruit crops (Lightfoot 1994).

The practice of lithic mulching has a positive influence on a number of soil properties that make the soils more beneficial for plant growth; an increase in infiltration, decrease in runoff, and decrease in evaporation all increase soil moisture, which allows the plant to take up more water. Alderfer and Merkel (1943) determined that for a 75 mm/hr rainfall, 40–60% runoff occurred on bare ground, whilst only 3–10% runoff occurred on pebble-mulch areas. Evaporation from bare ground was estimated to be 8.8 mm day, whilst under pebble mulch it was 1.8 mm/day; after two years, pebble mulch fields would have 30–40% more soil moisture (Choriki *et al* 1964). In addition, pebble-mulch areas stored 60% of winter moisture, compared to 40% for bare ground (Fairbourn 1973). This means that with a mean annual precipitation of 380 mm (similar to the 1,890 masl elevation discussed above), pebble mulch soils will retain an additional 38–58 mm of rainfall over bare ground. This increase would have been highly significant in low elevation, semiarid to arid areas, such as the northern portion of the US 89 project area and Wupatki National Monument, where the addition of just a few centimetres of precipitation could mean the difference between success or failure in crop production (Corey and Kemper 1968). For example, average yearly precipitation at the 100-room Wupatki Pueblo itself, at approximately 1,495 masl, is just over 200 mm, with only 100 mm falling during the growing season (Anderson 1990; Western Regional Climate Center 2002). Since this is below the threshold for corn agriculture, it is clear that the cinder mulch was an extremely critical variable in prehistoric settlement of this area.

This agricultural practice has additional benefits for soils. For example, with more water in the soil, nutrient availability increases. Lithic mulches also reduce rain splash erosion, runoff, and the formation of erosion rilling. The stony cover decreases wind speed across the ground surface, thereby diminishing the drying effects of wind, an important factor in the general Flagstaff and Wupatki areas that are seasonally buffeted by very high winds.

Soil temperature can also be regulated by the addition of lithic mulches, thereby lengthening the growing season and decreasing diurnal temperature fluctuations. Although naturally warm areas, such as sites in the lower elevations, may not need an extended growing season, the increase in soil temperature also increases the uptake rate of nutrients and water. Lengthening the growing season, however, would have been critical in the higher elevations. This is because most types of corn need 115–120 days to mature, although some short-season varieties grown in the pueblo Southwest mature in as few as 75–90 days (Meunchrath and Salvador 1995: 311). In the general Flagstaff area, at 70% probability (or in seven out of every 10 years), an elevation of 2,040 masl averages only 83 consecutive days with temperatures over $0°C$ (freezing) and 106 consecutive days with temperatures above $-2.0°C$ (a hard freeze). At an elevation of 2,195 masl, only 71 consecutive days are above freezing and 94 consecutive days above a hard freeze (Elson *et al* 2004: Table 2.2).

Experimental data from corn grown in pebble-mulch fields indicate that the combined effects of higher soil moisture, longer growing season, and higher rate of water and nutrient uptake allow plants to emerge two to three days earlier and begin tasseling four to seven days prior to corn grown in fields without a lithic mulch (Fairbourn 1973; Hakimi and Kachru 1978). This would have been extremely significant in the higher elevations, where use of a cinder mulch to extend the growing season by even a few days would have had a highly beneficial, if not critical, effect.

ARCHAEOLOGICAL IMPLICATIONS

Having considered the potential effects of the Sunset Crater eruption on the local environment and especially the impact of the resulting cinder mulch, we now consider the archaeological evidence. To begin with, an isopach map of cinder depth (Figure 6.7) was constructed with the software programme Surfer32, using measurements from about 100 sample units spread over an area of approximately 900 km² (Amos 1986; Hooten *et al* 2001). The map provides a minimum cinder depth after 900 years of erosional forces, although many of the measurements were taken from protected or buried areas, which more closely reflect the original deposition.

As noted above, experimental data combined with data from historic or modern volcanic eruptions indicate that crop yield is significantly reduced with cinder depths greater than 15–20 cm. The 30 cm isopach is

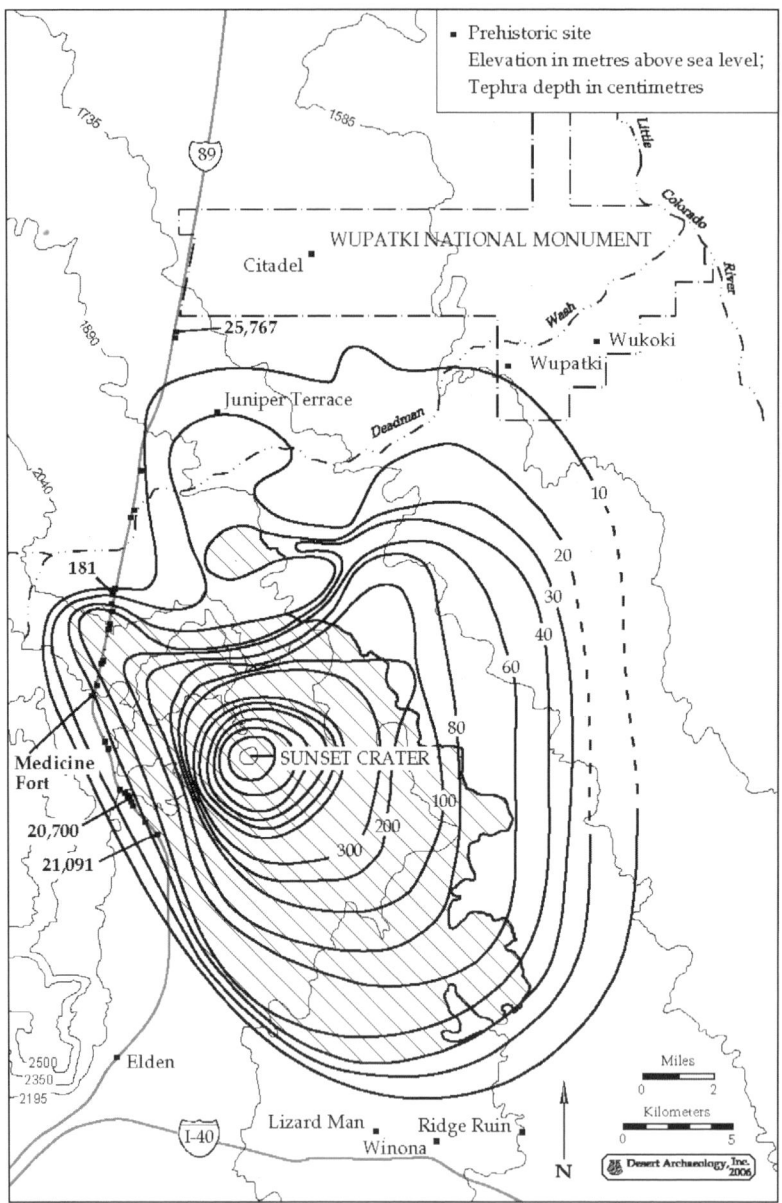

Figure 6.7 Isopach map of Sunset Crater tephra deposition. Cross-hatch indicates prime habitation areas above 1,890 m in elevation with greater than 30 cm of tephra. These had to be abandoned following the eruption.

used here as a very conservative measure of areas that had to be abandoned due to the loss of arable land, at least until erosion removed 15–20 cm of cinders, which was likely a slow process in this semi-arid landscape. The isopach map indicates that an area of approximately 400 km² was uninhabitable immediately following the eruption and likely abandoned. Of this, just over 265 km² was above 1,890 m in elevation and was, before the eruption, prime land for agriculture and settlement (cross-hatched area in Figure 6.8). Conversely, lower elevation areas with cinder fall between 3 cm and 10 cm would have become suitable for farming, because of the beneficial effects of mulching. This area includes the northern portion of the US 89 project area and almost all of Wupatki National Monument. The settlement of areas with 10 cm and 30 cm of cinder and ash fall would have been dependent on local conditions, and particularly local topography; it is likely that some of these 'in-between areas' stayed populated, whilst others were abandoned.

Groups that did move into newly fertile areas would have had to develop a new agricultural technology that necessarily entailed 'cinder management'. This strategy would involve keeping a consistent layer of cinders, 3–10 cm deep, over agricultural areas, which would not have been easy in a region buffeted by strong seasonal winds. Although some of the numerous linear rock alignments found throughout this area, and particularly in Wupatki National Monument, likely functioned as soil or water traps, or to protect young plants from the wind (Downum and Sullivan 1990; Travis 1990), we suspect that many were used to trap and manage cinders.

Figure 6.8 Sunset Crater lava with impressions of prehistoric corn ('corn rock') recovered from habitation site 4 km west of the Bonito lava flow (photograph by Helga Teiwes).

A comprehensive survey inventory of the 143 km^2 Wupatki National Monument (see Anderson 1990) revealed just under 2,400 prehistoric sites, or an average of approximately 17 sites per km^2. For the 977 sites that could be dated, only two, or around 0.2%, were definitively occupied prior to Sunset Crater, indicating enormous migration into this area following the eruption (Downum and Sullivan 1990: Table 5.2). The traditional interpretation, based primarily on ceramic type distinctions, is that the monument was occupied primarily by immigrant groups who joined with a small local population to settle this newly arable area and take advantage of the cinder mulch and increased crop yield (Colton 1936: 341, 1946; Downum 1988; Downum and Sullivan 1990; Wilcox 2002).

When the abandonment area of the Sunset Crater eruption is fully considered, the large-scale movement of non-local groups into this area can be questioned. Survey data from the Coconino National Forest site files indicate that an average of 10 sites per km^2 is a conservative estimate for the 265 km^2 that are above 1,890 m and experienced more than 30 cm of volcanic ash deposition. Excavation data from the US 89 project area, including sites both with and without architecture, also indicate an average of around four structures per site (Elson 2006a, 2006b). These data suggest that approximately 2,650 sites, containing over 10,000 structures, were abandoned following the eruption, which is comparable to the number of sites recorded in the Wupatki inventory survey. It is therefore not necessary to invoke migration by a large number of people from outside the immediate region (eg, the Cohonina and Kayenta Anasazi), as previous researchers have suggested, to populate Wupatki and other lower elevation areas following the eruption. These areas could have been settled by local groups displaced by the cinder and ash fall. Although we have no problems with including some outside migration into the lower elevations, the simplest explanation is that many, if not most, of these new settlements were made by the same people who had lived in the heavy ash fall or lava zones. This simple demographic exercise does not take into account site function or contemporaneity issues, but it does underscore the potential impact of the eruption on the local settlement, as well as the fact that the population displaced by heavy Sunset Crater cinder fall must be taken into account in all archaeological models.

Ceramics and Cultural Affiliation

The question that then remains is why do many settlements located north of the Coconino Divide and in the Wupatki area contain so much San Francisco Mountain Gray Ware pottery, long used as the basis for assigning cultural affiliation, if the people living there were *not* Cohonina migrants? First, petrographic temper provenance data indicate that people living north of the Coconino Divide were pottery consumers, not pottery producers. The ceramic temper resources in this area, labelled B, N, and Q in Figure 6.3, are not compatible with the temper observed in San Francisco Mountain Gray Ware ceramics (Miksa *et al* 2006). The presence of San Francisco Mountain Gray Ware at any given site therefore documents the distribution, consumption, and discard of pottery but *not* its production. Accordingly, because the ware was not locally made but was traded into the area, the abundance of San Francisco Mountain Gray Ware cannot be used, as it has commonly been, as a simple 'index' for a settlement's cultural affiliation.

Second, the eruption significantly disrupted Alameda Brown Ware production, but not the production of San Francisco Mountain Gray Ware. Petrographic data indicate that most of the Alameda Brown Ware recovered in the US 89 project area was made in a restricted geological zone, labelled K in Figure 6.3, located approximately 5–10 km west of Sunset Crater.

The isopach mapping suggests that about 50% of the ceramic production zone was covered with 20–40 cm of ash and cinders following the eruption, likely causing abandonment of numerous sites and disruption of ceramic production areas (see Figure 6.8). We think it is highly probable, then, that the eruption significantly reduced Alameda Brown Ware ceramic production, at least for a number of years until erosion had reduced the cinders to a farmable depth. San Francisco Mountain Gray Ware, on the other hand, was produced somewhere north-west of Flagstaff. We were not successful in locating the exact source area, but our work clearly indicated that the gray ware, which does not have volcanic grains in its temper, was made outside of the region affected by the Sunset Crater eruption and outside of the larger San Francisco volcanic field in general (Heidke *et al* 2006; Roberts 2001).

Currently, our best petrographic match is in the Cataract Creek area, some 50–60 km west of the US 89 project area. Although moving ceramic vessels 60 km may seem unreasonable, both Cataract Creek and Deadman Wash would have provided natural pathways, and this distance for movement of ceramics is ethnographically supportable (Beier 1980; Crossland and Posnansky 1978; Heidke *et al* 2006; Malville 2001). In this respect, San Francisco Mountain Gray Ware producers may have increased ceramic production to fill the gap left by the disruption in Alameda Brown Ware manufacture. Accordingly, the amount of San Francisco Mountain Gray Ware recovered from sites located in the Deadman Wash area increased almost immediately after the Sunset Crater eruption.

Ritual Behaviour

The Sunset Crater eruption may also have led to the initiation of volcano-related ritual behaviour. This is suggested by the recovery of over 50 pieces of Sunset Crater basalt with prehistoric corn cob and stalk impressions, termed 'corn rocks,' from the surface of a site 4 km from the nearest lava flow (Figure 6.8). Sunset Crater basalt is intrusive in this area; geochemical analyses indicate that, along with the rocks bearing corn impressions, around 950 additional pieces of Sunset Crater basalt, weighing more than 40 kg, were purposely brought to this location (Elson *et al* 2002: 125–29).

We have argued (Elson *et al* 2002) that the characteristics of the corn rocks, along with experimental data, suggest that the rocks were made deliberately through placement of corn cobs as an offering around a *hornito*, or small spattercone. Corn is a sacred plant to all Pueblo groups in the Southwest United States and it likely served a similar purpose in this region during prehistory. Why rocks with corn casts were transported to a habitation site 4 km from the lava flow is unknown, but it can be speculated that the corn rocks themselves may have been seen as a source of supernatural power or, given that one was found embedded in the wall of a structure, as protection from the malevolent forces of the volcano. Although both prehistoric and modern offerings are commonly associated with volcanoes in other parts of the world (Luhr and Simkin 1993; Plunket and Uruñuela 1998a, 1998b; Scarth 1999; Sigurdsson 1999), this is the first evidence from the south-western United States of possible ritual behaviour related to volcanism.

CONCLUSIONS

Volcanic eruptions are commonly considered to have a negative influence on local populations because they are an obvious direct threat to life and property. The explosive forces associated with eruptions – lava flows, tephra deposits, fires, and landslides – can devastate local environments and cause populations to seek other safer and more productive places to live. Longer term impacts, such as adverse effects on agricultural soils, can also be devastating to populations living in a subsistence economy dependent on domesticated crops. Psychological impacts are also possible, with the severity of the impact dependent on the nature and timing of the eruption, the extent of property destruction, and the overall loss of human life (Nolan 1993; Sheets and Grayson 1979: 4–5). Depending on rock types, climate, and other soil-forming factors, volcanic deposits may take hundreds to thousands of years to weather and become fertile again.

Not all volcanic eruptions have long-term negative impacts. As our research has shown, small-scale cinder-cone eruptions can have both negative and positive influences on the landscape and local populations. The Sunset Crater eruption buried villages, agricultural fields, and important landmarks, almost certainly causing immediate abandonment of some areas. In contrast, within a relatively short period after the eruption, other parts of the landscape became substantially more productive. The porous, highly permeable cinders that thinly covered the ground surface acted as a mulch over the barren desert soils. The mulching effect increased soil moisture and nutrient availability and also lengthened the growing season. Higher elevation areas covered by a thin cinder blanket became more productive because of the increase in the growing season, whereas lower elevation areas became more productive because of the increase in available moisture. As a direct result of these combined effects, the Wupatki National Monument area became a highly attractive region for human settlement and some of the largest pueblos ever built in the Flagstaff area were constructed there.

Our reconstruction of the processes that led to this result can now be summarised.

1. Prior to the eruption of Sunset Crater, areas below 1,890 masl in elevation were only opportunistically inhabited because they received too little precipitation for dry farming.
2. Sunset Crater erupted sometime in the mid-to-late 11th (and possibly the early 12th) century AD. The cinder mulch from the eruption opened elevations below 1,890 masl to permanent settlement.
3. An agriculturally productive and densely settled area of around 265 km^2 had to be abandoned because it contained greater than 30 cm of ash and cinder deposits. Many of the displaced groups moved into newly fertile areas below 1,890 masl, and it is therefore not necessary to invoke large numbers of outside immigrant groups to populate the lower elevations.
4. The cinder and ash fall significantly disrupted the zone of Alameda Brown Ware manufacture and likely reduced production of this ware.
5. In contrast, San Francisco Mountain Gray Ware was manufactured somewhere west of the San Francisco volcanic field and was not affected by the Sunset Crater eruption. Producers of San Francisco Mountain Gray Ware took advantage of the reduction in Alameda Brown Ware and increased their production. Consequently, additional ceramics were imported into areas north of Flagstaff, probably via a west to east route down Deadman Wash. Although additional research is needed, we propose that this increased ceramic production may represent community-level craft specialisation. What was being exchanged for the gray-ware ceramics is still unknown, although agricultural goods such as corn, and especially low-elevation goods like cotton, are possibilities.
6. Finally, ritual behaviour centred around the volcanic activity was initiated, suggesting some sort of alteration to existing belief systems. Although the exact nature and purpose of this ritual is unknown, it can be speculated that the corn rocks were made as an offering, perhaps to appease the forces responsible for the eruption or to serve as a source of protection from the volcano. Since ethnographic data indicate that ritual

behaviour related to volcanism is common in almost all areas with active (and often inactive) volcanoes, it would be surprising if ritual behaviour related to Sunset Crater were absent. In addition, studies of catastrophic events have shown that religious mechanisms for coping with a natural disaster are highly adaptive, enabling affected individuals and groups to more readily accept the event and begin the recovery process (Nolan 1979). Today, more than 900 years after Sunset Crater arose on the landscape, the Hopi, who are believed to be the descendents of the prehistoric Sinagua, still pass accounts of the eruption from generation to generation as a part of traditional clan knowledge, thereby underscoring the significance of this event (Colton 1932b; Ferguson and Loma'omvaya 2005; Malotki and Lomatuway'ma 1987).

ACKNOWLEDGMENTS

The research undertaken on the US 89-Wupatki to Fernwood Archaeological Project was funded primarily by the Arizona Department of Transportation (ADOT); financial support was also provided by research grants from the National Park Service (Flagstaff Area National Monuments) and Western National Parks Association. Assistance from Coconino National Forest, Desert Archaeology, and Northern Arizona University are also gratefully acknowledged. We would like to thank archaeologists from these agencies and institutions who facilitated our research in numerous ways, particularly Jeri DeYoung, William Doelle, Linda Farnsworth, Robert Gasser, Peter Pilles, and Bettina Rosenberg. The research was greatly helped by discussions with a number of people, and we particularly thank Diana Anderson, Jeffrey Dean, Christian Downum, Wendell Duffield, Susan Hall, Sarah Herr, Jeffery Homburg, Elizabeth Miksa, Nancy Riggs, Terry Samples, Paul Sheppard, Deborah Swartz, and Scott Van Keuren.

REFERENCES

Alderfer, R and Merkel, F (1943) 'The comparative effects of surface application versus incorporation of various mulching materials on structure, permeability, runoff, and other soil properties', *Soil Science Society of America Proceedings* 8, 79–86

Amos, R (1986) 'Sunset Crater, Arizona, evidence for a large magnitude Strombolian eruption', unpublished Master's thesis, Department of Geology, Arizona State University, Tempe

Anderson, B (compiler) (1990) *The Wupatki Archaeological Inventory Survey Project, Final Report*, Professional Paper No 35, Santa Fe, NM: Southwest Cultural Resource Center, Division of Anthropology, National Park Service

Beier, G (1980) 'Yoruba pottery', *Ceramics Monthly* 12, 48–53

Berlin, G, Ambler, J, Hevly, R and Schaber, G (1977) 'Identification of Sinagua agricultural field by aerial thermography, soil chemistry, pollen/plant analysis, and archaeology', *American Antiquity* 42, 588–600

Berlin, G, Salas, D and Geib, P (1990) 'A prehistoric Sinagua agricultural site in the ashfall zone of Sunset Crater, Arizona', *Journal of Field Archaeology* 17, 1–16

Blong, R (1984) *Volcanic Hazards, A Sourcebook on the Effects of Eruptions*, Orlando, FL: Academic Press

Boston, R (1995) 'Electron microprobe sourcing of volcanic ash temper in Sunset Red ceramics', unpublished Master's thesis, Department of Anthropology, Northern Arizona University, Flagstaff

Bullard, F (1984) *Volcanoes of the Earth*, Austin: University of Texas Press

Cameron, E (1999) 'Examining ethnicity through architecture, the Sunset Crater ash fall area- A.D. 1050 to 1150', unpublished Master's thesis, Department of Anthropology, Northern Arizona University, Flagstaff

Champion, D (1980) *Holocene Geomagnetic Secular Variation in the Western United States, Implications for the Global Geomagnetic Field*, Open-File Report 80-824, Washington, DC: US Department of the Interior Geological Survey

Choriki, R, Hide, J, Drall, L and Brown, B (1964) 'Rock and gravel mulch aid in moisture storage', *Crops and Soils* 16, 24

Colton, H (1932a) 'Sunset Crater, the effect of a volcanic eruption on the ancient Pueblo people', *Geographical Review* 22, 582–90

Colton, H (1932b) 'A possible Hopi tradition of the eruption of Sunset Crater', *Museum Notes* (Museum of Northern Arizona, Flagstaff) 5, 23

Colton, H (1936) 'The rise and fall of the prehistoric population of northern Arizona', *Science* 84, 337–43

Colton, H (1939) *Prehistoric Culture Units and Their Relationships in Northern Arizona*, Bulletin 17, Flagstaff: Museum of Northern Arizona

Colton, H (1946) *The Sinagua, A Summary of the Archaeology of the Region of Flagstaff, Arizona*, Bulletin 22, Flagstaff: Museum of Northern Arizona

Colton, H (1949) 'The prehistoric population of the Flagstaff area', *Plateau* 22, 21–25

Colton, H (1960) *Black Sand, Prehistory in Northern Arizona*, Albuquerque: University of New Mexico Press

Colton, H (1965) 'Experiments in raising corn in the Sunset Crater ashfall area east of Flagstaff, Arizona', *Plateau* 37, 77–79

Colton, H (1968) 'Frontiers of the Sinagua', in Schroeder, A (ed), *Collected Papers in Honor of Lyndon Lane Hargrave*, pp 9–15, Papers of the Archaeological Society of New Mexico, Santa Fe: Museum of New Mexico Press

Corey, A and Kemper, W (1968) *Conservation of Soil Water by Gravel Mulches*, Hydrology Paper No 30, Ft Collins: Colorado State University

Crossland, L and Posnansky, M (1978) 'Pottery, people and trade at Begho, Ghana', in Hodder, I (ed), *The Spatial Organization of Culture*, pp 77–89, Pittsburgh: University of Pittsburgh Press

Diehl, M (2006) 'Interassemblage macrobotanical variation from the U.S. 89 archaeological project', in Elson, M (ed), *Sunset Crater Archaeology, The History of a Volcanic Landscape. Environmental Analyses*, pp 27–46, Anthropological Papers No 33, Tucson, AZ: Center for Desert Archaeology

Downum, C (1988) '"One grand history", a critical review of Flagstaff archaeology, 1851–1988', unpublished PhD dissertation, Department of Anthropology, University of Arizona, Tucson

Downum, C and Gumerman, G (1998) *Archaeological Investigations at Sunset Crater Volcano National Monument*, Archaeological Report No 1159, Flagstaff: Northern Arizona University

Downum, C and Sullivan, A (1990) 'Settlement patterns', in Anderson, B (compiler), *The Wupatki Archeological Inventory Survey Project, Final Report*, pp 5.1–.90, Professional Paper No 35, Santa Fe, NM: Southwest Cultural Resource Center, Division of Anthropology, National Park Service

Edwards, J (2004) 'Soil fertility and prehistoric agriculture near Sunset Crater, Arizona', unpublished Master's thesis, Quaternary Sciences Program, Northern Arizona University, Flagstaff

Elson, M (ed) (2006a) *Sunset Crater Archaeology, The History of a Volcanic Landscape. Introduction and Site Descriptions*, Anthropological Papers No 30, Part 1, Tucson, AZ: Center for Desert Archaeology

Elson, M (ed) (2006b) *Sunset Crater Archaeology, The History of a Volcanic Landscape. Introduction and Site Descriptions*, Anthropological Papers No 30, Part 2, Tucson, AZ: Center for Desert Archaeology

Elson, M (ed) (2006c) *Sunset Crater Archaeology, The History of a Volcanic Landscape. Environmental Analyses*, Anthropological Papers No 33, Tucson, AZ: Center for Desert Archaeology

Elson, M, Ort, M, Hesse, S and Duffield, W (2002) 'Lava, corn, and ritual in the northern Southwest', *American Antiquity* 67, 119–35

Elson, M, Phillips, B and Ort, M (2004) 'Environmental Setting', in Elson, M (ed), *Sunset Crater Archaeology, The History of a Volcanic Landscape. Introduction and Site Descriptions*, pp 17–44, Anthropological Papers No 30, Part 1, Tucson, AZ: Center for Desert Archaeology

Fairbourn, M (1973) 'Effect of gravel mulch on crop yields', *Agronomy Journal* 65, 925–28

Ferguson, T and Loma'omvaya, M (2005) 'Nuvatukya'ovi, Palatsmo, Niqw Wupatki, Hopi history, culture, and landscape', in Elson, M (ed), *Sunset Crater Archaeology, The History of a Volcanic Landscape. Prehistoric Settlement in the Shadow of the Volcano* (unpaginated draft ms), Anthropological Papers No 37, Tucson, AZ: Center for Desert Archaeology

Fish, S, Fish, P, Miksicek, C and Madsen, J (1985) 'Prehistoric agave cultivation in southern Arizona', *Desert Plants* 7, 100, 107–12

Hakimi, A and Kachru, R (1978) 'Silage corn responses to different mulch tillage treatments under arid and semiarid climatic conditions', *Journal of Agronomy and Crop Science* 147, 15–23

Heidke, J, Roberts, S, Herr, S and Elson, M (2006) 'Alameda Brown Ware and San Francisco Mountain Gray Ware technology and economics', in Van Keuren, S and Elson, M (eds), *Sunset Crater Archaeology, The History of a Volcanic Landscape. Ceramic Technology, Distribution, and Use* (unpaginated draft ms), Anthropological Papers No 32, Tucson, AZ: Center for Desert Archaeology

Hevly, R, Kelly, R, Anderson, G and Olsen, S (1979) 'Comparative effects of climatic change, cultural impact, and volcanism in the paleoecology of the Flagstaff, Arizona, A.D. 900–1300', in Sheets, P and Grayson, K (eds), *Volcanic Activity and Human Ecology*, pp 487–523, New York: Academic Press

Holm, R and Moore, R (1987) 'Holocene scoria cone and lava flows at Sunset Crater, northern Arizona', in *Geological Society of America Centennial Field Guide-Rocky Mountain Section*, pp 393–97, Boulder, CO: Geological Society of America

Hooten, J, Ort, M and Elson, M (2001) *Origin of Cinders in Wupatki National Monument*, Technical Report No 2001–12, Tucson, AZ: Desert Archaeology, Inc

Lightfoot, D (1994) 'Morphology and ecology of lithic mulch agriculture', *Geographical Review* 84, 172–85

Lightfoot, D and Eddy F (1995) 'The construction and configuration of Anasazi pebble-mulch gardens in the northern Rio Grande', *American Antiquity* 60, 459–70

Luhr, J and Simkin, T (eds) (1993) *Parícutin, The Volcano Born in a Mexican Cornfield*, Phoenix: Geosciences Press, Inc

Malotki, E and Lomatuway'ma, M (1987) *Earth Fire, A Hopi Legend of the Sunset Crater Eruption*, Flagstaff, AZ: Northland Press

Malville, N (2001) 'Long-distance transport of bulk goods in the pre-Hispanic American Southwest', *Journal of Anthropological Archaeology* 20, 230–43

Maule, S (1963) 'Corn growing at Wupatki', *Plateau* 36, 29–32

Maxwell, T and Anschuetz, K (1992) 'The southwestern ethnographic record and prehistoric agricultural diversity', in Killion, T (ed), *Gardens of Prehistory, The Archaeology of Settlement Agriculture in Greater Mesoamerica*, pp 35–68, Tuscaloosa: University of Alabama Press

McGregor, J (1941) *Winona and Ridge Ruin, Part I, Architecture and Material Culture*, Bulletin No 18, Flagstaff: Museum of Northern Arizona and the Northern Arizona Society of Science and Art

McGregor, J (1951) *The Cohonina Culture of Northwestern Arizona*, Urbana: University of Illinois Press

Miksa, E, Montague-Judd, D and Heidke, J (2006) 'Petrographic analysis of tempering materials', in Van Keuren, S and Elson, M (eds), *Sunset Crater Archaeology, The History of a Volcanic Landscape. Ceramic Technology, Distribution, and Use* (unpaginated draft ms), Anthropological Papers No 32, Tucson, AZ: Center for Desert Archaeology

Muenchrath, D and Salvador, R (1995) 'Maize productivity and agroecology, effects of environment and agricultural practices on the biology of maize', in Toll, H (ed), *Soil, Water, Biology, and Belief in Prehistoric and Traditional Southwestern Agriculture*, pp 303–33, Special Publication 2, Albuquerque: New Mexico Archaeological Council

Nolan, M (1979) 'Impact of Parícutin on five communities', in Sheets, P and Grayson, D (eds), *Volcanic Activity and Human Ecology*, pp 293–338, New York: Academic Press

Nolan, M (1993) 'Human communities and their responses', in Luhr, J and Simkin, T (eds), *Parícutin, The Volcano Born in a Mexican Cornfield*, pp 189–206, Phoenix: Geosciences Press, Inc

Ort, M, Elson, M and Champion, D (2002) *A Paleomagnetic Dating Study of Sunset Crater Volcano*, Technical Report 2002–16, Tucson, AZ: Desert Archaeology, Inc

Pilles, P, Jr (1978) 'The field house and Sinagua demography', in Ward, A (ed), *Limited Activity and Occupation Sites, A Collection of Conference Papers*, pp 119–33, Contributions to Anthropological Studies No 1, Albuquerque: Center for Anthropological Studies

Pilles, P, Jr (1979) 'Sunset Crater and the Sinagua, a new interpretation', in Sheets, P and Grayson, D (eds), *Volcanic Activity and Human Ecology*, pp 459–85, New York: Academic Press

Pilles, P, Jr (1996) 'The Pueblo III period along the Mogollon Rim, the Honanki, Elden, and Turkey Hill phases of the Sinagua', in Adler, M (ed), *The Prehistoric Pueblo World, A.D. 1150–1350*, pp 59–72, Tucson: University of Arizona Press

Plunket, P and Uruñuela, G (1998a) 'Appeasing the volcano gods', *Archaeology* 51, 36–42

Plunket, P and Uruñuela, G (1998b) 'Preclassic household patterns preserved under volcanic ash at Tetimpa, Puebla, Mexico', *Latin American Antiquity* 94, 287–309

Rees, J (1979) 'Effects of the eruption of Parícutin volcano on landforms, vegetation, and human occupancy', in Sheets, P and Grayson, D (eds), *Volcanic Activity and Human Ecology*, pp 249–92, New York: Academic Press

Roberts, S (2001) 'Exploring prehistoric Cohonina social identity and interaction through San Francisco Mountain Gray Ware production', unpublished Master's thesis, Department of Anthropology, Northern Arizona University, Flagstaff

Robinson, W, Harrill, B and Warren, R (1975) *Tree-Ring Dates from Arizona H-I, Flagstaff Area*, Tucson: Laboratory of Tree-Ring Research, University of Arizona

Salzer, M and Dean, J (2006) 'Dendroclimatic reconstructions and paleoenvironmental analyses for the U.S. 89 project area', in Elson, M (ed), *Sunset Crater Archaeology, The History of a Volcanic Landscape. Environmental Analyses*, pp 125–56, Anthropological Papers No 33, Tucson, AZ: Center for Desert Archaeology

Salzer, M and Kipfmueller, K (2005) 'Reconstructed temperature and precipitation on a millennial timescale from tree-rings in the southern Colorado Plateau, U.S.A', *Climatic Change* 70, 465–87

Samples, T (1990) 'Cohonina archaeology, the view from Sitgreaves Mountain', unpublished Master's thesis, Department of Anthropology, Northern Arizona University, Flagstaff

Scarth, A (1999) *Vulcan's Fury, Man against the Volcano*, New Haven, CT: Yale University Press

Sheets, P and Grayson, D (1979) 'Introduction', in Sheets, P and Grayson D (eds), *Volcanic Activity and Human Ecology*, pp 1–8, New York: Academic Press

Sheppard, P, May, E, Ort, M, Anderson, K and Elson, M (2005) 'Dendrochronological responses to the 24 October 1992 tornado at Sunset Crater, northern Arizona', *Canadian Journal of Forestry Research* 35, 2911–19

Sheppard, P, Ort, M and Elson, M (2004) 'Collaborative research, dendrochronological, volcanological, and archaeological study of cinder cone eruptions of Parícutin and Sunset Crater', proposal funded by National Science Foundation, Earth Sciences Division, Petrology and Geochemistry, Washington, DC

Shoemaker, E and Champion, D (1977) 'Eruption history of Sunset Crater, Arizona', Investigator's annual report,unpublished manuscript on file, Area National Monuments Headquarters, Wupatki, Sunset Crater Volcano, and Walnut Canyon National Monuments, Flagstaff: Arizona

Sigurdsson, H (1999) *Melting the Earth, The History of Ideas on Volcanic Eruptions*, New York: Oxford University Press

Smiley, T (1958) 'The geology and dating of Sunset Crater, Flagstaff, Arizona', in Anderson, R and Harshbarger, J (eds), *Guidebook of the Black Mesa Basin, Northeastern Arizona*, pp 186–90, Albuquerque: New Mexico Geological Society

Smith, S (2006) 'The U.S. 89 pollen analysis and regional archaeobotanical overview', in Elson, M (ed), *Sunset Crater Archaeology, The History of a Volcanic Landscape. Environmental Analyses*, pp 1–25, Anthropological Papers No 33, Tucson, AZ: Center for Desert Archaeology

Stone, G and Downum, C (1999) 'Non-Boserupian ecology and agricultural risk, ethnic politics and land control in the arid southwest', *American Anthropologist* 101, 113–28

Sullivan, A III (1984) 'Sinagua agricultural strategies and Sunset Crater volcanism', in Fish, S and Fish, P (eds), *Prehistoric Agricultural Strategies in the Southwest*, pp 85–100, Anthropological Research Papers No 33, Tempe: Arizona State University

Sullivan, A III and Downum, C (1991) 'Aridity, activity, and volcanic ash agriculture, a study of short-term prehistoric cultural-ecological dynamics', *World Archaeology* 22, 271–87

Swartz, D and Elson, M (2006) 'The Lenox Park Site, NA 20,700 [AR-03-04-02-3352 (CNF)]', in Elson, M (ed), *Sunset Crater Archaeology, The History of a Volcanic Landscape. Introduction and Site Descriptions*, pp 467–518, Anthropological Papers No 30, Part 2, Tucson, AZ: Center for Desert Archaeology

Travis, S (1990) 'The prehistoric agricultural landscape of Wupatki National Monument', in Anderson, B (compiler), *The Wupatki Archeological Inventory Survey Project, Final Report*, pp 4.1–.54, Professional Paper No 35, Santa Fe, NM: Southwest Cultural Resource Center, Division of Anthropology, National Park Service

Vespermann, D, and Schmincke, H (2000) 'Scoria cones and tuff rings', in Sigurdsson, H (ed), *Encyclopedia of Volcanoes*, pp 683–94, New York: Academic Press

Waring, G (2006) 'Hopi corn and volcanic cinders: a test of the relationship between tephra and agriculture in northern Arizona', in Elson, M (ed), *Sunset Crater Archaeology, The History of a Volcanic Landscape. Environmental Analyses*, pp 71–84, Anthropological Papers No 33, Tucson, AZ: Center for Desert Archaeology

Western Regional Climate Center (2002) Available online at www.wrcc.dri.edu/index.html

Wilcox, D (2002) 'The Wupatki nexus, Chaco-Hohokam-Chumash Connectivity, A.D. 1150–1225', in Lesick, K, Kulle, B, Cluney C and Peuramaki-Brown, M (eds), *The Archaeology of Contact: Processes & Consequences*, pp 218–34, Calgary, Canada: The Archaeological Association of the University of Calgary

Yokoyama, I and De la Cruz-Reyna, S (1990) 'Precursory earthquakes of the 1943 eruption of the Parícutin volcano, Michoacán, Mexico', *Journal of Volcanology and Geothermal Research* 44, 265–81

CHAPTER 7

Ecological Roadblocks on a Constrained Landscape: The Cultural Effects of Catastrophic Holocene Volcanism on the Alaska Peninsula, Southwest Alaska

Richard Vanderhoek and RE Nelson

INTRODUCTION

This research addresses the impact of catastrophic volcanism on hunter-gatherer populations in Arctic environments. It focuses on the effect of 4,000–3,400 BP volcanism in the central Alaska Peninsula on regional human populations. This study is unique in both the size of the volcanically impacted landscape and the geographically constrained nature of the landscape affected by the volcanism.

This research began with the 1997–2000 National Park Service (NPS) archaeological survey of the Aniakchak National Monument and Preserve (Figure 7.1). This survey (VanderHoek and Myron 2004) was the first comprehensive archaeological ground-based survey of a section of the central Alaska Peninsula. Archaeological and geological survey work on the Aniakchak coast by NPS personnel produced evidence of numerous volcanic events (Dilley 2000b; VanderHoek 1999: 14; VanderHoek and Myron 2004), several of which were catastrophic in nature, mirroring evidence previously garnered by earlier US Geological Survey (USGS) researchers (Miller and Smith 1977, 1987). Pyroclastic flow deposits from one of the later of these catastrophic events, the 3,400 BP eruption of the Aniakchak volcano, were found over most of the survey region, emphasising the large-scale nature of the eruption.

The earliest archaeological evidence on the Aniakchak coast dates to approximately 2,100 BP, causing researchers to wonder if human colonisation (or recolonisation) in the region was delayed by a poorly productive, volcanically disturbed landscape (VanderHoek and Myron 2004: 148). In 2001, the authors travelled to the Alaska Peninsula, taking

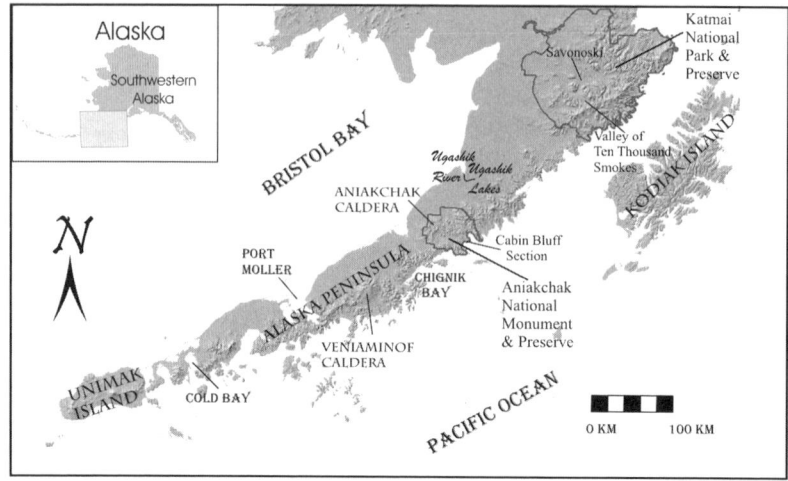

Figure 7.1 The Alaska Peninsula, south-west Alaska.

pollen and sediment samples in the Aniakchak region and sediment samples from the Katmai pyroclastic flow in the Valley of Ten Thousand Smokes (VTTS), in an attempt to address this question. All dates in this chapter are given in radiocarbon years.

SETTING

The Alaska Peninsula is a narrow geologic arm that reaches down from the south-west corner of the Alaskan mainland (at the 59th parallel) over 590 km to almost touch Unimak Island (between the 54th and 55th parallels), the first of the islands in the 1,770 km long Aleutian Island chain (Figure 7.1). The Alaska Peninsula ranges in width from over 170 km wide at its northern end to less than 10 km at its southern point. The peninsula, and the islands that are its continuation, make up the Aleutian volcanic arc, the result of active subduction of the Pacific Plate beneath the North American Plate (Kienle and Nye 1990).

The Alaska Peninsula has two distinctly different landscapes. The Pacific side of the peninsula is dominated by the Aleutian Range, with steep-sided mountains fronting on a coast indented by numerous bays. Rocky capes project to the south and east, with small sandy beaches often found at the heads of bays. The Bering Sea coast, by contrast, is a broad, flat plain composed largely of glacial and volcanigenic sediments deposited throughout the Pleistocene and Holocene, with straight sandy beaches only rarely interrupted by embayments. The seas on both sides of the peninsula are subject to severe storms, and the Aleutian low-pressure system regularly generates storm tracks that sweep across the region.

The peninsula provides a land bridge between the North American continent and the Aleutian Islands, allowing the dispersal of species between the two down the peninsula and along its coasts. The peninsula is a major flyway for a number of species of migratory waterfowl. All five species of salmon travel along its coasts as do whales and other marine mammals. The northern segment of the Alaska Peninsula caribou herd migrates south down the peninsula during the warm season, returning to the region between Lake Becharof and Lake Iliamna in the north in the winter (Hemming 1971).

Humans, too, have used the Alaska Peninsula as a travel corridor. Human populations moved onto the upper Alaska Peninsula more than 9,000 years ago (Henn 1978: 12), with populations reaching Umnak Island in the eastern Aleutians more than 8,000 BP (Aigner 1976; Dumond and Bland 1995). Occupation of mainland and island sites show human movement by this time was both terrestrial and along the coast by watercraft. Prehistoric travel by watercraft along the coast of southern Alaska shows that this travel was keyed to visible, known terrestrial landforms. Islands that could be seen from a mountain on the mainland or from another island were visited or colonised, whilst those that could not be seen from land (like the Pribilof Islands) were not. Early watercraft presumably consisted of small open or covered skin boats (kayaks or umiaks). Although maritime travel was present by 8,000 BP in the Aleutian Islands, it was tied to coastal geography, necessitating visible coastal references and generally sheltered terrestrial locations on which to land and camp. Thus, whilst humans could travel up and down the Alaska Peninsula in watercraft along the coasts, this travel was clearly tied to the land through the needs of navigation and shelter.

ALASKA PENINSULA VOLCANISM

The Aleutian volcanic arc and other subduction zone volcanoes generally have explosive eruptions fuelled by viscous, high-silica magma. These viscous magmas do not de-gas readily, so as they reach the surface, they often erupt explosively and may develop jet-like ash columns (Plinian eruptions). These eruptions frequently generate pyroclastic flows dominated by fine ash suspended in hot volcanic gases, but that may also include blocks up to 10 m or more in diametre, which can flow down slopes at speeds of 100 km/hr or more.

The Aleutian volcanic arc contains 44 of the 54 historically active volcanoes in the United States (Simkin *et al* 1981). At least 21 of these had major, caldera-forming eruptions in the Quaternary Period, with as many as 19 of these possibly forming in the Holocene (Kienle and Nye 1990: 10). At least five of the known Holocene eruptions were of bulk volumes greater than 50 km³ (Miller and Smith 1987). These eruptions were massive in scale, being over twice the size of the 1883 Krakatau eruption and almost 50 times the size of the 18 May 1980, Mount St Helens eruption (Sarna-Wojcicki *et al* 1981; Self and Rampino 1981).

The period between 4,000 and 3,400 BP was an especially active one for the volcanoes in the eastern Aleutian arc. Four of the six known Holocene caldera-forming eruptions on the Alaska Peninsula took place during this time, and at least four other volcanoes had sizeable eruptions during the same period (Riehle *et al* 1998).

The largest and most significant of these eruptions were the Veniaminof eruption (3,700–3,500 BP) and the 3,400 BP (also referred to in the literature as the 3,500 BP (Neal *et al* 2001; Waythomas and Neal 1998) eruption of the Aniakchak volcano, located approximately 110 km apart on the central Alaska Peninsula (Begét *et al* 1992; Miller and Smith 1987; Riehle *et al* 1998). Both of these eruptions produced more than 50 km³ of eruptive material and voluminous pyroclastic flows that blanketed the peninsula from one coast to the other (Miller and Smith 1987).

The Aniakchak volcano is the most active volcano in the eastern Aleutian arc, with at least 40 explosive eruptions in the last 10,000 years (Neal *et al* 2001: 4). The current caldera (Figure 7.2) formed during a massive eruption approximately 3,400 BP, ejecting ≥ 69 km³ of volcanic and lithic material (Dreher 2002). At least 24 km³ of material was injected into the stratosphere as ash (Dreher 2002: 28), with tephra deposits from this eruption, was found more than 1,500 km away on the northern Seward Peninsula (Begét *et al* 1992; Riehle *et al* 1987). Approximately 45 km³ of very mobile volcanic ash and rock, suspended in hot gasses, flowed across the landscape, covering an area of at least 2,500 km² (Dreher 2002; Miller and Smith 1977). Valleys close to the mountain were covered with more than 75 m of ash, with flow deposits found at least 60 km away (Miller and Smith 1977; Waythomas and Neal 1998). Ash flow exposures have been found along > 45 km of the Bering Sea coast north and west of the caldera in sea cliff exposures up to 15 m thick, and along sections of the Pacific coast in the south, showing flow movement over passes 260 m high in the Aleutian Range (Miller and Smith 1977, 1987; Waythomas and Neal 1998). The pyroclastic flow impacted Bristol Bay across a broad front, generating a tsunami that inundated low-lying coastal areas of northern Bristol Bay (Lea 1989a; Waythomas and Neal 1998).

If human populations were living in the Aniakchak region before the eruption, they may have had some warning of the eruption, but it is unclear if they could have travelled far enough away to not be impacted

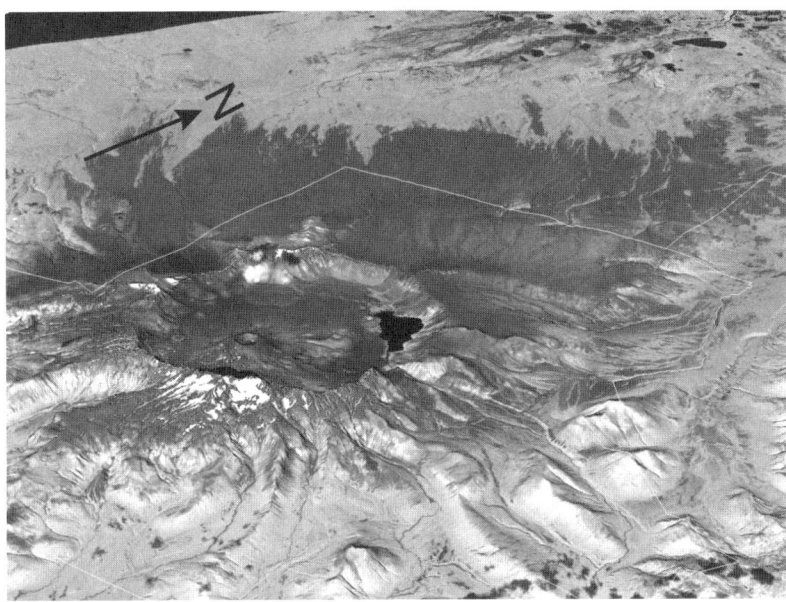

Figure 7.2 The Aniakchak caldera. All of the area in the image was covered by pyroclastic flow 3,400 years ago when the caldera was formed: the darker land north and west of the caldera remains unvegetated. The black region to the north-west is Bristol Bay. The white lines to the north and west are monument boundaries. The caldera is approximately 10 km in diameter. This three-dimensional image is a digital elevation model generated from a 2002 Landsat 7 coverage by Renaissance Remote Sensing.

by it. Historic accounts of the 1912 Katmai eruption report precursor earthquakes (Griggs 1922:19), alerting the local inhabitants that major geologic forces were at work. Katmai Village, the coastal village closest to the eruption, was abandoned several days before the eruption. Savonoski Village, 27 km inland from the erupting vent, was abandoned after the main eruption. American Pete, the chief of Savonoski Village, succinctly described the situation (from Griggs 1922: 17):

> The Katmai Mountain blew up with lots of fire, and fire came down trail from Katmai with lots of smoke. We go fast Savonoski. Everybody get in *bidarka* (skin boat). Helluva job. We come Naknek one day, dark, no could see. Hot ash fall. Work like hell.

The evidence from Katmai and other volcanic eruptions suggests that the Aniakchak eruption may also have generated ever-increasing earthquakes days before the main eruption, warning local populations. The Aniakchak eruption was much larger in scale than the Katmai eruption, though, with pyroclastic flows that spread more broadly across the landscape. If local populations didn't travel rapidly away from the Aniakchak volcano soon after the earthquakes started, they may not have been far enough away to survive the wide-ranging surge and pyroclastic flow effects from the eruption.

Researchers studying the effects of the Mount St Helens eruption have noted that eruptions that take place during the warm season have a greater effect on regional biota than eruptions that take place during the winter, when plants and small animals are protected by snow (del Moral and Bliss 1993: 36). USGS and NPS researchers have reported multiple lines of evidence that, when taken together, suggest a warm season eruption for the 3,400 BP Aniakchak event. The first of these include a pattern of tephra fallout to the north-north-west (Begét *et al* 1992; Riehle *et al* 1987), indicating a wind from the south, more common in the region during the summer months (Lea 1989b: 227, Figure 5.3). Additional support for a warm season eruption is that rip-up peat clasts were found entrained in the lower levels of the 3,400 BP pyroclastic flow (Dilley 2000a), suggesting flow over unfrozen ground. The next fact is that a tsunami was generated by the pyroclastic flow striking Bristol Bay (Waythomas and Neal 1998), suggesting the flow encountered an unfrozen bay. Last, the tsunami deposits on the northern Bristol Bay coast also contained rip-up peat clasts (Waythomas and Neal 1998: 112), which would only have happened if the tsunami swept across the northern coast during the time of year when the ground was unfrozen.

The tsunami itself would have had a profound effect on coastal populations in northern Bristol Bay, a conclusion that no one should doubt after watching the effects of the 26 December 2004 tsunami on the coastal populations of the Indian Ocean. Native settlements like those found historically near the mouths of major rivers would have been devastated by the tsunami, which left a layer of sand 8–14 m above the high-tide line along 50 km of the northern coast (Lea 1989a; Waythomas and Neal 1998: 119).

POLLEN DATA FROM THE ANIAKCHAK REGION

The postglacial vegetation of the Alaska Peninsula and Aleutian Islands has been subject to volcanic effects since its initial establishment. One researcher has already noted that periodic volcanic ash falls have left the Alaska Peninsula a 'patchwork of (biological) communities' (Heusser 1983: 292). The Aniakchak 3,400 BP volcanic eruption was a mix of voluminous ash falls and a far-reaching pyroclastic flow. These two, and especially the superheated pyroclastic flow, would have had a profound effect on the landscape.

To show that the Aniakchak flora had been severely affected by the 3,400 BP eruption, it was evident that close-interval pollen analysis was needed of the sediments capping the 3,400 BP pyroclastic flow. In August 2001, pollen sampling was conducted on both the Pacific and Bristol Bay coasts of the central Alaska Peninsula. The pollen samples collected on the Pacific coast of the Aniakchak National Monument and Preserve were taken from a geologic section ('Cabin Bluff') exposed in sea cliffs at Aniakchak Bay, approximately 42 km southeast of the Aniakchak caldera (Figure 7.3). Bulk samples were collected throughout the section for microfossil analysis, and samples were collected from peats and paleosols for radiocarbon dating.

Pollen samples from 2 and 5 cm sampling intervals through the 75 cm of sediments above the 3,400 BP pyroclastic flows, and from the top 2 cm below it, have produced a picture of the environment immediately before and after the eruption. Peats directly under the pyroclastic flow show vegetation here was rich grass-dominated tundra with only secondary sedge and minimal shrubs (Ericales, *Alnus* or *Betula*) just before the eruption, a product of an earlier volcanically disturbed landscape. Pollen concentrations in the samples are in the

Figure 7.3   Cabin Bluff Section. The geologic section is located on a bluff-top overlooking Aniakchak Bay. Note that the majority of Holocene sediments deposited here are volcanic products. Photo by R. VanderHoek.

range of 10^5 grains/cm^3. Grass was again dominant immediately after the eruption, but pollen concentrations in sediments were three to four orders of magnitude lower than in pre-eruption sediments. Plant taxa that prefer disturbed habitats (especially *Artemisia* and other Asteraceae) also became relatively common, implying unstable substrates with discontinuous vegetation (Nelson and VanderHoek 2002a). Low pollen accumulation rates suggest the Aniakchak landscape was discontinuously vegetated for a prolonged period of time after the 3,400 BP eruption, and the abraded nature of most pollen grains imply significant Aeolian transport with sharp mineral matter, again suggesting the biologically inhospitable nature of the environment (Nelson and VanderHoek 2002b).

Radiocarbon dates on bulk soil samples screened and picked to remove obvious modern contamination from the sediments above the 3,400 BP eruption nonetheless returned out-of-place and out-of sequence dates of $1,990 \pm 40/1930$ cal BP and $1,510 \pm 49/1390$ cal BP, with the younger date being from just above the 3,400 BP pyroclastic flow material and the older date from 55 cm above it, at approximately 20 cm below ground surface. These both exhibit contamination of soil by humic acids from decaying vegetation. Their dates show that by 2,000 BP, the soil surface had at least partially stabilised, vegetated, and the generation of organic acids had begun, making ~ 2,000 BP the minimum age for the revegetation of the area. These data correlate with revegetation studies carried out on volcanic deposits on the western end of the Aleutian arc, where researchers estimated that full recovery of the soil-vegetation system around Ksudach volcano in southern Kamchatka would take more than two thousand years (Grishin *et al* 1996).

REVEGETATION OF PYROCLASTIC FLOW DEPOSITS

The 1912 Katmai pyroclastic flow in the VTTS on the northern Alaska Peninsula is relatively close to Aniakchak, being only 230 km north of the Aniakchak caldera, and would appear to be a good geologic revegetation analogue to the 3,400 BP Aniakchak pyroclastic flow (Griggs 1922). That the valley was still steaming four years after the eruption highlights the prolonged high temperatures commonly found within thick pyroclastic deposits and their sterilising effect on the landscape. Although vegetation has come back to the region around the valley, the Katmai pyroclastic flow itself is still almost completely unvegetated over 90 years after its deposition, with only occasional mosses and low growing plants including *Oxyria digyna*, *Carex* sp, *Festuca* sp, and rare *Salix* and *Alnus* found in a few sheltered locations on the flow itself.

In 2001, soil samples were taken from two locations on the Katmai pyroclastic flow deposit, to be compared against samples taken from four locations on the Aniakchak 3,400 BP flow deposits. Field and lab data from these tests show that both the modern VTTS pyroclastic flow and the ancient Aniakchak pyroclastic flow samples lack many of the elements necessary for good plant growth. Deposits from silicic volcanic eruptions are typically depleted in nitrogen (Shoji *et al* 1993: 45). The amount of available nitrogen present on the surface of regional flow deposits ranged between 2 and 25% of that available at the surface of a vegetated control site. Available nitrogen at 5 cm below surface in the flow deposits was ~ 3% of the control site's values site. Total carbon content of the sediments exhibited similar patterns; both ancient and modern flow deposits contained between 0.8 and 6% of the concentrations at the control site. Potassium, calcium, and magnesium were also significantly lower in the flow deposits in comparison to the vegetated control site.

The cation exchange capacity is the sum total of the exchangeable cations that a soil can absorb (Brady and Weil 2002: 345) and is an indicator of nutrient storage capacity of the soil, for nutrients potentially available for plants. The cation exchange capacity of the surface flow deposits was between 10 and 2,000 times lower than samples from the upper 50 cm of the control site.

The mobility and abrasiveness of volcanic sediments are also important factors in the revegetation of pyroclastic flows and other volcanic landscapes. Volcanic sediments are often less compacted than other sediments, so are moved more freely by wind and water. Chronic erosion frequently eliminates establishing plants (del Moral and Bliss 1993: 59). The Aeolian process commonly strip the fines from the surface of volcanic deposits, leaving a lag pavement of larger material that does not hold enough water for plant growth (del Moral and Bliss 1993: 11). Volcanic material can also be very sharp edged and abrasive, cutting plant tissues (eg, see Riehle 2002: 80, Figure 22A).

Biological soil crusts, also called 'microbiotic' or 'cyano-bacterial-lichen' soil crusts, are sets of early colonisers of unvegetated landscapes throughout the world (Belnap and Gillette 1997). These crusts hold the soil together as it is vegetated by larger plants. Work on unvegetated landscapes in cold desert environments in the American west have shown that whereas biological soil crusts are moderately wind resistant, wind above a certain velocity will strip away these crusts, especially if they have been disturbed. Researchers have termed this wind speed the Threshold Friction Velocity (TFV), which they found to vary from 16 cm/sec (.3 miles/hr) on bare sand to 573 cm/sec (12.8 miles/hr) on crusts that had been undisturbed for 20 years or more (Belnap and Gillette 1997, 1998). The Alaska Peninsula is frequently subjected to winds many times maximum TFV required to strip away an established biological soil crust. The windblown plumes of these loose volcanic sediments are sometimes so voluminous they are even mistaken for volcanic eruptions (Miller *et al* 1998).

EFFECTS OF VOLCANIC PRODUCTS ON REGIONAL BIOTA

Pyroclastic flows are turbulent flowing mixtures of hot gas, ash, and volcanic debris that flow rapidly across the landscape. The Aniakchak pyroclastic material began flowing across the landscape at temperatures ranging between 890° and 1100°C (Dreher 2002: 39, 80). Even at considerable distances from the vent, these flows can maintain temperatures of 500°–650°C (Freundt *et al* 2000: 590). Temperatures this high sterilise the landscape covered by the pyroclastic flow, especially when this superheated flow material completely caps the ground surface. Massive ash deposits can maintain high temperatures for months, leaving virtually no chance for plant survival (Thornton 2000: 1060). Research from Mount St Helens and elsewhere has shown that pyroclastic flow landscapes require seed input for plant colonisation (primary succession) from outside the flow (del Moral and Bliss 1993; Thornton 2000). The vast area covered by the 3,400 BP deposits means that propagules must have come in from considerable distances.

Revegetation of the Aniakchak 3,400 BP pyroclastic flow in the subarctic maritime environment of the Alaska Peninsula would have been slower than pyroclastic deposits in other, lower latitude environments. Flow deposits from tropical environments like those from Krakatau, or even temperate ones like Mount St Helens, exist in more moderate climates that both increase the speed of soil weathering and favour plant survival.

Volcanic ash has a considerable effect on plants and animals, depending on the thickness of tephra and the particular species. These effects were observed first hand after the 1912 eruption of Novarupta in the Katmai region. After the 1912 Katmai eruption, virtually all animals died or left the areas of heaviest tephra fall on the Alaska Peninsula (Riehle *et al* 2000: 262). Anadromous fish were also affected, with most streams in the Katmai region empty of fish five years after the eruption (Griggs 1922: 161). Salmon were found to either have suffocated or returned to the sea in streams where more than 10 cm had fallen on the region, with eroding tephra and unstable beds keeping salmon out of many streams for years (Riehle *et al* 2000: 262).

The ash fallout from the 3,400 BP Aniakchak eruption stretched northwards for over 1,500 km, covering many thousands of square kilometres of the northern Alaska Peninsula and central and western Alaska (Begét *et al* 1992; Riehle *et al* 1987). Although this ash fall may have had a considerable effect on all mammals, birds and fish, grazing mammals like caribou may have been the large mammal most seriously affected. These animals would have been negatively impacted by the ash-caused increase in tooth wear (Trowbridge 1976), by the ingestion or inhalation of ash (Gregory and Neall 1996), from volcanic acid poisoning (see below and also Grattan *et al*, Chapter 8 this volume), and from their loss of grazing habitat. Studies in Kamchatka have shown that with at least 10 cm of tephra 'the moss layer is eliminated' (Grishin *et al* 1996: 150), and deposits of between 10 and 20 cm would eliminate all moss and lichen (Grishin *et al* 1996: 129). This suggests that lichens and other low-growing taxa would be adversely affected or eliminated by a heavy ash fall. This is critical for caribou survival, because this low-growing plant matter makes up a high percentage of the winter caribou diet, a season that lasts for half the year (Skoog 1968).

Another series of volcanic products harmful to animals are acids containing fluorine (as HF), chlorine (as HCl), and sulfur (as SO_2, which converts to H_2SO_4), which are present in the volcanic gases and may also adhere to tephra during an eruption. These acids are then ingested by animals along with the tephra during grazing, in acid-contaminated water, or are absorbed into lichen and ingested when the lichen is eaten (Gregory and Neall 1996). In historic tephra falls, fluorine has been the most common cause of livestock poisoning (Cronin *et al* 1998: 28). Fluorine poisoning was responsible for high sheep mortality in New Zealand in 1995, after the dusting of fields with between 2 and 3.5 mm of fluorine-rich tephra (Cronin *et al* 1998: 22, Figure 1). Distal, fine-grained tephra can carry a greater load of volcanic acid than coarser proximal material because of the greater surface area. Thus, a volcanic eruption with a high fluorine component could have led to a high mortality for medium-sized, low-grazing mammals like caribou even many hundreds of kilometres away from the volcano.

Summer eruptions cause most plants and wildlife to be more strongly affected by the effects of a tephra fall, happening as it does during the active time for most species. This especially impacts plants, with the abrasive (and often acid-carrying) tephra landing on open leaves. A winter ash fall has a lesser effect, coming when plants and small animals are dormant or sheltered under a blanket of snow (del Moral and Bliss 1993: 37; Halpern *et al* 1990; Thornton 2000: 1060).

POSSIBLE ARCHAEOLOGICAL EVIDENCE FOR 3,400–3,700 BP VOLCANISM

If volcanic activity on the central Alaska Peninsula was drastic enough around 3,400 BP to dramatically affect human populations in the region, evidence for this should be found in the archaeological record. Workman (1979) hypothesised a series of expected cultural effects of catastrophic volcanism. They include (1) abandonment of a region, with breaks in the local archaeological sequence; (2) intrusive appearance of cultural complexes or traits derived from the affected area in the peripheries; and (3) evidence of intensified regional contacts synchronous with the volcanic event in question. Evidence against substantial cultural impact from volcanism would include immediate reoccupation of affected sites without recognisable cultural breaks (Workman 1979: 366). We see no evidence of immediate reoccupation of sites in this region and evidence for all three of Workman's proposed cultural effects. See a summary of archaeological evidence in Table 7.1.

(1) There is no discernable human occupation on the central Alaska Peninsula for hundreds of years after 3,400 BP, with little regional occupation until ~2,100 BP. Four years of archaeological surveys in the Aniakchak National Monument and Preserve have identified 27 prehistoric sites, with the earliest sites dating to approximately 2,100 BP. No sites were found that dated to between 3,400 and 2,100 BP, a time when regions on the northern and southern ends of the peninsula had thriving human populations. Archaeological surveys in Aniakchak have focused not only on desirable later prehistoric locations but also those that would have been appealing during the Middle and Mid- to Late Holocene. These sites were tested by probe, auger, examination of eroding bluff faces, and 1×1 m and 1×2 m test excavations, several of them over 2 m deep. Field knowledge of the physical characteristics of regional tephras helped excavators determine when they encountered Middle and Early Holocene strata. No cultural occupations were found within these strata.

Regions directly adjacent to Aniakchak also show an occupational hiatus after 3,400 BP. The Ugashik region to the north of Aniakchak shows occupations around the Ugashik Narrows that date between ~3900 and 3,400 BP. These occupations are capped by a tephra (Ash IV), and available evidence suggests the region is then vacated until 2,100 BP (Henn 1978). Similarly, preliminary archaeological evidence from the salmon-rich Chignik region to the south of Aniakchak shows no sites until ~2800 BP (Corbett 1995), almost a thousand years after the massive eruption of the nearby Mount Veniaminof. The region sees little population development until almost 2,000 BP (Corbett 1993, personal communication; Dumond 1992).

(2) Intrusive cultural complexes are a common occurrence around or shortly after 3,400 BP. The maritime Old Whaling culture, and the similar Devil's Gorge occupation, both located north of the Bering Strait more than 1,170 km north-west of Aniakchak, have recently been suggested to have an origin on the Alaska Peninsula. Dumond sees these cultures deriving from a homeland on the lower Alaska Peninsula, which was left to move northwards about 3,400 BP (2000: 99).

Table 7.1 Available evidence by region for mid-Holocene human occupation on the Alaska Peninsula. Heavy dashed line (▬ ▬ ▬) marks Aniakchak and Veniaminof volcanic eruptions. Dashed line with diamonds (♦------♦) denotes time range for stylistically dated occupation

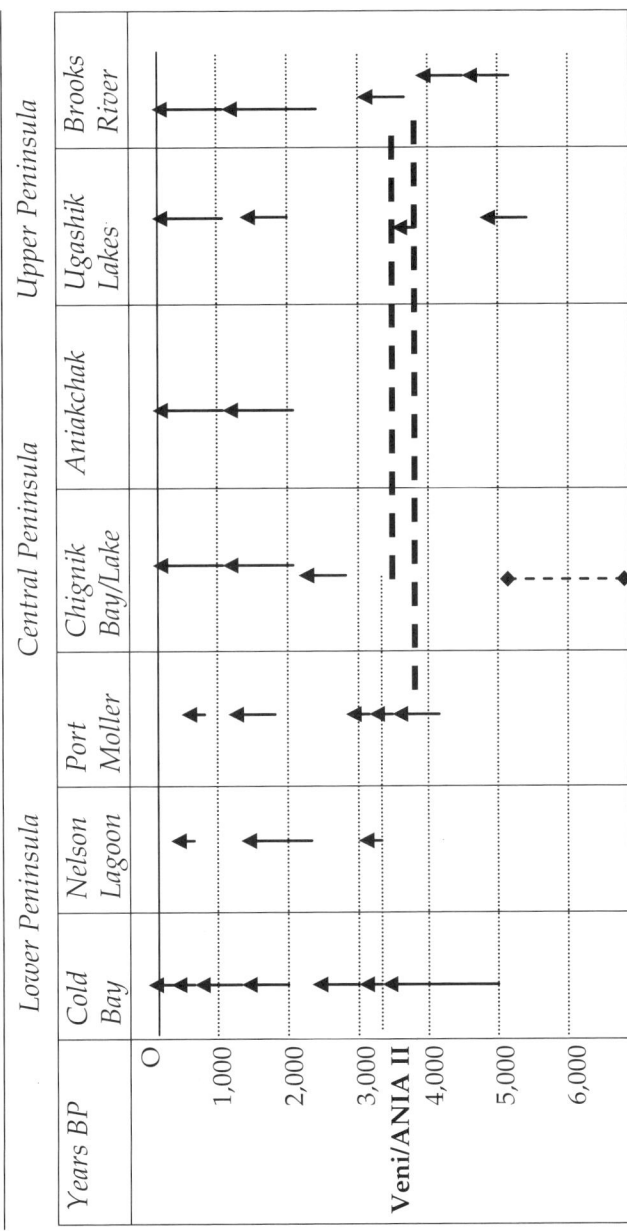

Port Moller and the Cold Bay region on the southern Alaska Peninsula also have intrusive populations at this time. The Russell Creek phase at XCB-022 in Cold Bay is a short-lived occupation with slab-lined box hearths and small, finely flaked stone points. Its closest precursor is seen by its excavators (Jordan and Maschner 2000; Maschner 1999) in the earlier Arctic Small Tool tradition (ASTt) on the northern peninsula (for a differing opinion, see Dumond 2001). A recent publication by these researchers places this occupation originating shortly after 3,400 BP (Maschner and Jordan 2001: 153).

A similarly related occupation is found at the same time at the Hot Springs Site in Port Moller (Maschner and Jordan 2001; Maschner 2004). Both of these manifestations last a short time and are very dissimilar from the cultures that come before and after them.

The first known occupation of the Nelson Lagoon region, located between Cold Bay and Port Moller, takes place just after 3,400 BP. Recent survey in the Nelson Lagoon watershed found the earliest site to be a large village (> 150 dwelling depressions) with an occupation date of 3,310 ± 40 BP (Knudsen 2005).

(3) Intensified regional contacts may be harder to quantify. Researchers have commented on the apparent movement of people or ideas into the Aleutian Islands from the lower Alaska Peninsula at this time (Dumond 2000, 2001: 305).

Kodiak Island sees a transition from the Ocean Bay Culture to the Kachemak Culture that occurs about or just before this time, with early Kachemak cultural horizons often dating to the mid-4th millennium BP (Clark 1997; Mills 1994). Although some see the Ocean Bay to Kachemak transition on Kodiak as *in situ* development (Clark 1997), some influx of people or ideas seems apparent.

The Brooks River Gravels manifestation of ASTt, at the salmon-rich Brooks River on the northern Alaska Peninsula, may also reflect some effect from the Aniakchak 3,400 BP eruption. Except for the oldest of the C-14 dates for this occupation (3,610 ± 85), the rest occur during and after the time of the Aniakchak 3,400 BP eruption (Dumond 1981, 1998: 73), making it possible that they reflect a population pushed north from the Aniakchak and Ugashik regions by volcanic effects. Alternatively, considering the high (possibly > 200) number of ASTt dwellings at Brooks River (Dumond 1981: 128) they may reflect the coalescing of an existing regional population in a resource-rich area after a large region to the south became untenantable.

Archaeological evidence has even caused one researcher to suggest the Aniakchak region as a cultural boundary shortly after the time of the 3,400 BP eruption. In discussing the marine-adapted Margaret Bay culture of the Aleutians and lower Alaska Peninsula, and the contemporary Brooks River Gravels phase on the northern Alaska Peninsula, both of which exhibited ASTt characteristics, Dumond (2001: 298) noted that this evidence

> points to the presence in the first millennium BC of a fairly definite cultural boundary on the Alaska Peninsula, a boundary between Brooks River Gravels and Macro Margaret Bay, that must have been located somewhere between the Ugashik River system and Port Moller.

The Aniakchak and Veniaminof volcanoes are the dominant geographic features between the Ugashik River area and the Port Moller embayment (Figure 7.1).

ANALYSIS

The scale and frequency of natural disasters affect their impact on regional human populations. Most active explosive volcanoes, like those in the Pacific 'Ring of Fire', produce periodic small eruptions with little effect on the biota or human populations of the region. Very large eruptions, like the 3,700–3,400 BP eruptions of Veniaminof and Aniakchak volcanoes on the Alaska Peninsula are another matter, as they affect all parts of the ecosystem and are outside the understanding or memory of local populations.

Large volcanic eruptions that affect virtually all the regional biota push hunter-gatherer populations out of a region. Agrarian populations in temperate or tropical climates, especially those with developed sociopolitical systems, can sometimes put large numbers of people to work on redeveloping an area's agricultural productivity, depending on the scale of the eruption and other ecological and social factors (Elson *et al*, Chapter 6 this volume; Sheets 1979: 536 and Chapter 4 this volume; Shimoyama 2002; Torrence 2002). Hunter-gatherer populations are tied to the natural renewal cycle of the landscape, and must wait for the biota to regenerate, clearly illustrated by the 7,280 BP Kikai Akahoya eruption in southern Japan (Machida and Sugiyama 2002).

Multiple lines of evidence suggest that the 3,700–3,400 BP volcanism from the Aniakchak and Veniaminof volcanoes on the central Alaska Peninsula appears to have created a zone of very low biological productivity that lasted a substantial period of time. Vast areas covered by pyroclastic flow deposits produced gigantic 'dead zones'. The Aniakchak pollen evidence shows a prolonged period of discontinuous vegetation, with soils dominated by abrasive aeolian sediments. The Katmai VTTS pyroclastic flow provides a good geologic analogue for the ecological effects of the 3,400 BP Aniakchak eruption, showing how little the region would have revegetated in over 90 years. Research on the Katmai pyroclastic flow and in other volcanic regions has highlighted the role factors like particle abrasion and ablation (both Aeolian and fluvial) play on vegetation attempting to colonise the landscape.

Large explosive volcanic eruptions, with far-ranging pyroclastic flows and thick (>0.5 m) tephra-fall deposits, take a long time to revegetate, especially in higher latitudes and altitudes, because soil-weathering processes are slower. The locking up of available phosphorus (Ping and Michaelson 1986), and the absence of available nitrogen and other needed elements in the volcanic substrates would also make the survival of almost any sprouting windblown seeds exceptionally difficult, contributing further to the preservation of the barren landscape.

The effects of volcanism on a landscape are intensified if the eruption takes place in the summer months, as plants are not shielded by snow and are in their growth period. This is especially true for the Aniakchak region, for not only would a large summer eruption affect the regional flora, but it would also affect any fauna within its area of effect, having a drastic effect on the Alaska Peninsula caribou herd that summer in the region. The 3,400 BP eruption would have eradicated those caribou in the area covered by the pyroclastic flow and isolated those caribou in the south from winter forage areas in the north (Dumond 2004: 121). Other faunal resources, like migratory waterfowl, shellfish, anadromous fish and near-shore fish, bird, and mammal species, would also have been massively affected.

Terrestrial, riverine, and marine resources recover from an influx of volcanic sediments at different rates. Deep sea and near-shore resources should recover first as the volcanic products in these systems are soon either diluted or flushed out. Intertidal and riverine resources may also rapidly rebound unless constant pulses of sediment, from ash washing off hillsides, continue to impact these environments. Massive influxes of sediment can change the nature of intertidal environments from rocky to sandy shores, requiring the colonisation by different intertidal bivalves for those areas to be biologically useful for humans. Terrestrial plants, especially low-growing species, and the mammals dependent on them may also be impacted for a considerable period of time.

The effects of catastrophic volcanism would appear to have an even greater cultural effect on a region if it is geographically circumscribed, as with an island or peninsula. This may differentiate the cultural effects of major continental eruptions like the 6,800 BP Mount Mazama eruption (Bacon 1983) and the 3,500 BP Mount St Helens eruptions (Begét et al 1992: 55; Mullineaux 1986:20, Table 2) from the 6,300 BP Kikai-Akahoya eruption in Japan (Machida and Sugiyama 2002) or the 3,400 BP Aniakchak eruption on the Alaska Peninsula. In a continental eruption, people will be pushed away from the eruption and eruptive cloud, but can still move on foot to another terrestrial location, limited only by terrain, resources, and other cultural groups. Later human movement around the region can circumnavigate the impacted area. A major volcanic event on an island or peninsula will either push out or eradicate the original population and create a 'dead zone' with distinctly bounded sides that resists reoccupation. This region can act as a roadblock that, whilst with some effort can be circumvented, nonetheless restricts the steady flow of cultural contact and can function to isolate a region. This circumscribed effect is exacerbated if the landform is located in higher latitudes like the Alaska Peninsula, where climate limits soil weathering and the revegetation process. Here, prehistoric maritime technology before 2,000 BP was limited to relatively small skin boats, with navigation that was largely tied to destinations that could be seen from the mainland, or from some island that was visible from the mainland, with sleep and other bodily functions still generally requiring a landfall. In this case, a large dead zone of low biological productivity with no terrestrial resources and limited marine resources for several hundred kilometres would cause a literal 'no-man's-land' to occur. The archaeological evidence shows a large area vacated around the time of this volcanism, with populations 'pushed' away from the region and not returning for many hundreds of years. This geographic and temporal barrier appears to have led to a separation and eventual divergence of human populations, especially because one population (Aleut) was more isolated and the other (Eskimo) was more subject to outside influences.

ACKNOWLEDGMENTS

The authors gratefully acknowledge grants from the following agencies: National Science Foundation (Dissertation Improvement Grant #1116325), National Park Service (National Natural Landmark Grant), and the Natural Sciences Research Fund of Colby College.

REFERENCES

Aigner, JS (1976) 'Dating of the early Holocene maritime village of Anangula', 18 *Anthropological Papers of the University of Alaska*, 18, 51–62

Bacon, CA (1983) 'Eruptive history of Mount Mazama and Crater Lake caldera, Cascade Range, USA', *Journal of Volcanology and Geothermal Research* 18, 57–115

Begét, J, Mason, O and Anderson, P (1992) 'Age, extent and climatic significance of the c. 3,400 BP Aniakchak tephra, western Alaska, USA', *The Holocene* 2, 51–56

Belnap, J and Gillette, DA (1997) 'Disturbance of biological soil crusts: impacts on potential wind erodibility of sandy desert soils in southeastern Utah', *Land Degradation and Development* 8, 355–62

Belnap, J and Gillette, DA (1998) 'Vulnerability of desert biological soil crusts to wind erosion: the influence of crust development, soil texture and disturbance', *Journal of Arid Environments* 39, 133–42

Brady, NC and Weil, RR (2002) *The Nature and Properties of Soils*, Upper Saddle River, NJ: Prentice Hall

Clark, DW (1997) *The Early Kachemak Phase on Kodiak Island at Old Kiavak*, Mercury Series, Archaeological Survey of Canada Paper 155, Hull: Canadian Museum of Civilization

Corbett, DG (1993) 'Investigation at Mud Bay, Site CHK-042', manuscript on file, National Park Service, Anchorage: Lake Clark Katmai Studies Center

Corbett, DG (1995) 'Chignik Lake Village, CHK-031, Excavation', manuscript on file, National Park Service, Anchorage: Lake Clark Katmai Studies Center

Cronin, SJ, Hedley, MJ, Neall, VE and Smith, RJ (1998) 'Agronomic impact of ash fallout from the 1995 and 1996 Ruapehu volcanic eruptions, New Zealand', *Environmental Geology* 34, 21–30

del Moral, R and Bliss, LC (1993) 'Mechanisms of primary succession: insights resulting from the eruption of Mount St Helens', in Begon, M and Fitter, AH (eds), *Advances in Ecological Research*, pp 1–66, London: Academic Press

Dilley, TE (2000a) Unpublished field notes, on file, National Park Service, Anchorage: Lake Clark Katmai Studies Center

Dilley, TE (2000b) 'Geology and tephrochronology of Aniakchak National Monument, Alaska', manuscript on file, National Park Service, Anchorage: Lake Clark Katmai Studies Center

Dreher, ST (2002) 'The physical volcanology and petrology of the 3,400 YBP caldera-forming eruption of Aniakchak volcano, Alaska', unpublished PhD thesis, Department of Geology and Geosciences, Fairbanks, University of Alaska

Dumond, DE (1981) *Archaeology on the Alaska Peninsula: The Naknek Region, 1960–1975*, Eugene: University of Oregon Anthropological Papers

Dumond, DE (1992) 'Archaeological reconnaissance in the Chignik-Port Heiden region of the Alaska Peninsula', *Anthropological Papers of the University of Alaska* 24, 89–108

Dumond, DE (1998) 'The archaeology of migrations: following the fainter footprints', *Arctic Anthropology* 35, 59–76

Dumond, DE (2000) 'A southern origin for Norton Culture?', *Anthropological Papers of the University of Alaska* 25, 87–102

Dumond, DE (2001) 'Toward a (yet) newer view of the (pre)history of the Aleutians', in Dumond, DE (ed), *Archaeology in the Aleut Zone of Alaska: Some Recent Research*, pp 289–309, Eugene: University of Oregon Anthropological Papers

Dumond, DE (2004) 'Volcanism and history on the northern Alaska Peninsula', *Arctic Anthropology* 42, 112–25

Dumond, DE and Bland, RL (1995) 'Holocene prehistory of the northernmost North Pacific', *Journal of World Prehistory* 9, 401–51

Freundt, A, Wilson, CFN and Carey, SN (2000) 'Ignimbrites and block-and-ash flow deposits', in Sigurdsson, J (ed), *Encyclopedia of Volcanoes*, pp 581–99, San Diego: Academic Press

Gregory, NG and Neall, VE (1996) 'Volcanic hazards for livestock.' *Outlook on Agriculture* 25, 123–29

Griggs, RF (1922) *The Valley of Ten Thousand Smokes*, Washington, DC: National Geographic Society

Grishin, SY, del Moral, R, Krestov, PV and Verkholat, VP (1996) 'Succession following the catastrophic eruption of Ksudach volcano (Kamchatka, 1907)', *Vegetatio*

Halpern, CB, Frenzen, PM, Means, JE and Franklin, JF (1990) 'Plant succession in areas of scorched and blown-down forest after the 1980 eruption of Mount St. Helens, Washington', *Journal of Vegetative Science* 1, 181–94

Hemming, JE (1971) *The Distribution and Movement Patterns of Caribou in Alaska*, Game Technical Bulletin 1, Alaska Department of Fish and Game, Fairbanks

Henn, W (1978) *Archaeology of the Alaska Peninsula: The Ugashik Drainage*, Eugene: University of Oregon Anthropological Papers

Heusser, CJ (1983) 'Pollen diagrams from the Shumagin Islands and adjacent Alaska Peninsula, southwestern Alaska', *Boreas* 12, 279–95

Jordan, JW and Maschner, HDG (2000) 'Coastal paleogeography and human occupation of the western Alaska Peninsula', *Geoarchaeology* 15, 385–414

Kienle, J and Nye, CJ (1990) 'Volcano tectonics of Alaska', in Wood, CA and Kienle, J (eds), *Volcanoes of North America: United States and Canada*, pp 9–16, Cambridge: Cambridge University Press

Knudsen, G (2005) 'Archaeological survey of the Nelson Lagoon drainage', paper presented at the Alaska Anthropological Association 32nd annual meeting, Anchorage

Lea, PD (1989a) 'Holocene tsunami deposits in coastal peatlands, northeastern Bristol Bay, SW Alaska', *Geological Society of America Abstracts with Programs* 21, 344

Lea, PD (1989b) 'Quaternary environments and depositional systems of the Nushagak lowland, southwestern Alaska', unpublished PhD thesis, Department of Geological Sciences, Boulder, University of Colorado

Machida, H and Sugiyama, S (2002) 'The impact of the Kikai-Akahoya explosive eruptions on human societies', in Torrence, R and Grattan, JP (eds), *Natural Disasters and Cultural Change*, pp 313–25, London: Routledge

Maschner, HDG (1999) 'Prologue to the prehistory of the lower Alaska Peninsula', *Arctic Anthropology* 36, 84–102

Maschner, HDG (2004) 'Redating the Hot Springs village site in Port Moller, Alaska', *Alaska Journal of Anthropology* 2, 100–16

Maschner, HDG and Jordan, JW (2001) 'The Russell Creek manifestation of the Arctic small tool tradition on the western Alaska Peninsula', in Dumond, DE (ed) *Archaeology in the Aleut Zone of Alaska: Some Recent Research*, pp 151–71, Eugene: University of Oregon Anthropological Papers

Miller, TP, McGimsey, G, Richter, DH, Riehle, JR, Nye, CJ, Yount, ME and Dumoulin, JA (1998) *Catalog of Historically Active Volcanoes in Alaska*, U.S. Geological Survey, 98–582 Open File Report, Anchorage: Alaska Volcano Observatory

Miller, TP and Smith, RL (1977) 'Spectacular mobility of ash flows around Aniakchak and Fisher calderas, Alaska', *Geology*, 5, 173–76

Miller, TP and Smith, RL (1987) 'Late Quaternary caldera-forming eruptions in the eastern Aleutian arc, Alaska', *Geology* 15, 434–38

Mills, RO (1994) 'Radiocarbon calibration of archaeological dates from the central Gulf of Alaska', *Arctic Anthropology* 31, 126–49

Mullineaux, DR (1986) 'Summary of pre-1980 tephra-fall deposits erupted from Mount St. Helens, Washington State, USA', *Bulletin of Volcanology* 48, 17–26

Neal, CA, McGimsey, RG, Miller, TP, Riehle, JR and Waythomas, CF (2001) *Preliminary Volcano-Hazard Assessment for Aniakchak Volcano, Alaska*, U.S. Geological Survey 00-519 Open-File Report, Anchorage: Alaska Volcano Observatory

Nelson, RE, and VanderHoek, R (2002a) 'Impact of the Aniakchak 3,500 BP pyroclastic flow on the environments of the lower Alaska Peninsula', American Quaternary Association Program and Abstracts of the 17th Biennial Meeting, Anchorage, Alaska, 8–11 August

Nelson, RE and VanderHoek, R (2002b) 'A postglacial volcanically dominated pollen record from the Pacific coast of the Central Alaska Peninsula', *Geological Society of America Abstracts with Programs*, vol 34, no 6, September

Ping, CL and Michaelson, GJ (1986) 'Phosphorus sorption by major agricultural soils in Alaska', *Communications in Soil Science and Plant Analysis*, 17, 299–320

Riehle, JR (2002) *The Geology of Katmai National Park and Preserve, Alaska*, Anchorage: Publication Consultants

Riehle, JR, Dumond, DE, Meyer, CE and Schaaf, JM (2000) 'Tephrochronology of the Brooks River Archaeological District, Katmai National Park and Preserve, Alaska: what can and cannot be done with tephra deposits', in McGuire, WJ Griffiths, DR Hancock, PI.

and Stewart, IS (eds), *The Archaeology of Geological Catastrophes*, pp 245–66, Geological Society Special Publication, London: The Geological Society

Riehle, JR, Meyer, CE, Ager, TA, Kaufman, DS and Ackerman, RE (1987) 'The Aniakchak Tephra deposit, a Late Holocene Marker horizon in western Alaska', in Hamilton, TD and Galloway, JP (eds), *Geologic Studies in Alaska by the U.S. Geological Survey, 1986*, pp 19–87, Washington, DC: US Geological Survey

Riehle, JR, Waitt, RB, Meyer, CE and Calk, LC (1998) 'Age of formation of Kaguyak caldera, eastern Aleutian arc, Alaska, estimated by tephrochronology', in Gray, JE and Riehle, JR (eds), *Geological Studies in Alaska by the U.S. Geological Survey, 1996*, pp 161–68, Washington, DC: US Geological Survey

Sarna-Wojcicki, A, Shipley, S, Waitt, TB, Dzurisin, D and Wood, SH (1981) 'Areal distribution, thickness, mass, volume, and grain size of air-fall ash from the six major eruptions of 1980', in Lippman, PW and Mullineaux, DR (eds), *The 1980 Eruptions of Mount St. Helens, Washington*, pp 577–600, Washington, DC: US Geological Survey

Self, S and Rampino, MR (1981) 'The 1883 eruption of Krakatau', *Nature* 294, 699–704

Sheets, PD (1979) 'Environmental and cultural effects of the Ilopango eruption on Central America', in Sheets, PD and Grayson, DK (eds), *Volcanic Activity and Human Ecology*, pp 525–64, New York: Academic Press

Shimoyama, S (2002) 'Volcanic disasters and archaeological sites in southern Kyushu, Japan', in Torrence, R and Grattan, JP (eds), *Natural Disasters and Cultural Change*, pp 326–42, London: Routledge

Shoji, S, Nanzyo, M and Dahlgren, RA (1993) 'Genesis of Volcanic Ash Soils', in Shoji, S, Nanzyo, M, and Dahlgren, R A (eds), *Volcanic Ash Soils: Genesis, Properties and Utilization*, pp 37–71, Amsterdam: Elsevier

Simkin, T, Siebert, L, McClelland, L, Bridge, D, Newhall, C and Latter, JH (1981) *Volcanoes of the World*, Smithsonian Institution, Stroudsburg, PA: Hutchinson Ross Publishing Company

Skoog, RO (1968) 'Ecology of the caribou *(Rangifer tarandus granti)* in Alaska', unpublished PhD thesis, Department of Biology, University of California, Berkeley

Thornton, IWB (2000) 'The ecology of volcanoes: recovery and reassembly of living communities', in Sigurdsson, H (ed), *Encyclopedia of Volcanoes*, pp 1057–81, San Diego: Academic Press

Torrence, R (2002) 'What makes a disaster? A long-term view of volcanic eruptions and human responses in Papua New Guinea', in Torrence, R and Grattan, JP (eds), *Natural Disasters and Cultural Change*, pp 292–312, London: Routledge

Trowbridge, T (1976) 'Aniakchak crater: Alaska's volcanoes, northern link in the ring of fire', in Rennik, G and Perry, H (eds), *Alaska Geographic*, pp 71–73, Anchorage: Alaska Geographic Society

VanderHoek, R (1999) 'The 1999 NPS Archaeological Survey of Aniakchak National Monument and Preserve: (or, the curse of Aniakchak)', manuscript on file, National Park Service, Anchorage: Lake Clark Katmai Studies Center

VanderHoek, R and Myron, R (2004) *Cultural Remains from a Catastrophic Landscape: An Archaeological Overview and Assessment of Aniakchak National Monument and Preserve*, US Department of Interior, Anchorage: Aniakchak National Monument and Preserve

Waythomas, CF and Neal, CA (1998) 'Tsunami generation by pyroclastic flow during the 3,500-year BP caldera-forming eruption of Aniakchak Volcano, Alaska', *Bulletin of Volcanology* 60, 110–24

Workman, WE (1979) 'The significance of volcanism in the prehistory of subarctic northwest\North America', in Sheets, PD and Grayson, DK (eds), *Volcanic Activity and Human Ecology*, pp 339–72, New York: Academic Press

CHAPTER 8

The Long Shadow: Understanding the Influence of the Laki Fissure Eruption on Human Mortality in Europe

John Grattan, Sabina Michnowicz, and
Roland Rabartin

INTRODUCTION

Many of the chapters in this book have considered the impact of volcanic eruptions on cultures located relatively close to a volcano. These peoples may have heard the eruption and seen the tephra fall or lava flow and may have had to respond directly to the resultant environmental pressures and support communities subsequently displaced by these phenomena. In many cases, knowledge of how to cope with these situations may have been embedded in personal memory or folklore. This chapter will consider alternative scenarios, in which a volcanic eruption may wield an influence on communities far from the volcano who would have had no knowledge of the event or the phenomena associated with it and may not even have recognised that a volcanic eruption was the cause of their suffering at all.

Developing such a scenario presents a considerable challenge (Grattan *et al* 2002); how does the researcher estimate the impact of the eruption and assess the sensitivity of people and environments at great distances from the volcano? This scenario will be explored and illustrated through in the context of the Laki fissure eruption of 1783–84, the European mortality crisis of the same period, and the worldwide climate disruption and famine in the years that followed.

Unlike many volcanic eruptions, which are of quite short duration, this event persisted for eight months, from June 1783 to February 1784 and emitted ~122 million tonnes of sulphur dioxide, 15 million tonnes of fluorine, and seven million tonnes of chlorine into the atmosphere (Thordarson and Self 2003). Until Thorarinsson's seminal paper 'Greetings from Iceland' (1981), the eruption was little known outside

Iceland and a small community of volcanologists, and any impact that this huge eruption may have had outside Iceland was largely unknown.

Wider interest in Icelandic volcanism and any environmental and perhaps cultural consequences were aroused in the geographical and archaeological research communities in the early 1990s when several complimentary strands of research developed simultaneously. These were the recognition of the ice core record of volcanic sulphate (Hammer et al 1981); the inception of a tephrochronology in Europe of Icelandic micro tephras that had travelled long distances (Dugmore 1989); long tree-ring chronologies and pollen histories, both of which indicated periods of intense environmental stress (Baillie and Munro 1998, Blackford et al 1992; Hall 2003; Pilcher et al 1995); and settlement histories that were interpreted deterministically as indicating abandonment forced by external pressures (Burgess 1989). Prompted by these developments, the Laki fissure eruption was studied in an attempt to construct robust paradigms that the archaeological community could use to assess the potential impact of Icelandic volcanism on the wider world (Grattan and Brayshay 1995; Grattan and Charman 1994; Grattan and Gilbertson 1994; Grattan and Pyatt 1994, 1999; Thordarson and Self 2003). The eruption had the greatest known climatic impact of any eruption of the past 12,000 years; in Europe the summer of 1783 was intensely hot, whilst the winter of 1783–84 was notoriously cold and was followed by several cold years (Sadler and Grattan 1999). Amongst major eruptions Laki is unique because its impact was not limited merely to climate change. The eruption also pumped millions of tonnes of sulphur, fluorine, and chlorine into the atmosphere, which people thousands of miles away could see, taste, and smell. Throughout the summer of 1783, a poisonous volcanic fog spread across Europe, shrivelling the leaves on the trees and withering the crops. Needless to say, such a concentration of acid gasses and strange weather had serious consequences for human health.

VOLCANOGENIC CLIMATE CHANGE: A BRIEF REVIEW

Until recently, climatic change has been the most frequently invoked mechanism by which eruptions have been assumed to influence people and environments far from volcanic regions (Burgess 1989; Gunn 2000; Maddox 1984; Peiser et al 1998; Rampino and Ambrose 2000; Velikovsky 1950; White and Humphreys 1994). Major volcanic eruptions, which introduce sulphur into the stratosphere, undoubtedly disrupt the Earth's weather (Rampino and Self 1993; Robock 2000), but perhaps not to the extent that archaeologists may hope or imagine. What may be more important is the sensitivity of the culture and environment affected to disturbance by the eruption, a factor discussed by Sheets in Chapter 4.

Temperature reduction alone is a crude mechanism by which to assess the influence of an eruption. The great eruption of Tambora reduced hemispheric average temperature by 0.7°C, whilst the Laki fissure eruption reduced temperatures worldwide by an average of 1°C. In any event, all such averaged temperature reductions are well within normal annual variation (Sadler and Grattan 1999). It is now clear that any volcanic influence on climate has a variable seasonal and regional impact; for instance, continents may in fact be warmer than normal in the first winter after an eruption (Robock 2000). This focus on temperature reduction tends to mask the potential impact of volcanic eruptions on global weather.

Eruptions that are powerful enough to inject sulphur gas into the stratosphere can disrupt the delicately balanced exchange of heat energy that drives the world's weather. In the stratosphere, volcanic gasses react with water vapour and form a sulphuric acid aerosol that inhibits the passage of solar energy to the Earth's surface. Powerful winds in the stratosphere then quickly spread the aerosol veil around the world, first as a narrow belt then, as the months pass, as a blanket that can encompass the entire Earth. In the skies above Iceland, Laki's eruption columns easily penetrated the stratosphere – there a mere 9 km from the surface – and each episode of the eruption injected millions of tonnes of sulphur gas into the stratosphere, where it was swept eastwards around the planet by high-speed altitude winds.

The world's climate is driven by the redistribution of energy received from the sun at the Earth's surface; essentially, heat is redistributed from the equator towards the poles and between the oceans and continents. The seasonal arrival of monsoon rains is one feature of this great global engine, as is the latitude at which low-pressure systems track across the Atlantic, delivering rain to Europe. People worldwide rely on the seasonal weather these systems deliver and are vulnerable to unexpected changes to them. Instrumental records, weather diaries, tree-ring studies, and modern computer modelling all tell us that 1783 and the years that followed was a time of great disruption to the world's weather. Climatic anomalies that affected millions of people occurred in Europe, Africa, the Middle East, India, Japan, Central America, and North America. The European situation will be presented in detail below.

Volcano Climate Synergies in 1783

The climatic influence of the eruption appears to have been felt almost immediately by the Alaskan Inuit. The Alaskan spring was normal, but because of the volcanic aerosols gasses in the atmosphere summer did not arrive and instead winter returned. The consequences for the Inuit would be dramatic, with little food reserves they relied on the seasonal arrival of migratory birds and animals, the thawing of rivers, and the ripening of berries; none of these would happen. Tree-ring evidence shows that during the summer of 1783 in Alaska, temperatures remained below freezing. Starvation and death followed and there were few survivors (Jacoby *et al* 1999).

In Ethiopia, the summer rains failed and the Nile floods of 1783 were amongst the lowest ever recorded; millions died in the ensuing famine. From Egypt, European travellers reported that the streets had been 'swept clean of beggars'. Though Oman *et al* (in press) have modelled the failure of these rains as a consequence of a Laki-style eruption, the years of political turmoil, civil war where control of grain transport was a major weapon, rapacious taxation that drove thousands from the land, and outright brigandage had rendered Egyptian society vulnerable to a low flood of the Nile. The country was poorly administered and there were no reserves, thus the climatic influence of the eruption was intensified by the disorganisation of the state.

At exactly the same time, the monsoon rains failed in India and millions are again reported to have died. In a parallel to the situation in Egypt, the subcontinent was wracked by political conflict, war, and rapacious taxation, not helped by the destabilising influence of British expansion. When the monsoon rains failed in 1783, millions died in the 'Chalisa Famine' that ensued. Warren Hastings, who was the first British governor-general, was touched by the plight of the famine-stricken poor:

> From the confines of Buxar to Benares I was followed and fatigued by the clamours of the discontented inhabitants. ... I have seen nothing but traces of complete devastation in every village. The administration of the province is misconducted, and the people oppressed.

The political situation in Japan was rather different, but millions were to die in 1783. The Japanese population had grown considerably under the stable government of the Tokugawa dynasty, to the extent that it is considered that the carrying capacity of the land was at its limit. This large population was the vulnerable feature of Japanese society. In 1783, an anomalous blocking high-pressure cell was established to the north, and cold wind and rains blighted the country throughout the summer. The rice failed to ripen and the human consequences were dramatic. So many people died that their bodies littered the roads to the extent that they became a hazard to traffic in the years that followed. In the early 19th century, population in the north of Japan was still so low that the government had to adopt and encourage a deliberate policy of migration into those regions.

These brief examples illustrate that the climatic impact of the Laki fissure was undoubtedly global. With a death toll that ran into millions, it must rank as one of the greatest natural disasters in human history, but as with all famines of the 20th century, human society and its failings was at least as culpable in the inception of the famine as the eruption itself.

THE LONG REACH OF VOLCANOGENIC ACID FOGS

A powerful additional mechanism by which volcanic eruptions may impact on remote environments has received fresh attention in the last decade: this is the direct impact of volcanic volatiles, typically compounds of sulphur and often accompanied by fluorine and chlorine, on people and the environment. In 1783, the acid volatiles released by the eruption devastated the environment throughout Iceland. Contemporary descriptions of their impact on plants, animals, and humans resemble the aftermath of a serious industrial disaster and makes harrowing reading (Steingrimsson 1998; Thordarson and Self 2003). However, it is now clear that these same volatiles were transported through the atmosphere to Europe, and in the summer of 1783 a dense acid fog, derived from the volcanic gasses, formed throughout Europe. Many people could smell and taste the sulphur in the cloud, and acid concentrations were high enough to burn throats and eyes, damage plants, and kill insects and fish (Demarée and Ogilvie 2001; Van Swinden 2002). People fell seriously ill, often exhibiting symptoms that are today associated with chronic air pollution (Durand and Grattan 1999, 2001), and died in numbers and in patterns that prompted contemporary observers to fear they were witnessing the return of the plague (Grattan *et al* 2002; Grattan, Durand, Gilbertson *et al* 2003; Grattan, Durand, and Taylor 2003; Witham and Oppenheimer 2005).

Volcanogenic Mortality Crisis in Europe?

Although the broad picture of the eruption's impact on the environment is now understood, the full story of its impact on mortality remains the subject of ongoing research. What is known to date can be simply summarised: as a result of the environmental devastation and the famine and disease that followed, up to a third of Iceland's entire population perished (Jackson 1982). In England, the national death rate doubled, with as many as 30,000 extra deaths recorded (Grattan *et al* 2002; Witham and Oppenheimer 2005). In France, mortality at least doubled, and initial estimates suggest figures well in excess of 30,000 and perhaps as high as 200,000 extra deaths (Grattan *et al* 2005). The precise cause of these deaths remains unknown, but the emerging pattern of crisis mortality across a vast area does suggest the operation of an overarching vector. The data presented below are from a variety of sources. In the case of England, the *Population History of England Database* (Schofield 1998) contains monthly mortality data for 404 parishes, in some cases going back into the 16th century. Further data have been compiled from other sources, mainly from the *Bills of Mortality* published in summary form each month in newspapers and other serials such as '*The Gentleman's Magazine*'. In France, the mortality data have not been so conveniently compiled on a national basis, and the data presented here are the result of our research in departmental archives of the length and breadth of France.

Contemporary observers in both countries were certain that people were dying in unusual numbers and that the hot sulphurous fog had something to do with it (Table 8.1). A palpable sense of concern, even panic, was apparent in many accounts (Grattan *et al* 2005; Rabartin and Rocher 1993), which is quite understandable when one considers the actual rates of mortality being experienced. Figure 8.1 compares patterns of mortality in parishes from Bedfordshire, England, and three French departments – the Loiret, Seine Maritime, and Eure et Loire. All show simultaneously anomalous levels of high mortality in the summer and early autumn of 1783.

In rural Europe at this period of the 18th century, the summer was usually a period of low mortality, guaranteed by abundant food and fair weather, a pattern interrupted only by occasional outbreaks of disease or famine. The trends apparent in Figure 8.1 imply the operation of a common vector that was acting on vulnerabilities that these communities may have had in common. Three potential vectors originated with the eruption and can be identified: chronic air pollution, high summer air temperatures, and, in the winter, persistently low temperatures (Grattan and Sadler 1998; Oman *et al* in press). However, we must also consider whether the communities affected by the eruption had common areas of weakness, perhaps illness and disease, the spread of which may have been facilitated by the eruption's modification of the environment.

Patterns of Vulnerability

Until recently, we have only had vague hints as to who was dying and why. The English poet, Cowper, suggested that conditions in Bedfordshire were so bad it was feared to be the return of the plague. He also noted that 'such multitudes are indisposed by fevers in this country, that farmers have with difficulty gathered in their harvest, the labourers having been almost every day carried out of the field incapable of work

Table 8.1 Contemporary associations of mortality and the volcanic fog in 1783

Pendant cette obscurité du soleil, on n'entendait que maladie et morts très innombrable.
While the sun was obscured there was a sickness which caused innumerable deaths.
Curé of Broué

Les brouillards ont été suivis de grands orages et de maladies qui ont mis au tombeau le tiers des hommes dans plusieurs paroisses.
The fogs have been followed by great storms and sicknesses which have driven a third of the men in many parishes to their tombs.
Curé of Landelles

Au commencement de ce dégel, la paroisse de Champseru a été affligée d'une maladie pestilentielle; les malades se sentaient pris a la gorge, quelques ignorants de chirurgiens ont commencé par la saignée et l'émétique; depuis dix-septs jours en voilà quatorze mors sur dix-huit. On prétend que les brouillards de mai, juin, juillet et août, qui offusquèrent le soleil qui paraissait rouge comme du sang, nous pronostiquaient ce fléau. Dieu en préserve ma paroisse!
Until the beginning of the thaw the parish of Champseru has been afflicted by a pestilential sickness. Patients were afflicted by a sickness of the throat. Many ignorant doctors treated it by bleeding and applying emetics and after 18 days there were 40 dead. One pretends that the fogs of May, June, July and August that offended the sun and turned it red as blood forecast this curse. May God preserve my parish.
Curé of Umpeau

Ces gens avec les coffres faibles ont éprouvé une sensation semblable à cela éprouvée une fois exposés brûlant au soufre.
Those people with weak chests experienced a similar sensation to that experienced when exposed to burning sulphur.
Van Swinden

Plusieurs personnes ont éprouvé le 24 après midi à l'air libre une pression incommode, mal de tête, une difficulté dans la respiration exactement semblable à celle qu'on éprouve quand on hume l'air imprégné d'une vapeur de soufre brûlant les asthmatiques ont éprouvé des récidives.
After the 24th, many people in the open air experienced an uncomfortable pressure, headaches and experienced a difficulty breathing exactly like that encountered when the air is full of burning sulphur, asthmatics suffered to an even greater degree.
Brugmans

(For further details, refer to Grattan *et al* 2005.)

and many die' (Cowper Letters 1981). In France, correspondents also reported large numbers of men being 'swept to their tombs', which implies that a similar process was operating in both countries.

In fact, parish registers suggest a more egalitarian pattern of death. A snapshot of this period can be gathered from the parish register of the Church of St Peter and St Paul, in Flitwick, Bedfordshire. The first entry, on 3 August, records the baptism of Robert, the son of Thomas Lincoln, a labourer; Thomas Lincoln's own burial was subsequently noted on 17 October. Another baptism, that of John, the son of Christopher Cox, a labourer, and Mary, his wife, took place on 23 August; sadly, the child's burial was entered in the register on 28 October, just five weeks later. Other entries list the burial of Thomas Butcher, a farmer, on 3 September,

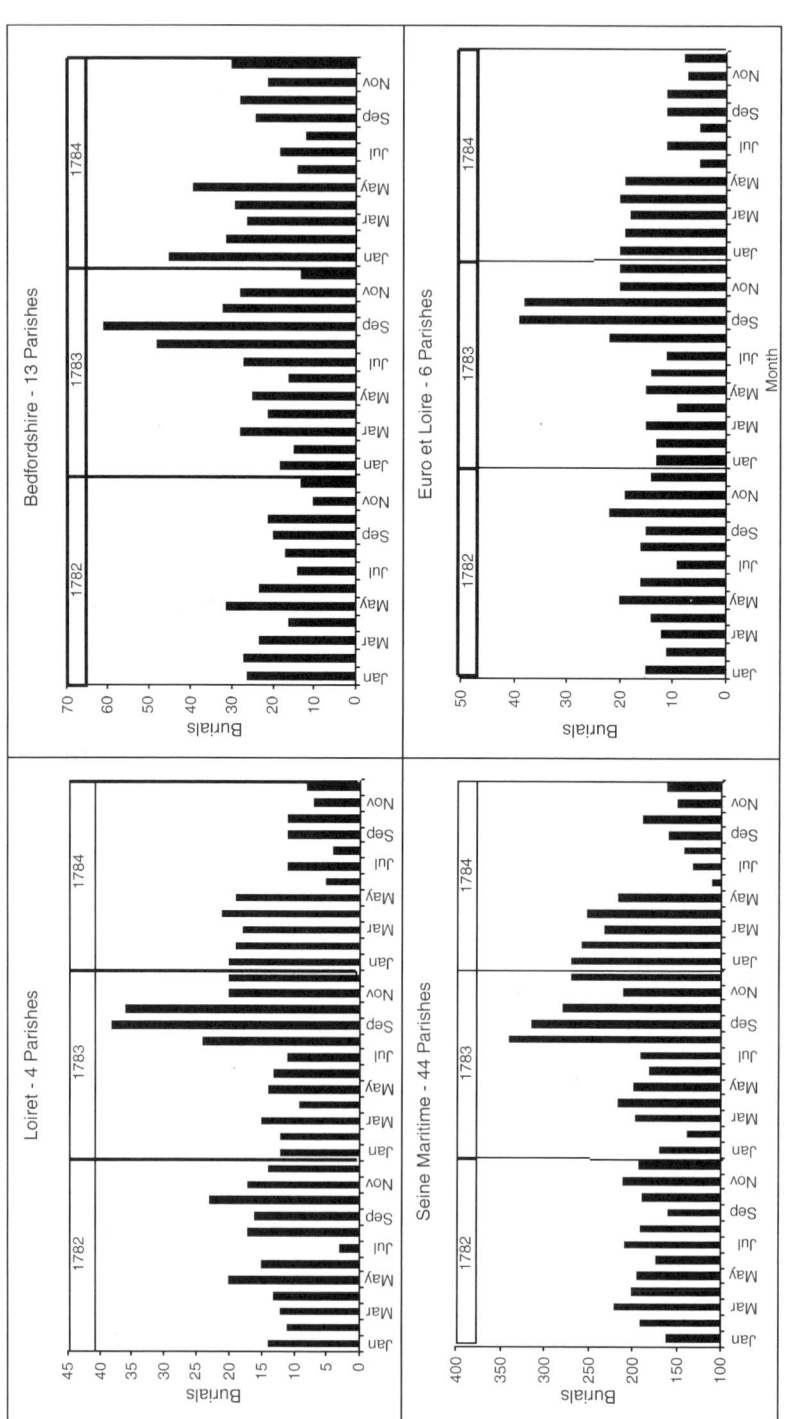

Figure 8.1 Comparable patterns of mortality in England and France.

and Samuel Flinders and Joseph Williams, both labourers, on 10 and 27 September, respectively. The other deaths in September – there were two burials on 28 and 30 September alone – were all the children of labourers. A particularly poignant entry was that for Joseph Smithfield, whose baptism on 31 October and subsequent burial on 4 November was recorded on subsequent lines.

Deaths were not confined to the hard-working poor. The register of St Mary the Virgin in Maulden, Bedfordshire, records the deaths of James Hesse, a gentleman, his wife, and newborn daughter between August and November 1783. The numbers involved in a single parish may seem of little note, but the cumulative disruption to the normal pattern of mortality was significant. In England, the period July 1783–June 1784 was classified as a One Star mortality crisis (Wrigley and Schofield 1989), indicating an annual mortality rate ~20% above the 51-year moving mean, which describes the state of the English nation's health at this time as 'unhealthy'.

Although the cycles of mortality in the countryside were usually predictable and governed by the stresses associated with passing of the seasons, mortality in the overcrowded and unsanitary cities was unpredictable and chaotic. Nevertheless, data reconstructed from the Bills of Mortality published in the *Gentleman's Magazine* suggest that in London, too, a mortality crisis raged in the summer and autumn of 1783 (Figure 8.2a). Typically, the *Gentleman's Magazine* reported that one-third of children born in London died before they reached the age of two, but between July and September 1783, over a thousand more children under the age of six died than might have been anticipated. Deaths were below average in October, but this can be explained as a shift in the normal pattern; those children who were ill or weak and might normally have survived until October appear to have died earlier in the summer.

In France, data compiled from 145 rural parishes confirm the trends seen in London, with the highest death toll amongst the young. Patterns of mortality in 1783 reflect pre-existing vulnerabilities; hence, in the below-one-year age group, where mortality was normally high, deaths rose from 298 in 1782 to 705 in 1783 before falling back to 302, which presumably reflects the normal rate of mortality. The greatest percentage increase came in the six–10-year age group, in which deaths rose from 37 to 134. The death toll for adolescents and young adults was also striking, rising from 51 to 128, but mature adults in all age groups demonstrated an increase in mortality. These mortality figures appear to confirm the comments of most contemporary observers, which were that people were dying in great numbers and in unusual patterns in the summer and autumn of 1783. The demographic patterns, presented here for the first time, indicate that the modifications to the environment caused by the eruption affected most age groups, but in particular the young. Why were they so vulnerable? (Figure 8.2b).

Spatial Patterns of the Mortality Crisis

We may gain further insight into the events of 1783 by studying the development of the mortality crisis, month by month, in a relatively small geographical area. Mortality indices were calculated for parishes in the English county of Bedfordshire, which are contained in the *Population History of England Database* following the procedures laid

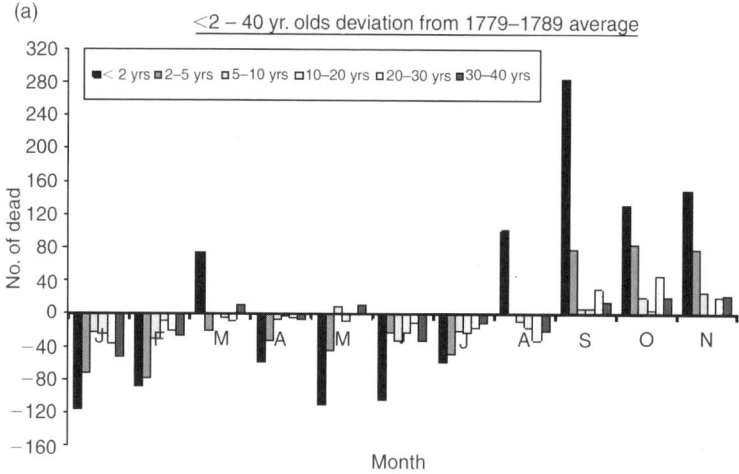

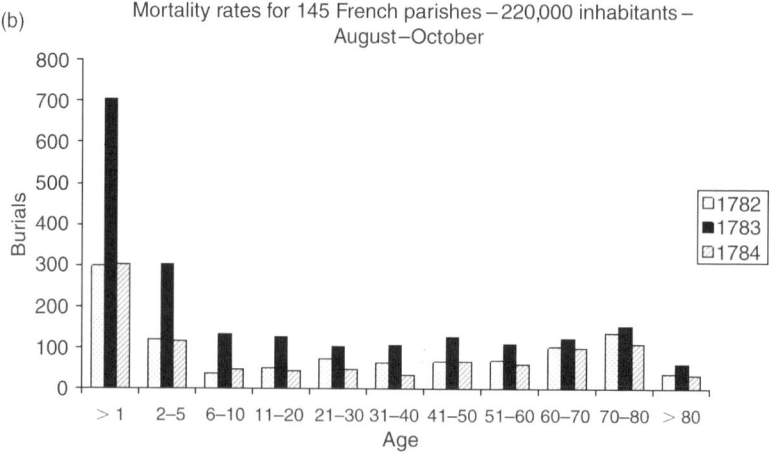

Figure 8.2 (a) 1783 age group deaths in London: variation from 1779–1789 mean; (b) mortality in France, late summer–autumn 1783.

down in Dobson (1980, 1997); these were plotted onto a series of base maps (Figure 8.3a–e).

In July (Figure 8.3a), eight parishes began to experience crisis mortality; in contrast, 12 were healthy, with low mortality. Parishes in both categories are closely juxtaposed; for instance, the parishes of Bolnhurst and Pavenham, which were in crisis, were separated by Thurleigh, Felmersham, and Milton Ernest, which were healthy at that time. In A1ugust (Figure 8.3b), the picture changed dramatically and the gaps had been filled in. By September (Figure 8.3c), 22 parishes were in crisis. By October (Figure 8.3d), the number of parishes considered to be in crisis fell to 14, but this may be partly because of the nature of the event: the weaker members of the community were culled early in the crisis, and those who remained were healthier and more robust. Hence, the parish of Woolton, which was in crisis in July, August, and September, appeared healthy again by October; the vulnerable inhabitants of the parish had probably already died by then. In November, as winter approached, different vectors may have come into play, the number of parishes in crisis had again risen to 18 (Figure 8.3e). The only parish of those for which data are presented here that remained healthy throughout this period, was Riseley in the north of the county. Riseley had suffered from an unspecified epidemic in 1781 and it may be that the vulnerable members of the community died at that time and that in 1783 the remaining parishioners were comparatively healthy. These patterns of mortality do suggest a rather more subtle mechanism is responsible for these events than volcanic death raining from the sky; to understand them, we need to understand the landscape of health in 18th-century Europe.

The Health Context in 1783

In combination, high air temperatures and poor air quality is a modern killer, most recently seen in Europe in 2003, when over 30,000 extra deaths were caused by this lethal environmental cocktail (WHO 2004). In 1783, summer air temperatures, particularly in July, were amongst the hottest ever recorded, and it is thought that the volcanic gasses released may have been responsible (Grattan and Sadler 1998; Oman et al in press). In addition, poor air quality was widespread as millions of tonnes of sulphuric acid were added to the European air mass each day, accompanied by widely reported human illness, especially symptoms that today are usually associated with chronic air pollution. However, in the literature of the day, it is not possible to identify the operation of a single environmental vector or outbreak of a single specific illness that can be unequivocally blamed on the eruption. Authors talked variously of difficulty breathing, of sicknesses of the throat, but

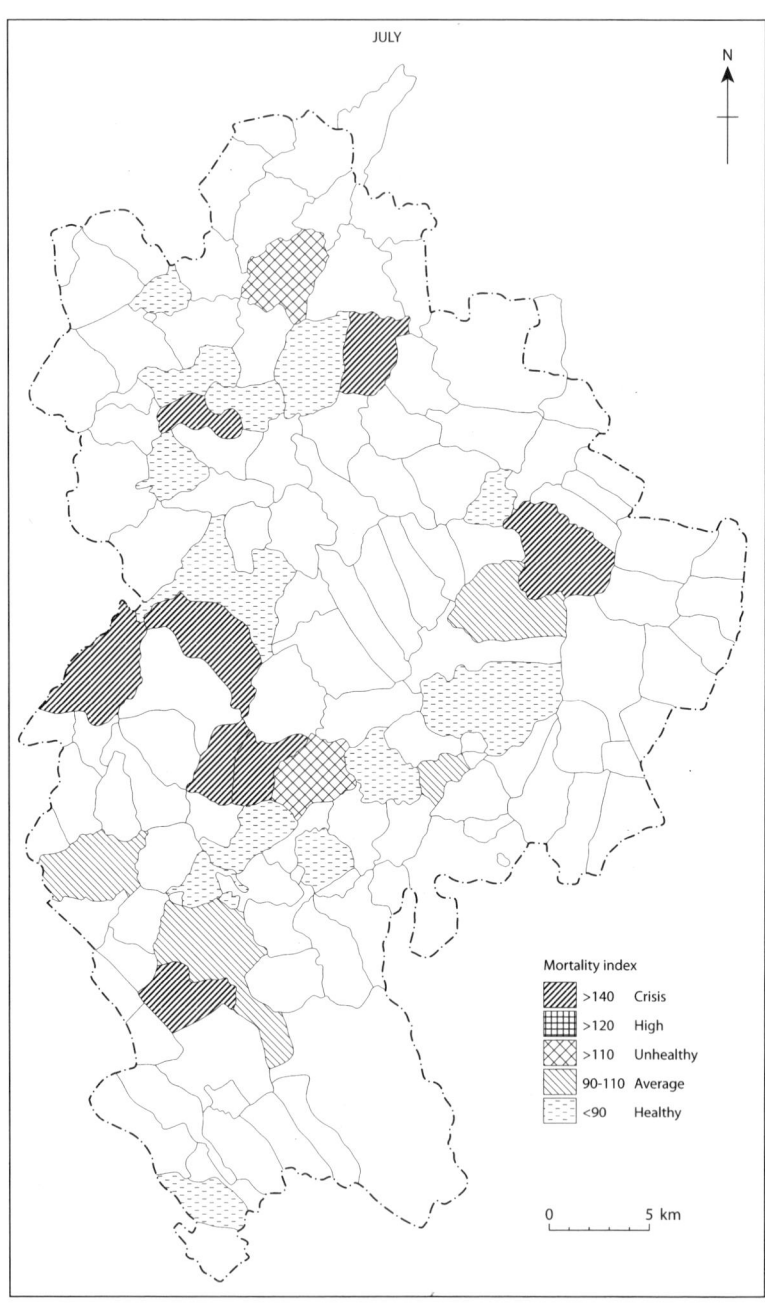

Figure 8.3(a)

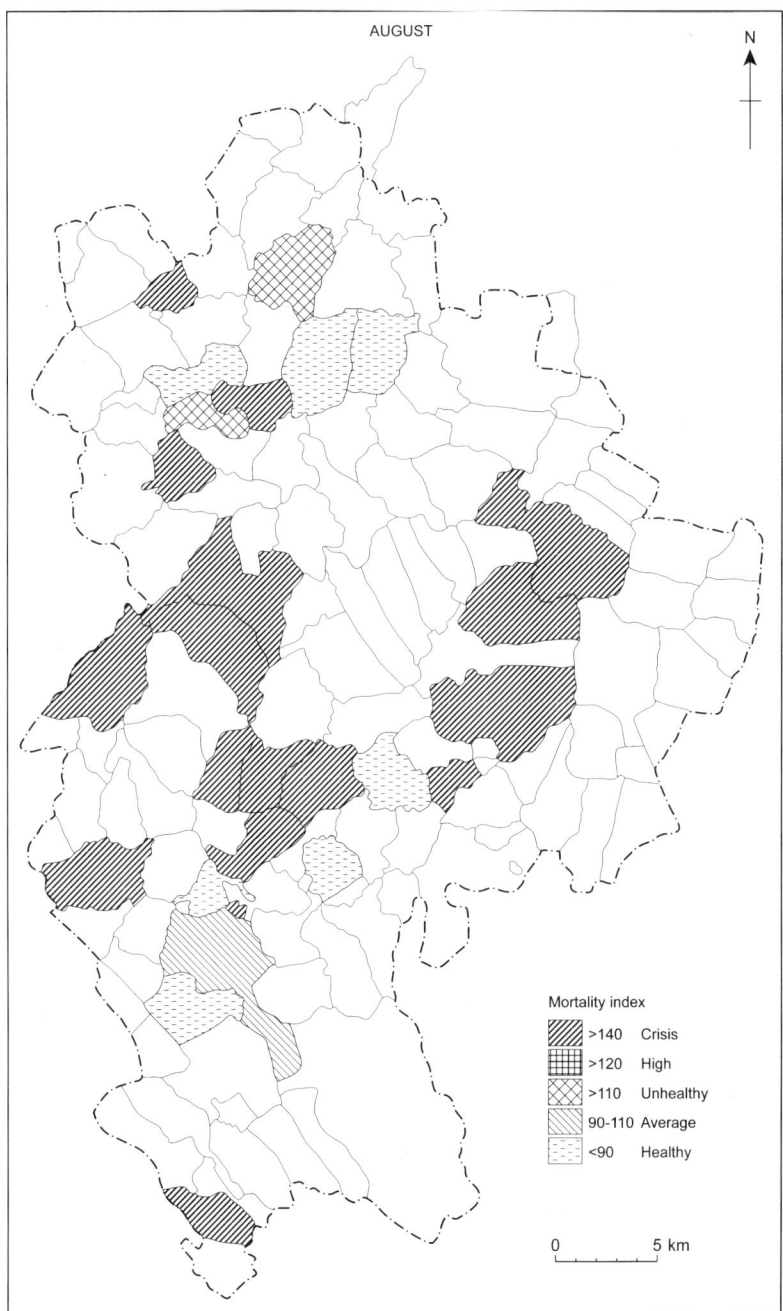

Figure 8.3(b)

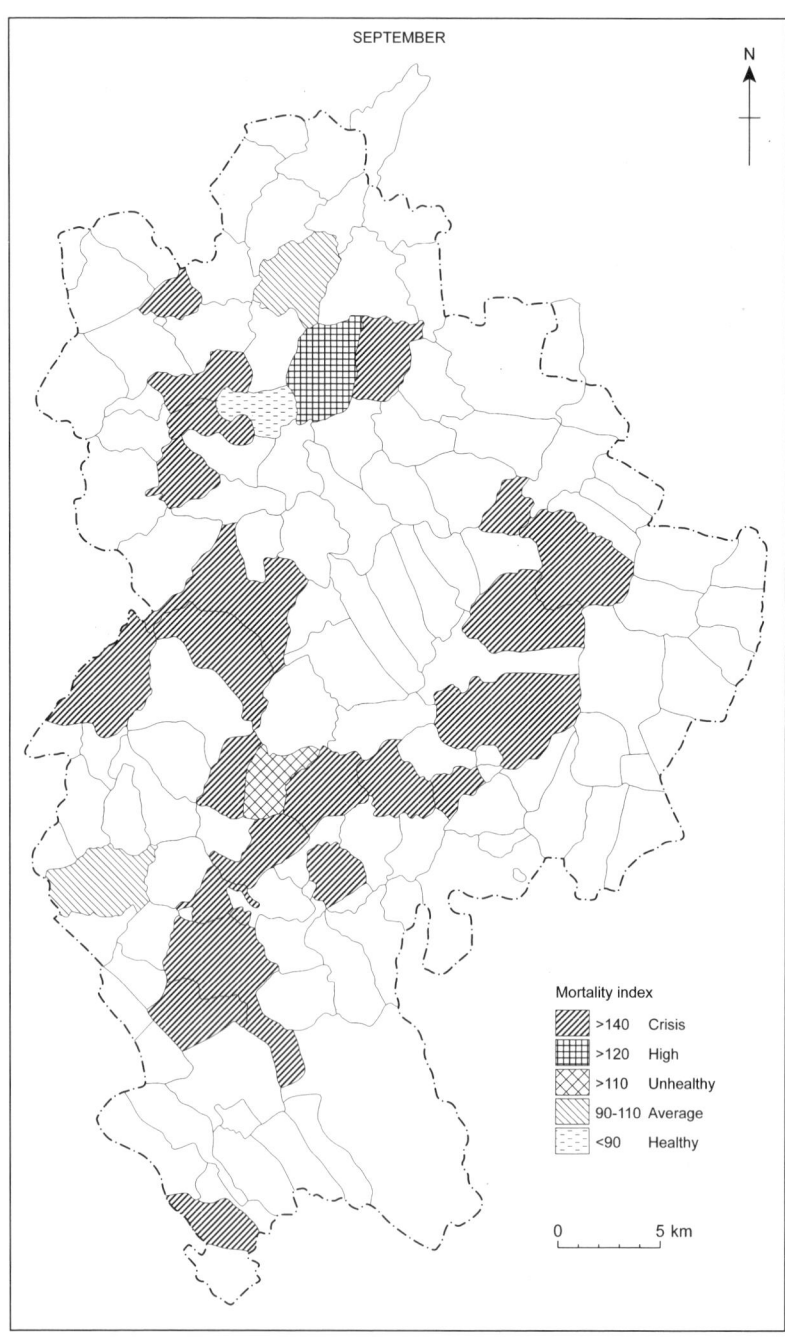

Figure 8.3(c)

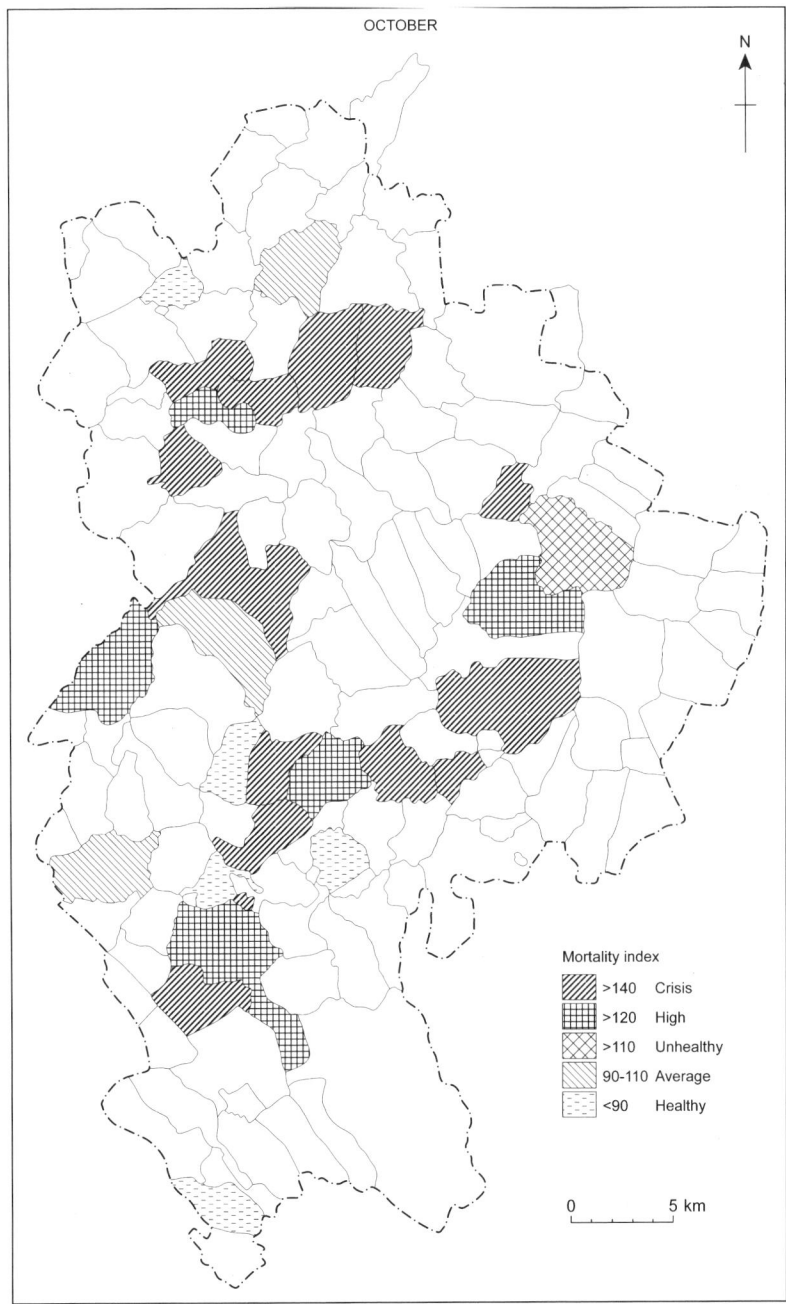

OCTOBER

N

Mortality index

▨ >140 Crisis
▦ >120 High
▨ >110 Unhealthy
▨ 90-110 Average
▨ <90 Healthy

0   5 km

Figure 8.3(d)

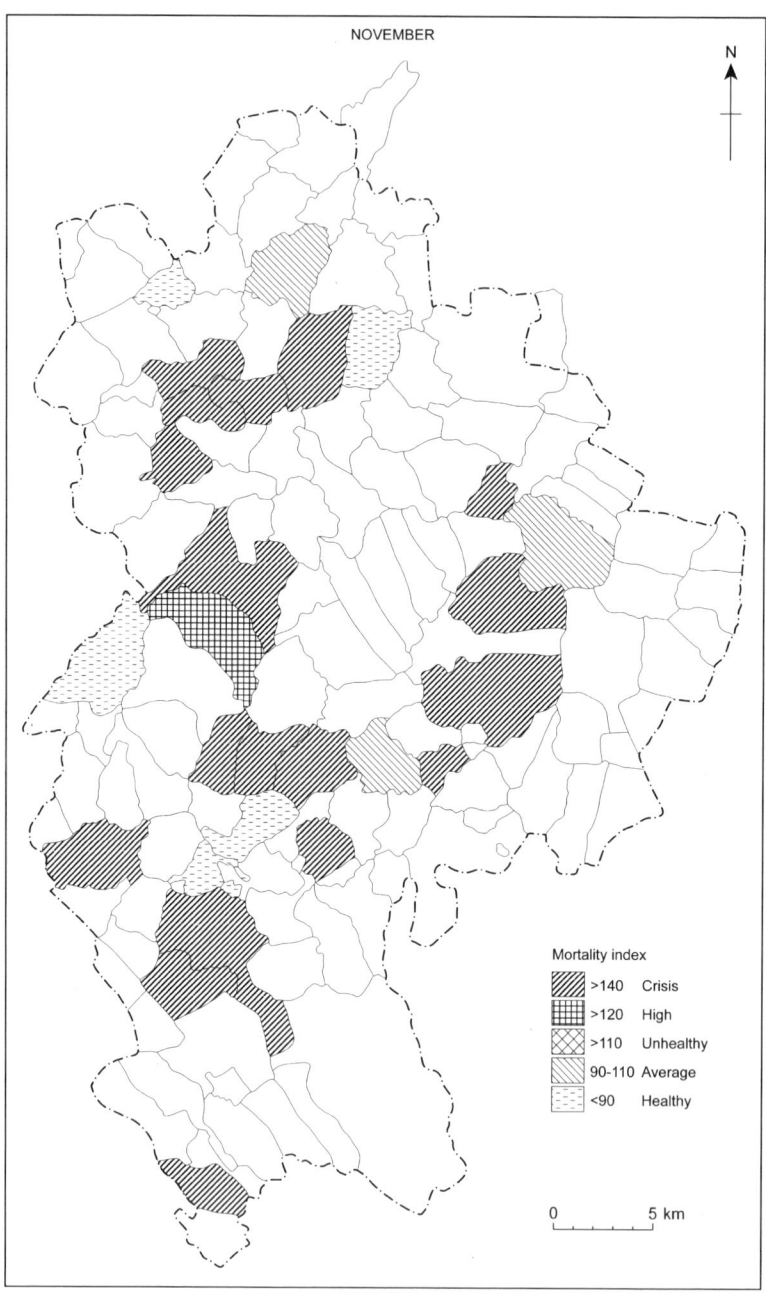

Figure 8.3(e)

Figure 8.3 Mortality indices for Bedfordshire parishes: (a) July 1783, (b) August 1783, (c) September 1783, (d) October 1783, (e) November 1783.

also of fevers and of epidemics; people collapsed and died suddenly or contracted illnesses from which they endured a lingering death. Can we blame the influence of the volcano for all of these?

That year, the abbot of Bazinghem, in France, wrote that there had been a larger number of deaths than births and blamed the 'great heat' of the summer for an outbreak of dysentery. In Italy, Rafael d'Amat de Cortada, Baron de Maldà, recognised that more than one illness was occurring: 'At that time, the great heat and extreme droughts turned into a constellation of illnesses, which were malignant fevers, from which no few persons died' (Demarée and Ogilvie 2001).

The use of the term 'constellation', which is surely deliberate, is very revealing. Amongst the constellation of illnesses always present, though not always in epidemic proportions, were plague, malaria, smallpox, typhus, typhoid, influenza, scarlet fever, measles, and tuberculosis; in addition, water-borne enteric illness posed an ever-present threat to the young, the elderly, and the weak. It is known that some diseases, such as malaria, responded to seasonal conditions, and were endemic in many areas (Dobson 1980, 1997). It was a killer particularly of the young and of those newly arrived in malarial districts (Riley 1987), but many of the mortality crises that took place in 1783 occurred outside the areas traditionally blighted by malaria.

Both epidemic typhus, transmitted by the human body louse, and endemic typhus, transmitted by rat and cat fleas, was an ever-present threat. In 1783, typhus reached out from its secure heartland in the unsanitary overpopulated industrial hovels and struck at rural counties such as Oxfordshire, Wiltshire, Gloucestershire, Worcestershire, and Buckinghamshire.

Enteric dysentery due to typhoid was a constant menace in the 18th century, and was so common that Howe (1997) described it as an 'almost 'natural' cause of death'. Children appear to have been particularly prone to summer and early autumn mortality crises when environmental conditions facilitated the transmission of air-borne infection and enteric disease (Bradley 1971), which would help make possible the contamination of water and food by insect, animal, or human vectors. In this respect, it is interesting that Gilbert White (1993) linked several of these vectors in his famous account of 1783: 'the heat was so intense that the butchers' meat could hardly be eaten on the day after it was killed and the flies swarmed so in the lanes and hedges'. Creighton (1965) considered that dysentery was epidemic in Britain in the period 1779–85, and several notable episodes of the diseases were reported in 1783. One of these occurred in Carlisle, where 17 children under the age of three but only two over the age of five died. In Newcastle, dysentery was described as having 'attacked great numbers of the poor in the autumn of 1783' (Creighton 1965), and an incredible 154 deaths occurred between September and November 1783 in neighbouring Tynemouth. Dysentery also caused many deaths in the Netherlands in the summer and autumn of 1783 (Van Westeringh 2004).

The persistent background level of infection by the smallpox virus could flare up into an epidemic when the population of unaffected children became high enough (Creighton 1965) or when the vectors that transmitted the disease were particularly favourable, as they were in the summer of 1783. Dr Heysham described how in Whitehaven, Cumbria, an epidemic of smallpox in 1783 was mainly confined to the labouring classes who largely lived in cellars: 'incredible numbers were attacked of whom scarcely one in three survived' (cited in Creighton 1965: 538). In the parish of All Saints, Loughborough, 114 people died in 1783, nearly double the average. In an unusual note added at the end of the register for 1783, the clerk recorded that 44 of the deaths that year were because of an outbreak of smallpox during the summer.

The events of 1783 fell in the midst of what Creighton (1965) described as 'The Epidemic Agues of 1780–1785'. Barker (1796) noted that these agues did not respond to treatment by Peruvian bark, and a correspondent in the *Gentleman's Magazine* sought to blame the enclosure of common land and the building of canals for the unexplained appearance of ague 'in the most healthy situations in it [the country]'. Barker wrote of 1783 that 'the influenza also began to appear again; and those who had coughs last year began now to be afflicted with them again, the disorder at length frequently ending in a consumption' and death shortly afterwards. Barker, writing of Northamptonshire, also confirmed Cowper's observation of the difficulty getting in the harvest due to illness amongst the labourers. Today, except amongst the very elderly, influenza is no more than an inconvenience. In the 18th century, it was a killer; an influenza epidemic began in 1782 and appears to have intensified in 1783 (Creighton 1965).

DISCUSSION

We should at least consider a null hypothesis in the discussion, one in which the eruption had no influence on climate, the environment, and human health. To do so requires that we ignore, or at the least dismiss, the vast volume of toxic volatiles emitted by the eruption and to deny any subsequent climatic modification. We would also need to ignore the well-established research that has demonstrated Laki's impact on Europe's environment and world climate.

These examples of the climatic disasters that occurred in 1783 and that encompassed the deaths of millions reinforces one of the key themes of this book: that it is the organisation and inherent resilience of a society that determines its ability or lack thereof to weather disasters. Of the examples cited above, only the Alaskan Inuit were vulnerable because of their society's exploitation of a highly marginal environment and specialised, but ecologically fragile, resource base. In the cases of Egypt and India, the climatic crisis of 1783 was intensified by social disorganisation and economic breakdown. In Japan, overpopulation made the society vulnerable to the unanticipated failure of the rice crop. With the possible exception of the Alaskan Inuit, disaster was not inevitable in these cases.

To assess the impact of the Laki fissure eruption on human health and mortality requires a subtle understanding of the interactions between the eruption and the 18th-century environment, including a consideration of how the eruption could wield so much influence on such a diverse range of locations and environments. Stothers (1999) suggested that increased wetness in Europe and the Middle East, which is a modelled climatic disruption following significant volcanic eruptions, facilitated the spread of disease, with the result that significant plague pandemics occurred. Similarly, Oppenheimer (2003) noted that increased wetness in Ireland following the eruption of Tambora in 1816 appears to have facilitated a devastating outbreak of typhus, which caused the deaths of over 44,000 people. England in 1783 was in the throes of at least three ongoing epidemics of agues, including influenza, typhus, and dysentery. In addition, the 18th-century population contained a suite of diseases that were more or less virulent, depending on the environmental and demographic vectors as well as the individual health of each person. Thus, the severity of the mortality crisis and its geographical distribution were intimately related to the demographic structure of the population in the towns and villages, the living conditions of particular socioeconomic groups, and the prevalent environmental vectors. The eruption appears to have been able to influence all of these; in locations in which mortality vectors were close to a critical threshold, it was able to influence these decisively.

For example, in the 18th-century, summer diarrhoea killed many children, particularly in 1783, yet few adults died from this condition as their physical condition was robust enough to enable them to endure and recover. The critical influence of the eruption in this instance was that it intensified the heat of the summer and promoted the vectors responsible for both spreading the disease – insects, bad food, and poor water – and intensifying the physiological stress of the season.

CONCLUSIONS

Recently developed paradigms, many proposed in our previous book (Shimoyama 2002; Torrence 2002; Torrence and Grattan 2002), seek to emphasise the vulnerability of cultures and environments rather than the magnitude of the eruption itself, an approach that requires a consideration of cultures or environments in detail rather than in abstract (Driessen 2002; Grattan and Gilbertson 2000) and allows there to be both victims and beneficiaries when volcanic eruptions occur. This study confirms that volcanic eruptions can have a profound influence of populations far from an eruption. The magnitude of the event and its longevity clearly were factors in the phenomena described in this chapter, but at least as important was that both the population and environment were sensitive to the volcanic influences on several levels.

From an archaeological standpoint, this is the critical paradigm: the eruption on its own, whatever influence it yielded, is irrelevant unless the receptor community and/or environment is sensitive or vulnerable to those influences. For archaeologists or geoarchaeologists to invoke a distant volcanic eruption as the cause of phenomena noticed during research they must first classify the eruption and understand the range of impacts it could have. They also need to establish the extent to which the environment, culture, or people were sensitive or vulnerable to the volcanic mechanisms; mere coincidence is not enough.

REFERENCES

Barker, J (1796) *Epidemicks*, Birmingham, T Longman

Bradley, L (1971) 'An enquiry into seasonality in baptisms, marriages and burials', *Local Population Studies* 6, 15–31

Baillie, MGL and Munro, MAR (1988) 'Irish tree rings, Santorini and volcanic dust veils', *Nature* 322, 344–46

Blackford, JJ, Edwards, KJ, Dugmore, AJ, Cook, GT and Buckland, P (1992) 'Hekla-4: Icelandic volcanic ash and the mid-Holocene Scots Pine decline in northern Scotland', *The Holocene* 2, 260–65

Burgess, C (1989) 'Volcanoes, catastrophe and the global crisis of the late second millenium BC', *Current Archaeology* 117, 325–29

Cowper, W 1981 (1783) in King, J & Ryskamp, C (eds). 1981. *The letters and prose writings of William Cowper*, vol 2, Oxford: Clarendon Press,

Creighton, C (1965) *A History of Epidemics in Britain*, vol 2, Cambridge, Cambridge University Press

Demarée, GR and Ogilvie, AEJ (2001) 'Bons Baisers d'Islande: climatic, environmental and human dimensions impacts of the Lakágigar eruption (1783–1784) in Iceland', in Jones, PD, Ogilvie, AEJ, Davies, TD and Briffa, KR (eds) *History and Climate: Memories of the Future*, pp 219–46, Dordrecht, the Netherlands: Kluwer

Dobson, MJ (1980) 'Marsh fever – the geography of malaria in England', *Journal of Historical Geography* 6, 357–89

Dobson, MJ (1997) *Contours of Death and Disease in Early Modern England*, Cambridge: Cambridge University Press

Driessen, J (2002) 'Towards an archaeology of crisis: defining the long term impact of the Bronze Age Santorini eruption', in Torrence, R and Grattan, JP (eds), *Natural Disasters and Cultural Change*, pp 252–63, London: Routledge

Dugmore, AJ (1989) 'Icelandic volcanic ash in Scotland', *Scottish Geographical Magazine* 105, 168–72

Durand, M and Grattan, JP (1999) 'Extensive respiratory health effects of volcanogenic dry fog in 1783 inferred from European documentary sources', *Environmental Geochemistry and Health* 21, 371–76

Durand, M and Grattan, JP (2001) 'Volcanoes, air pollution and health', *The Lancet* 357, 164

Gentleman's Magazine, The. A monthly serial published 1731–1914

Grattan, JP and Brayshay, MB (1995) 'An amazing and portentous summer: environmental and social responses in Britain to the 1783 eruption of an Iceland volcano', *The Geographical Journal* 161, 125–34

Grattan, JP and Charman, DJ (1994) 'Non-climatic factors and the environmental impact of volcanic volatiles: implications of the Laki Fissure eruption of AD 1783', *The Holocene* 4, 101–16

Grattan, JP, Durand, M, Gilbertson, DD, Pyatt, FB and Taylor, S (2003) 'Long range transport of volcanic gases, human health and mortality: a case study from eighteenth century Europe', in *Geology and Health: Closing the Gap*, Skinner, HCW and Anthony Berger, A (eds), pp 19–31, Oxford: Oxford University Press

Grattan, JP, Durand, M and Taylor, S (2003) 'Illness and elevated human mortality coincident with volcanic eruptions', *Volcanic Degassing, Geological Society Special Publication* 213, 401–14

Grattan, JP and Gilbertson, DD (1994) 'Acid-loading from Icelandic tephra falling on acidified ecosystems as a key to understanding archaeological and environmental stress in northern and western Britain', *Journal of Archaeological Science* 21, 851–59

Grattan, JP and Gilbertson, DD (2000) 'Prehistoric "settlement crisis", environmental changes in the British Isles and volcanic eruptions in Iceland: an exploration of plausible linkages', in McCoy, F and Heiken, G (eds), *Volcanic Hazards and Disasters in Human Antiquity, Geological Society of America Special Paper* 345, 33–42

Grattan, JP and Pyatt, FB (1994) 'Acid damage in Europe caused by the Laki Fissure eruption – an historical review', *The Science of the Total Environment* 151, 241–47

Grattan, JP and Pyatt, FB (1999) 'Volcanic eruptions, dust veils, dry fogs and the European Palaeoenvironmental record: localised phenomena or hemispheric impacts?', *Global and Planetary Change* 21, 173–79

Grattan JP, Rabartin, R, Self, S and Thordarson, TH (2005) 'Volcanic air pollution and mortality in France 1783–84', *Comptes Rendu Geosciences* 7, 641–51

Grattan, JP and Sadler, J (1998) 'Regional warming of the lower atmosphere in the wake of volcanic eruptions: the role of the Laki fissure eruption in the hot summer of 1783', *Geological Society Special Publication* 161 Volcanoes in the Quaternary, 161, 161–72

Grattan, JP, Schuttenhelm, R and Brayshay, M (2002) 'The end is nigh? Severe social and environmental responses to volcanic pollution of the lower atmosphere: an alternative forcing mechanism for archaeologists and historians', in Torrence, R and Grattan, JP (eds), *Natural Disasters and Cultural Change*, pp 87–106, London: Routledge

Gunn, JD (2000) *The Years without a Summer: Tracing AD 536 and Its Aftermath. British Archaeological Reports International Series* 872, Oxford: Archaeopress

Hall, VA (2003) 'Assessing the impact of Icelandic volcanism on vegetation systems in the north of Ireland in the fifth and sixth centuries', *The Holocene* 13, 131–38

Hammer, CU, Clausen, HB and Dansgaard, W (1981) 'Past volcanism and climate revealed by Greenland ice cores', *Journal of Volcanology and Geothermal Research* 11, 3–10

Howe, GM (1997) *People, environment, disease and death*, Cardiff: University of Wales Press

Jackson, EL (1982) 'The Laki eruption of 1783: impacts on population and settlement in Iceland', *Geography* 67, 42–50

Jacoby, GC, Workman, KW and D'Arrigo, RD (1999) Laki eruption of 1783, tree-rings and disaster for the NW Alaskan Inuit, *Quaternary Science Reviews* 18, 1365–71

Maddox, J (1984) 'From Santorini to Armageddon', *Nature* 307, 107

Oman, L, Robock, A, Stenchikov, GL, Thordarson, T, Koch, D, Shindell, DT and Gao, C (in press) 'Modelling the distribution of the volcanic aerosol cloud from the 1783–1784 Laki eruption', *Journal of Geophysical Research*

Oppenheimer, C (2003) 'Climatic, environmental and human consequences of the largest known historic eruption: Tambora volcano (Indonesia) 1815', *Progress in Physical Geography*, 27, 230–59

Peiser, BJ, Palmer, T and Bailey, ME (1998) *Natural Catastrophes during Bronze Age Civilizations*. British Archaeological Reports, International Series 728, Oxford: Archaeopress

Pilcher, JR, Hall, VA, and McCormac, FG (1995) 'Dates of Holocene Icelandic volcanic eruptions from tephra layers in Irish peats', *The Holocene* 5, 103–10

Rabartin, R and Rocher, P (1993) 'Les volcans, le climate et la révolution française', *Memoire de l'Association Volcanologique Européenne* 1, 1–56

Rampino, MR and Ambrose, S (2000) 'Volcanic winter in the Garden of Eden: the Toba super-eruption and the late Pleistocene human population crash', in McCoy, F and Heiken, G (eds), *Volcanic Hazards and Disasters in Human Antiquity, Geological Society of America Special Paper* 345, 71–82

Rampino, MR and Self, S (1993) 'Climate-volcanism feedback and the Toba eruption of ~74000 years ago', *Quaternary Research* 40, 269–80

Riley, JC (1987) *The Eighteenth Century Campaign to Avoid Disease*, London: Macmillan

Robock, A (2000) 'Volcanic eruptions and climate', *Reviews of Geophysics* 38, 191–219

Sadler, J and Grattan, JP (1999) 'Volcanoes as agents of past environmental change', *Global and Planetary Change* 21, 181–96

Schofield, R (1998) 'Parish register aggregate analyses: the "population history of England" database' *Local Population Studies*, Supplement

Shimoyama, S (2002) 'Basic characteristics of disasters', in Torrence, R and Grattan, JP (eds), *Natural Disasters and Cultural Change*, pp 19–27, London: Routledge

Steingrimsson, J (1998) *Fires of the Earth: The Laki Eruption 1783–1784, Reykjavik*: University of Iceland Press

Stothers, RB (1999) 'Volcanic dry fogs, climate cooling and plague pandemics in Europe and the Middle East', *Climatic Change* 42, 713–23

Thorarinsson, S (1981) 'Greetings from Iceland. Ash falls and volcanic aerosols in Scandinavia', *Geografiska Annaler* 63A, 109–18

Thordarson, T and Self, S (2003) 'Atmospheric and environmental effects of the 1783–1784 Laki eruption: a review and reassessment', *Journal of Geophysical Research-Atmospheres* 108, 33–54

Torrence, R (2002) 'What makes a disaster? A long-term view of volcanic eruptions and human responses in Papua New Guinea', in Torrence, R and Grattan, JP (eds), *Natural Disasters and Cultural Change*, 292–312, London: Routledge

Torrence, R and Grattan, JP (2002) *Natural Disasters and Cultural Change*, London: Routledge

Van Swinden, SP (2002) 'Observations of the cloud (dry fog) which appeared in June 1783', *Jökull* 50, 73–80

Van Westeringh, W (2004) 'De rode loop in Andelst 1783', *Aqua Vitae* 7, 22–25

Velikovsky, I (1950) *Worlds in Collision*, London: Gollancz

White, GW (1993) *The Natural History of Selbourne*, London: Thames and Hudson

White, RS and Humphreys, CJ (1994) 'Famines and cataclysmic volcanism', *Geology Today*, 94, 181–85

Witham, CS and Oppenheimer, C (2005) 'Mortality in England during the 1783–4 Laki Craters eruption', *Bulletin of Volcanology* 67, 15–26

World Health Organisation (2004) *Heat waves, risks and responses, Geneva WHO*

Wrigley, EA and Schofield, RS (1989) *The Population History of England 1541–1871*, London: Edward Arnold

CHAPTER 9

Volcanic Oral Traditions in Hazard Assessment and Mitigation

Shane J Cronin and Katharine V Cashman

INTRODUCTION

In preliterate societies, momentous natural events such as volcanic
catastrophes are often incorporated into cultural histories as stories
that are repeated over generations, thereby forming 'oral traditions'
(Vansina 1985). Worldwide, many such accounts have been proven by
geological study to be more than just simple legends or myths (eg,
Blong 1982). To date, these stories have been largely neglected in modern volcanic hazard mitigation, that is, in helping local communities
make themselves more resistant to volcanic hazards posed by their
environment (exceptions include Cronin, Ferland *et al* 2004; Cronin,
Gaylord *et al* 2004). Here we address a number of questions about the
integration of oral traditions into modern hazard analysis. How accurately do oral traditions record volcanic events? How can oral traditions be used to further the understanding of local hazards? Could
they help develop hazard prevention, response, and recovery plans?
How can these records be used to supply a physical (natural hazard)
context for archaeological research on cultural histories, particularly to
explain sudden migrations or changes in agricultural practices?

Modern societies assess potential volcanic hazards from several lines
of physical and descriptive evidence. Of these, eyewitness accounts
are fundamental to the characterisation of volcanic processes and form
the basis for defining almost all of the known 'types' of eruption. For
example, 'Plinian' eruptions are named for Pliny the Younger, who
provided the first scientific descriptions of an eruption, the 79 AD eruption of Vesuvius (Schmincke 2000). Even in an age of instrumentation
and remote sensing, eyewitness observations of scientists and non-
experts are vital to fully reconstructing eruptive sequences and processes (Rosenbaum and Waitt 1981). However, such accounts normally

require careful analysis to understand the perspectives and language of the observers, who are often under extreme stress and who have different levels of understanding of the processes being described (Loughlan *et al* 2002).

If one assumes that oral traditions have their roots in eyewitness accounts of eruptions, then not only aspects of viewer background knowledge, but also processes that modify initial eyewitness descriptions through time, should be considered in story interpretation. These modifiers may include religious interpretation, performance, cultural conventions, personal creativity, and world view (Finnegan 1992), in addition to the simple retelling of the tale over time by people who have not witnessed the event. Given such an array of filters applied to already imperfect eyewitness accounts, is there any scientific information that can be salvaged from oral traditions of volcanic eruptions? We propose that there is, but that the information must be very carefully verified with supporting geological evidence (cf Blong 1982).

MULTIPLE ROLES OF ORAL TRADITIONS

The term 'oral tradition' is used here to describe a range of accounts, stories, or sagas that are repeated over generations, generally within a non-literate society, until they are eventually committed to writing, either in their native tongue or, more frequently, in another (Vansina 1985). They are normally regarded by their tellers as descriptions of real events. Not all oral traditions are true records of historical fact; some may be better considered as local artistic expression or works of human imagination, and hence prone to speculative interpretations (Finnegan 1995). On rare occasions, they may even be a hoax or practical joke (Grattan *et al* 2000). In many traditional societies, however, oral traditions may be as important for the maintenance of the social order as are kinship systems and social organisation (Gossen 1974).

As History

There is a wide range of opinion regarding the reliability of oral traditions as historical records, with the debate particularly sharp in the Pacific north-west of the United States (eg, Echo-Hawk 2000; Mason 2000). The present consensus is that many oral traditions can be used, with care, as historical data (Whiteley 2002). However, the main difficulty in their use is identifying their 'time depth', the age of the events they describe. The durability of oral traditions is also hotly debated, with long-entrenched views claiming that they cannot persist over even short periods with any reliability (Lowie 1915), whilst more recent studies infer descriptions of events that may be up to 40,000 years old (Echo-Hawk 2000). Regardless of their age, however, establishing a chronology in calendar years may be impossible without secure relationships to other more broadly known events in a culture's history (eg, the first arrival of Europeans in the New World) or direct linkage to a datable event, such as a volcanic eruption (Blong 1982).

Oral traditions relating to volcanism probably begin as eyewitness descriptions of dramatic or overwhelming events where, as in many contemporary eyewitness accounts, the observers have variable scientific understanding of volcanic phenomena. From these beginnings, such accounts are progressively modified by a range of processes (Vansina 1985):

(1) Retelling over generations and modifications to fit the narrator's views and life experiences that may lead to omission of specific observations that the narrator has not personally witnessed and therefore does not understand;

(2) Embellishment of stories by retellers to improve their entertainment value (Finnegan 1992);

(3) Putting accounts into modern contexts, particularly when no physical reminders of the event remains (eg, reforestation of volcanic vent areas);

(4) Cultural/religious modifications, which may involve either personification of momentous eruptions as conflicts between deities within a belief system, or use of catastrophic events as vehicles to educate others in aspects of religion, culture, and correct religious and moral practice. These modifications may include new religious teachings, such as the arrival of Christianity in an area (Moodie *et al* 1992; and see Chester and Duncan, Chapter 10 this volume);

(5) Relationship to intertribal conflict, so that different versions arise through different tribal affiliations of the groups retelling the traditions;

(6) Incorporation into origin/migration histories. Because many catastrophic eruptions involve people fleeing from an area of danger, they also involve new beginnings for tribes or families. Hence, aspects of tribal origins are often woven into the accounts;

(7) Relationship to landscape, where changes resulting from eruptions are not only incorporated into stories but may also be extrapolated to explain the origins of similar landscape features in the local area;

(8) Layering and amalgamation of multiple events, particularly when eruptions are relatively frequent but of low impact;

(9) Translation and writing, although not artefacts of the oral tradition itself, the recording of oral traditions may often introduce fundamental changes, particularly as the result of bias or ulterior motives of the recorder. Examples include using traditions to aid religious conversion (Moodie *et al* 1992), register moral offences towards sexual or violent themes, or misrepresent or even romanticise local cultures and practices;

(10) Adaptation into other forms or media, such as song, dance, or artwork

These potential modifications inherent to the transmission of oral traditions pose challenges to the extraction of scientific data. However, below we illustrate ways in which specific and accurate information relating to volcanic events and processes can be obtained from traditional stories, particularly when verified by independent geological research.

As Aids to Volcanic Hazard Assessment

Evaluation of the risk posed by volcanic activity requires both knowledge of the specific types of volcanic processes likely to affect a community (ash fall, pyroclastic flows, mudflows, etc) and the degree to which the community is vulnerable. Community vulnerability includes both its ability to anticipate eruptive activity (ie, to recognise and respond to precursors of impending eruptions) and to survive and recover from eruptive processes. Under ideal conditions, several types of information can be extracted from eyewitness or oral tradition accounts that can improve community knowledge of, and vulnerability to, volcanic hazards:

(1) Precursory signs of volcanic activity, such as duration, changes over time, nature, and spatial distribution of surface manifestations (eg, hydrothermal features and ground cracks);

(2) Detailed eruption descriptions, including the location of vent areas, ash column heights, wind directions, common paths of lava flows, pyroclastic flows (incandescent gas and ash flows) and mud flows (lahars), timing and duration of various eruption phases, and the nature and spatial distribution of impacts on local populations;

(3) Post-eruption changes in landscape, both destructive and constructive;

(4) Community hazard mitigation information such as past areas of danger, safe areas, evacuation routes taken, protective measures, and other response measures and recovery strategies that have been adopted.

Although no one account addresses all aspects of a particular eruption, the collective body of traditions in a society often provides a fairly complete picture of volcanic phenomena that have affected a given region in the past, and thus are events that are likely to occur in the future.

Cultural History and Disaster Recovery

Local or subregional volcanic catastrophes are unlikely to have generated pan-Pacific cultural change or migration, however they could have been major drivers of change in places such as Tonga (Cronin, Gaylord *et al* 2004) or Vanuatu (Galipaud 2002). In these cases, oral traditions recording the volcanic events may provide clues to the causes of prehistoric cultural changes. For example, Shanklin (1989) documents oral traditions that describe migration history that is specifically linked to a volcanic disaster.

Oral traditions may have also played a major role in societal recovery from volcanic disasters. Physical recovery (crops, houses, tools, and materials) is now increasingly seen as a relatively minor issue in comparison to the long-lasting process of psychosocial recovery (Paton 1996; Tierney 1989). Oral traditions arise during the time immediately after an event, a time when the event is discussed and contextualised within a framework that can be understood by the culture (ie, in terms of its own heroes, gods, or perceptions of the natural world). Discussion and assimilation are important steps in coping with a traumatic and overwhelming event, helping facilitate the return to normal cultural practices or adaptation to new ones.

ORIGINS OF ORAL TRADITION: A CONTEMPORARY EXAMPLE

How do volcanic oral traditions originate and why? Here it is helpful to consider the case of Vanuatu, a south-west Pacific archipelago because it possesses the unique combination of many active volcanoes (Figure 9.1) and many rural communities based on subsistence farming that have only been moderately influenced by external contact. Aspects of contemporary oral history have been gathered from community education activities on the islands of Ambae, Ambrym, Tongoa, and Paama (Cronin, Gaylord *et al* 2004) undertaken with the able assistance and experience of Mr Douglas Charley (Department of Geology, Mines and Water Resources, Vanuatu). Through these education activities, we have found that many traditional beliefs and practices (*kastom*) are directly related to volcanic phenomena.

Beginning with the theme of origins or causes, the local view is that many eruptions are initiated by a sorcerer who calls the volcano into life by ritual, song, or incantation for purposes of revenge, demonstration of power, or simply ceremonial requirements. For example, it is widely believed that large eruptions on the flanks of Ambrym in 1894 (Purey-Cust 1896) and 1913 (Frater 1917) were started by a sorcerer. In both accounts, the sorcerer initiated an eruption by digging a hole in the ground from which hot water emerged; a subsequent eruption was encouraged with special calls and songs. Such changes in local groundwater systems are common precursors to volcanic eruptions, as illustrated by hot floods that emerged from the summit crater of Mt Pelée, Martinique, in the week prior to a catastrophic eruption of that volcano in 1902 (Tanguy 1994). Eruptions may also be halted by a sorcerer if he makes certain incantations or sings particular *kastom* songs. Icons and

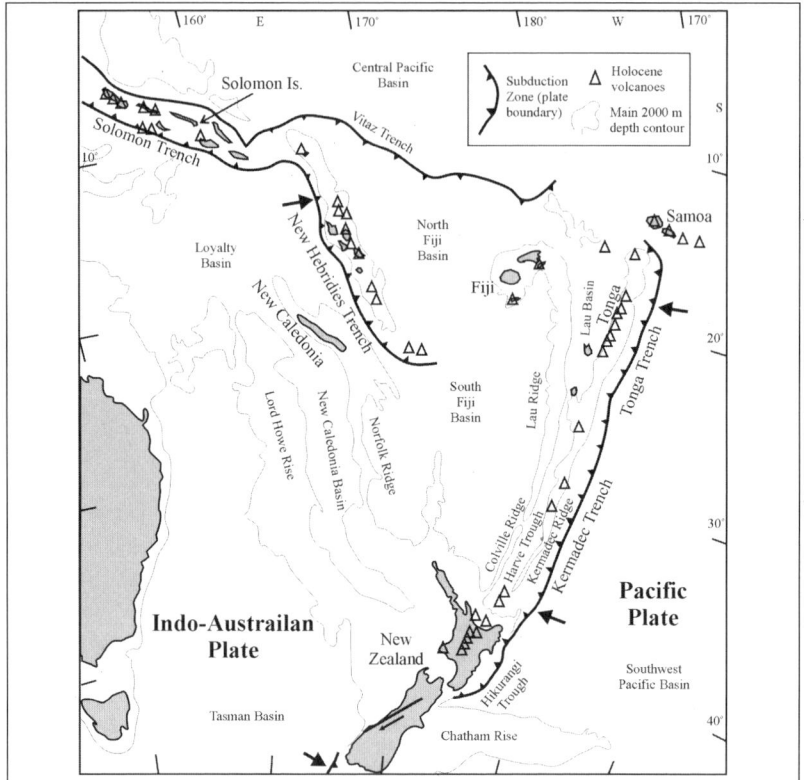

Figure 9.1   Active volcanoes described in the text from the southwest Pacific.

ritual locations, such as the ceremonial site shown in Figure 9.2, centre in many sorcery activities relating to volcanism.

A second important point is that there are rules that must be observed with respect to volcanoes. Should these be broken, eruptions, and hence punishment of people, will result. For example, the summit caldera of Ambae contains two lakes, one of which is located over the active crater. This area is *tambu* (forbidden), because it is believed that if a person were to go there, an eruption would ensue. In fact, a small phreatic eruption on Ambae in 1995 is widely believed to have been caused by transgression of this *tambu*. Application of *tambu* has clear hazard mitigation implications, as it serves to protect local people from unexpected explosions from the crater lake. However, *tambu* also hampers scientific monitoring efforts on the volcano, because it limits visits for scientific observation and installation of monitoring equipment to those sanctioned by a series of ritual ceremonies required to warn and placate the volcano.

Figure 9.2 A *kastom* ceremonial site lined with carved *tamtams* (wooden slit drums) in an upland village of western Ambrym is used at times for rituals relating to hindering volcanic hazards (along with other cultural ceremonies). This area is showered by ash fall on a yearly basis and lava flows have inundated nearby locations on two occasions since the 1950s.

In other cases, spirits are believed to occupy past centres of volcanic activity (eg, within water-filled volcanic craters on the low flanks of east Ambae). Respecting these spirits by conducting animal sacrifices and other annual ceremonies is essential to keep the volcano pacified. Thus, until recently annual sacrifices were carried out by people who were forced to abandon Lopevi Island and relocate on eastern Paama during large eruptions in 1963. Practices regarding the placation of spirits then faltered with adoption of new Christian practices and weakening of old ways. However, after a period of relative quiet since 1963, Lopevi has recently come to life with eruptions in 1999, 2001, 2002, and 2003. Many elders in the eastern Paama communities attribute this activity to the lapse in annual pig sacrifices and *kastom* ceremonies, which has raised the ire of local sorcerers, some of whom have been accused of starting the latest eruptions. A more recent example of volcano worship in Vanuatu involves a 'cargo cult' that arose during World War II activities on Tanna Island (Lindstrom 1995) and incorporated Yasur volcano as a central feature in its ceremonies and beliefs (Figure 9.3). This volcano, which has been in semi-constant eruption over at least the last 300 years (Wiart 1995), has consistently affected the cult villages living around its flanks.

The Vanuatu contemporary belief systems provide examples of how volcanism can be but is also incorporated into cultural beliefs and ritual. Here, oral traditions provide coherent scientific descriptions of volcanism and establish measures (particularly *tambu*) to protect people from volcanic hazards. The Lopevi example also demonstrates affirmation of an oral tradition by subsequent volcanic events. Importantly,

Figure 9.3   View of Yasur volcano in July 2003. Cargo cult villages that surround this constantly active volcano are affected on a daily basis primarily by minor volcanic gas and ash-fall hazards.

these 'everyday' stories offer a context for tales or descriptions of more devastating eruptions, which then survive to become oral traditions.

ASSESSING THE HAZARD

Volcanologists need to comprehend the range of possible eruptive behaviour of a particular volcano to assess, or forecast, potential future hazards. Despite seemingly broad similarities between forms and compositions, all volcanoes are different, and comprehensive individual assessment of their hazards must be carried out requiring detailed geologic study and examination of historic/recorded evidence of past eruptions. We believe that oral traditions should be relied on to supplement this record, as demonstrated by the examples provided below.

Precursor Signals

Records of volcanic precursors are found in oral traditions, even though the details may be stylised to suit a story or lesson. For example, New Zealand Maori and European observers recorded both physical and spiritual phenomena prior to the 1886 eruption of Tarawera (Cowan 1930; Izett 1904). The former included changes in water levels of the lake at the foot of the volcano, another example of precursory hydrologic activity. The latter involve an enduring tale, purportedly witnessed also by European tourists, of a phantom canoe in the lake that bore a row of standing figures, their heads bowed and hair plumed with *Huia* and white heron feathers as if prepared for death.

A more explicit description of precursory activity is provided by the oral tradition of 'Tombuk' from Vanuatu, which records activity related to a voluminous (c 30 km³) caldera-forming eruption at Kuwae in central Vanuatu that occurred in 1452 AD (Robin *et al* 1994). The following version lacks detailed eruption descriptions, but outlines associated landscape changes and strongly emphasises pre-eruptive phenomena.

> The magic was ready. Tombuk took all six pig bladders and climbed the large tree at the Nakamal, tying bladders to branches at progressively higher levels within the tree. Then he began to sing a special *kastom* song whilst perched at the very top of the tree. After the song, he broke the first and uppermost bladder, causing the ground to shake and move. The young sorcerer then dropped down to the second level and resumed singing, before breaking the second bladder. This caused even more ground shaking. By this time all the others from the village had gathered around the tree and began teasing and jeering at Tombuk for singing in the tree. He did not take any notice and descended to the third level. Here the young sorcerer sang again and broke the third bladder. Now ground motions and shakes were much stronger and the gathered villagers began to realise what Tombuk was planning. … The young man then dropped

down to the next level and began again to sing. When he broke the fourth bladder, the ground did not stop heaving and pitching people and houses to the ground. No one was able to stand up, and all were rolling on the ground. The people began to cry and plead with Tombuk, screaming apologies and begging him to stop his magic. But the young sorcerer dropped down to the next level in the tree and continued singing. Once he had broken the fifth bladder, the ground began to open up under the people below. Chiefs broke down and pleaded with Tombuk, while some villagers were completely swallowed up and burnt within the huge cracks. ... He jumped down to the base of the tree where his mother stood. ... As he broke the sixth and last bladder, the volcano roared into life, throwing him and his mother high in the air. Tombuk fell down in Ambrym and his mother in Lopevi and the land of Kuwae was destroyed, leaving only its edges on Epi and Tongoa. (translated from a Bislama interview recorded by Douglas Charley, Department of Geology and Mines, Vanuatu)

The strong earthquake activity alluded to in this story is a common precursor to large eruptions. For example, extensive ground shaking and deformation preceded a smaller 1991 caldera-forming eruption of Pinatubo volcano in the Philippines (Ewert *et al* 1996; Harlow *et al* 1996), and the climactic eruption of Mount St Helens in 1980 (Endo *et al* 1981). Strong earthquakes preceding the Bronze Age eruption of Santorini volcano (Greece) provided sufficient warning of the impending eruption to enable evacuation before the climactic event. In contrast, the absence of strong seismic activity before the 79 AD eruption of Vesuvius left people in the towns of Pompeii and Herculaneum vulnerable to the devastating pyroclastic flows and lahars that accompanied the eruption (Cioni *et al* 2000).

Precursory details may also be recorded with less stylistic embellishment. In contrast to the Vanuatu tradition noted above, oral traditions collected on Savo Island, Solomon Islands (Figure 9.1), are more pragmatic in their summary of events that heralded three brief but powerful (vulcanian) eruptions on their island since 1567 AD (Cronin, Petterson *et al* 2004; Petterson *et al* 2003; Toba 1993). Here, long-term precursors, including gradual 'shrinking' of the island through coastal erosion of volcaniclastic fan sediments, are distinguished from those manifested shortly before eruptive activity. According to tradition, once the coastline had reached the low hills of the island, the next eruption was due. A more specific long-term precursor is an increase in the aerial extent and vigour of surface geothermal activity.

Changes in fumarole and hot spring areas are common in some areas during the years before renewal of volcanic activity, an example being a 13-year period of changing geothermal activity noted prior to the devastating 1902 eruption of Mt Pelée (Smith and Roobol 1990). Short-term precursors on Savo include hydrologic changes, specifically infilling of the usually dry central crater with water (which also occurred at Mt Pelée). Also recorded are earthquakes and tremors that increase in frequency until the volcano comes to life, local tsunamis, increasing storminess, landslides in and around the crater area, and vegetation die-off around the crater.

Studies from Ambae volcano, a large basaltic shield with an active caldera-hosted crater lake in Vanuatu (Figure 9.1), also illustrate precursors related to migration of volcanic gasses and changes to local groundwater systems that provide early evidence of magma migration to the surface. Ambae has experienced significant eruptions on a similar scale as Savo, with events at 1575, 1670, and 1870 AD (Blot and Priam 1962; Warden 1970). As at Savo, warning signs of impending eruptions are plentiful in oral traditions (Cronin, Gaylord *et al* 2004). Evidence of such changes are recorded at Ambae as the occurrence of pervasive gas smells, die-off of trees around the active crater lake, unusually active bubbling and other lake colour disturbances, rumbling and booming from the crater, rapid rotting of taro roots in the ground, and taro leaves burnt by acid rain. Unusual animal behaviour, such as bird migration or swarming of ants, may also provide clues of local environmental change. Oral traditions from Ambae also include recognition of earthquake activity as precursory to volcanic eruptions. Like Savo, bad weather is also in the precursor arsenal, as are dreams and unusual natural appearances to impending eruptive activity.

Together, these examples from Savo and Ambae give practical information that could be applied directly to improved community hazard mitigation plans. All of the precursors noted (with the possible exception of premonitory dreams) are used by modern volcanologists to monitor restless volcanoes, thus demonstrating the accuracy of local observers and oral traditions in communities where volcanic activity is a fairly frequent phenomenon. In particular, for subsistence agriculturalists, the effects of changing hydrologic and hydrothermal systems on plant and animal behaviour are the most easily observed signs of unusual happenings on a volcano. Interestingly, both examples also refer to stormy weather preceding and accompanying eruptive activity. The latter is easily explained, as large ash eruptions tend to produce widespread atmospheric disturbance, including lightning and rainfall (Blong 1984). However, there is no good explanation for the occurrence of rough seas and high winds before eruptions. We suggest that reporting this activity may be serendipitous, or may even act as Shakespearean-type dramatic aids to impress on listeners the magnitude of the catastrophic event.

Eruption Processes

Descriptions of specific events in oral traditions may provide critical inputs to understanding the types of eruption processes that are possible at any given volcano, especially in parts of the world where geological exposures are poor or geologic investigations are scanty.

Ash Fall

Ash fall from towering volcanic eruption columns is the most widespread of volcanic phenomena. Importantly, ash fall commonly affects regions sufficiently far from a volcano that the recipients may be completely unaware of its origin. In oral traditions, ash fall is most commonly recorded as the blocking out of sunlight, effectively turning night into day. This is a common eyewitness description by those who have been enveloped in an ash plume. At greater distances from the volcano, ash fall may be described as snow-like (see Ray 1980 for the Pacific north-west) or soot/sand-like (see Blong 1982 for the Pacific south-west) material falling from the sky. Because ash fall is experienced differently by communities at different distances from a volcano, traditional explanations also vary widely.

An excellent example of oral traditions relating to ash fall is the eruption of Tibito Tephra over much of central Papua New Guinea at around 1630–80 AD (Blong 1982). Most of the oral traditions are from areas where the source of the eruption was not witnessed. All stories describe a period of several days of darkness, in cases more impenetrable than night (*bingi*), when dust or sand covered everything, a time scale consistent with modern experience of large eruptions. From 54 accounts, Blong (1982) reported that despite stylistic variations, the legends record essentially accurate data that agree with the physical evidence for the Tibito Tephra fall, but that the accounts vary in the level of detail recorded and in the nature of the physical phenomena described. These differences are probably a consequence of diversity in memory, culture, house construction, elevation, vegetation/crops, and ash-fall thickness. Embellishments include confusion with Christian teachings, use of pig sacrifices to make the darkness lift, and other emphases on 'proper' behaviour to reduce the impact of future eruptions.

In another important example, ancient ash-fall events and major eruptions of El Chichon in Mexico were also described primarily as sudden darkness.

> The earth darkened for five days. The sun was still bright and clear. Then at midday it got dark. 'Now we will surely die', said the people. … And when it darkened, the demons rushed forth from the broken pots. Lions came forth! Snakes came forth! Jaguars came forth! So it was that the poor people perished in the jaws of the demons. How the people screamed and shrieked. … Well when dawn came upon the earth, there was no longer a single person left. Now there were only birds. … (Gossen 2002: 171–82)

The animals mentioned may represent those unsettled by the ash-fall event or earthquakes associated with it. Similar disturbances to the animal population were noted before and during the 1902 eruption of

Mt Pelée; snakes plagued the population during the early eruptions (Tanguy 1994), probably as a consequence of intense seismic and groundwater disturbances. Alternatively, considering that all the animals mentioned are frightening creatures, they could equally be dramatic aids to impress on the listener the horror of the situation.

A more recent oral tradition (perhaps only 60–80 years old) from the same region describes a less destructive eruption and ash-fall event.

> There was a roar of the earth gods, just like thunder, and the ash rose out of the mountain like clouds and came down as rain and hail in small and large pieces. The sky was black and the sun invisible and people could not see their way on the paths. Sheep and cattle, horses and rabbits died because of the ash that they ate with the grass. The ash polluted the water supply and caused the meat in the market in San Cristóbal to rot. On the second day the sky was clear and everyone was happy. (Gossen 2002: 280)

Other similar tales describe ground so white with ash that animals could not eat and that when the ash began to fall, many thought that the end of the world had arrived (Gossen 2002: 286). This account concurs with modern examples of destructive ash and acid rain impacts on plants and animals during recent volcanic eruptions in New Zealand (Cronin *et al* 1998).

Mythological memories of ash fall permeate the oral traditions of Fiji and Tonga, such as those recording an eruption of Nabukelevu volcano (Kadavu, Fiji; Figure 9.1) that occurred at some time between 220–420 AD and 1630–80 AD, based on radiocarbon dating of a recent major ashfall deposit in the area (Cronin, Ferland *et al* 2004). Eruptive activity is stylised as a battle between two rival regional deities (Beauclerc 1909; Deane 1909). One of these protagonists blocked the view of the sunset for the other by building a giant yam heap (Nabukelevu), probably recording the growth of a lava dome, and ash columns that raised the 'height' of the volcano (Cronin, Ferland *et al* 2004). During the night, when one of the gods transported earth from the volcano summit in woven baskets towards the east (the direction of high-level prevailing winds) (Reid and Penney 1982), the earth dropped out of the baskets, thus generating the ash fall.

The same metaphor for ash fall (ie, foreign deities stealing part of a mountain – and sometimes leaving a crater behind – and transporting it to the east), is also found in examples from Tonga (Gifford 1924; Mahony 1915; Taylor 1995). Such traditions were used to describe large caldera-forming eruptions of Niuafo'ou and Tofua (Figure 9.1). Here, the marauding Samoan *teevolos* were sometimes confronted by the local deity as they fled to the east (and over the inhabited islands of Tonga), which caused the interlopers to drop their spoils. In other cases, the invaders were scared off by the invocation of an impromptu 'sunrise' that occurred not over the horizon in the east, but instead rose out of the sea to the west of the inhabited Tongan islands (ie, where the volcanoes lie in relation to the observers). These stories may refer to the latest large Tofua eruption, which deposited >10 cm of tephra over inhabited islands in Central Tonga around 1440–1640 AD (Cronin and Taylor unpublished radiocarbon data).

Together, these oral traditions provide a rich metaphorical description of the effect of ash falls on local communities. When the source of the ash is known, ash fall is attributed directly to the action of deities that inhabit the relevant volcanoes. When communities experience ash fall from distant (unseen) volcanoes, memories are recorded as darkness, particularly startling when it blocks the sun's light, and terrifying when it persists for several days (eg, *bingi*).

Volcanic Flows

Traditions that relate to more disastrous proximal (near-source) events produced by large eruptions (eg, volcanic blasts, pyroclastic flows, and mud flows) are less frequent. The direct impact of the eruption of Long Island (Arop) that produced the Papua New Guinea *bingi* is recorded in oral traditions from regions near the volcano (Blong 1982), where a man from West New Britain was wounded in a struggle with Arop people and later invokes the eruption as revenge.

> Straight away big tidal waves came and the slopes collapsed. There were continuous earthquakes and a big rain – many mosquitoes came afterwards and then a bush fire burnt all the houses. All of the houses and all of the bush were burnt. No one knows how the fire started, but it wasn't fire from the mountain. The people didn't go away. Malala and Bok as well as Matapun were wiped out by the fire. There the people stayed away for three days.
>
> Two men came in a canoe. … They found everybody dead at Malala and everywhere else. … The two men called upon the mountain to cover all the bodies up. …
>
> Before the eruption there was a big mountain where the lake is now. … Bodies were buried, houses were buried, coconuts were buried by the ash. The island was smaller before the eruption. The erupted material went in all directions. All the animals and birds were killed by the fire; by the fire not by the eruption. … After the eruption there was no water in the lake. Water comes up in the wet season and covers the place where the boat is now. …
>
> Ash fell on Tolokiwa during the big eruption, but it didn't break down houses or destroy gardens. The skin of the people turned black because of the dust. Stones fell too. It was a white dust and it happened only once. (Blong 1982: 218–19)

The first part of this story describes an eruption that had sufficient energy to kill everyone and burn everything in its path, but not to bury the bodies, suggesting a low-particle density, hot pyroclastic surge (Wohletz and Sheridan 1979) during early stages of activity. An historical analogy is the 1902 eruption of Mt Pelée, where dramatic seismic and hydrologic disturbances presaged a volcanic blast that killed >26 000 people. A medical doctor who witnessed the aftermath of the Pelée event (Will 1903) describes victims strewn in the streets, devoid of clothing, many showing no physical signs of having been burnt, just as in the Papua New Guinea oral tradition. By comparing what he saw in St Pierre with survivors' reports of a 1902 eruption of Soufriére, St Vincent, Will (1903) concluded that in both pyroclastic surge events most people died through asphyxiation rather than burning. A similar conclusion was reached by Capasso (2000) concerning the cause of death in the town of Herculaneum as a result of the 79 AD eruption of Vesuvius.

The second part of the Arop eruption story describes a very different type of activity, which went in all directions, created a central crater, and buried bodies and houses. This matches observations of denser pyroclastic flows that would be expected to accompany caldera formation. Also important is the statement that the island was *larger* after the eruption, possibly as a consequence of the deposition of pyroclastic flows. A similar oral tradition exists at Savo in the Solomon Islands (Petterson *et al* 2003), in which syn-eruptive pyroclastic flow deposits formed prograding deltas that caused the island to *grow*. At Savo, oral traditions describe moving pyroclastic surges/flows quite precisely as red hot gaseous materials that descend valleys, emitting *foam* (Toba 1993), an apt description of flows of volcanic pumice.

Events associated with the 1982 El Chichon, Mexico, eruption gave an explanation for additional components of these traditions. Although pyroclastic flows and surges caused fatalities and destruction (Duffield 2001), the volcano also produced boiling mud flows when a body of water ponded behind still-hot pyroclastic debris attained near-boiling temperatures and then broke through, inundating the valleys and villages below. Chamula oral traditions of ancient events in this area describe similar pyroclastic flow and surge events as 'boiling rain' that destroyed the first people (Gossen 1974).

> Now the first people did not have houses. ... They still did not have corn, either. ... They simply ate the grass, along with the fruit of the trees, along with the fruit of the vines. Moreover, they could not talk well. They knew nothing of singing.
> Now, while Our Lord Sun/Christ was living on earth, he saw that the people could not speak. So he destroyed them with boiling rain. ... The boiling rain just poured out of the mountains and rose to the sky. It rose just like clouds. Then when the rain fell, it was just like ordinary rain, but it was boiling hot. As for these people, then, all of these people perished in the rain of boiling water. (Gossen 2002: 199–201)

Another account of this destruction records that the boiling rain killed half of the people already in their houses (note the presence of houses; Gossen 2002: 223–31). Even those who sheltered in caves were affected, although another account states that people had already been scalded by the time they arrived at the caves, whilst others burned to death by the unbearably hot rain as they were sowing their cornfields (Gossen 2002: 255–59). In other versions, the boiling rain is reported to have lasted three days (Gossen 2002: 336), or was replaced by a great flood (Gossen 2002: 240–53), or a boiling flood (Gossen 2002: 336). It seems probable that most of these stories record a previous eruption of El Chichon, most likely in the 13th century. The inconsistencies among the different accounts reflect the long time depth of these stories and may reflect some combination of (1) the effects of retelling in differing family groups; (2) variable influences of missionaries, as all of the stories have Christian overtones; or (3) repeated events of a similar nature, such that more than one eruption has been condensed into a single tradition.

The 'boiling flood' in the Chumula oral traditions is a special type of volcanic mudflow, or lahar. On Savo, Solomon Islands, oral traditions record lahars as distinct from pyroclastic flows, with their most distinctive feature being that they appear to be solid (either whilst flowing or in the deposits left behind), but people sink into them (Cronin, Petterson *et al* 2004; Toba 1993).

Oral traditions of lahars may also include the source of the destructive event, often a lake high on the volcano. In an example from the Cascade Range in the United States, Clark (1953: 32) reports a legend describing how a man climbed Mt Rainier (Takobed) in search of spirit power. Power was granted to him, but the gift was accompanied by the warning that once he grew old and died, the head of Mt Rainier would burst open and water from the lake would flow down the mountain to the valleys below. This eventually came to pass, with the water sweeping the trees from the current location of the town of Orting, leaving the prairie covered in stones. Geologists have since determined that this location was engulfed by a large lahar about 500 years ago (the Electron mudflow). For this reason, Orting is an area identified by modern geological studies as being at particularly high risk from lahars (eg, Scott *et al* 1995).

Debris Avalanches and Volcanogenic Tsunamis

Debris avalanches are formed from catastrophic failure of volcanic edifices. The 1980 eruption of Mount St Helens, United States, highlighted the importance of such volcano collapse in eruptive cycles. In the Nabukelevu (Kadavu, Fiji) example, the collapse of volcano flanks has been placed in the context of a battle between rival demi-gods

(Beauclerc 1909, Deane 1909). One of the deities tore down the other's hill over several nights, taking away the earth in coconut baskets. The hill's owner caught the interloper in the act and chased him. To escape, the fugitive dived down and hid beneath the sea, whilst his pursuer 'stooped down and drank the sea at that place dry' (Beauclerc 1909: 23). Cronin, Ferland *et al* (2004) describe at least two debris avalanche deposits on the lower flanks of Nabukelevu, radiocarbon dated at 850–50 BC and 120–430 AD, which contain pottery fragments, human-bone fragments, and charcoal. Along with several others from this volcano, they entered the surrounding seas and undoubtedly generated locally destructive tsunamis (cf Kienle *et al* 1987). The sea being sucked dry (Beauclerc 1909; Deane 1909) is consistent with contemporary eyewitness reports of tsunamis, where the sea retreats well beyond its normal tidal range before the first of the wave train arrives.

Volcanic Gas

Impacts of volcanic gas may be quite easily overlooked in eyewitness accounts because they are not normally catastrophic. An exception is the 1986 carbon dioxide released from Lake Nyos, Cameroon, which killed 1,700 people and all animals within 14 km of the lake (eg, Kling *et al* 1987). Shanklin (1989) provides a comprehensive overview of oral traditions related to similar events in this region. They describe explosive gas releases or violent lake overturns in terms of revenge between competing tribes and designate some lakes as 'maleficent'. As a consequence, the Grasslands people native to the region do not settle next these lakes. In contrast, occupation of the land close to Lake Nyos by newly arriving settlers led to the 1986 tragedy. This is an interesting example, not only for the application of land-use planning through banned areas, but also because the relationship between carbon dioxide release and volcanic activity has been extensively debated by the scientific community (Sigvaldason 1989).

More commonly, volcanic gas release is most damaging when gas emissions persist for days, months, or years. A particularly graphic description of the deleterious effects of volcanic gasses is provided by an eyewitness account of the 1783–1784 eruption of Laki volcano, Iceland.

> More poison fell from the air than words can describe: ash, volcanic hairs, rain full of sulphur and saltpeter, all of it mixed with sand. The snouts, nostrils and feet of livestock grazing or walking on the grass turned bright yellow and raw. All water went tepid and light blue in color and rocks and gravel slides turned grey. All the earth's plants burned, withered and turned grey, one after another, as the fire increased and neared the settlements … the foul smell of the air, bitter as seaweed and reeking of rot for days on end, was such that many people,

especially those with chest ailments, could no more than half-fill their lungs with air. (Steingrimsson and Knuz 1998: 41)

In fact, volcanic gasses were the primary cause of both human and livestock fatalities related to this eruption.

Lava Flows

Lava flows are not normally catastrophic from the perspective of human lives lost, but they may wreak enormous hardship on people whose land is inundated and rendered infertile. Oral traditions related to lava flow emplacement are numerous in Hawaii (Westervelt 1963), where they involve various manifestations of *Pele* (a word used for the volcano goddess as well as for the eruptive activity). One story that can be directly related to an eruption of Hualalai volcano in c 1800 AD is used to provide a moral lesson: the home of one girl who does not pay homage to *Pele* is destroyed, whilst that of another girl faithful to the goddess is saved (Maguire 1966). The story also records the locations and sequence of vent activity related to this eruption. Information provided by this story, together with second-hand accounts of the eruption provided to missionaries a few decades later, were used to construct a detailed interpretation of the eruption chronology (Kauahikaua *et al* 2002). Moreover, careful evaluation of this story led to the discovery of a previously unknown vent for this activity.

Taken as a whole, these examples demonstrate ways in which all major volcanic phenomena have been documented and incorporated into oral traditions. Thus, they provide accurate information on the types of volcanic phenomena likely to affect a given community.

HAZARD MANAGEMENT

In the same way that oral traditions can be a mine of information about volcanic hazards, they may also describe aspects of hazard or emergency management that were practiced at the time or recommended for the future. They may, therefore, function as teachings for subsequent generations about how to survive and recover from destructive volcanic events.

Disaster Response

Oral traditions describing a community's response to a hazard may include direct response-related advice or descriptions about how people responded at the time. An explicit example can be found in some of the oral traditions of Papua New Guinea, which contain practical and ritual precautions for survival (Glasse 1963). In these stories, specific instructions are given for how to respond to *bingi*. Members of kin groups are told to build communal houses at the first sign of ash fall, to lay in food and water to last for four days, and to gather pigs and dogs inside. Some gardens are to be covered with grass, and all wives are to return to their natal groups. During the ash fall, the only people to leave the communal house are men who are the last surviving sons. If these responses are not carried out, the *bingi* turns into a holocaust, destroying all life. These comprehensively prescribed responses imply that the *bingi* traditions represent several ash-fall events experienced by the communities (Blong 1982).

Other traditions may indirectly describe community responses, such as defining areas that were 'safe' during large eruptions. Safe areas may be local, such as areas of high ground in regions prone to valley lahars (Toba 1993), or they may be distant, as in descriptions of populations retreating to the safety of Klamath Lake, around 40 km from the vent area involved in the climactic eruption of Mt Mazama, United States, about 7,700 years ago (Clark 1953). Refuge may also be sought by a particular type of shelter, such as the caves referred to by many Chamula traditions, that offered better protection than the houses (Gossen 1974).

Hazard Mitigation

The Polynesian/Melanesian concept of *tabu/tambu/tapu/kapu*, all meaning 'forbidden', as applied to a place, may have been a traditional equivalent to a scientifically based hazard exclusion zone or safety zone in a volcanic area. Although the local usage may apply to other sacred places, such as burial grounds and sacred religious sites, it is also commonly applied to the summit area of active volcanoes and 'crater' lakes. As mentioned previously, the summit area of Ambae in Vanuatu is still regarded as *tambu*, with the access of even foreign scientists being highly restricted (Cronin, Gaylord *et al* 2004). Oral traditions from Cameroon protected local populations by discouraging settlement in areas affected by deadly volcanic gas releases from lakes (Shanklin 1989). In New Zealand, the highest volcanoes in the North Island, Ruapehu, Tongariro, Ngauruhoe, and Taranaki, have all erupted in the time frame of Maori occupation of the country (eg, Lowe *et al* 2000) and were all considered *tapu* by the Maori who lived around them (Izett 1904). Similarly, in the Pacific north-west of the United States, Spirit Lake north of Mount St Helens was considered to be the home of evil spirits and hence was avoided by Native Americans who lived in the area (Clark 1953: 63). Some accounts describe a demon living in the lake that is so huge it can reach out across the entire lake to snatch people and canoes. The hazardous reputation of Spirit Lake was likely based on the protracted volcanic activity on Mount St Helen's north flank during the first half of the 19th century. It proved to be uncannily accurate in 1980 when the north side of the mountain collapsed catastrophically, sending a debris avalanche into the southern end of Spirit Lake.

In modern settings, however, oral tradition alone may not be an effective hazard mitigation tool, particularly where overseas aid following disasters benefits the community. Davies (2002) reports that oral histories of tsunami on the weather coast of Papua New Guinea are held only by some of the older men within the tribe, implying that events that occurred more than ~50 years before the present may have no impact on preparedness of communities. Of more concern is his observation that even intervals less than 10 years may permit sufficient loss of communal memory to allow resettlement in areas where many deaths were caused by tsunami. However, it is possible that community memories of volcanic eruptions have longer durations, in part related to the identifiable unrest that commonly precedes these events. For example, Davies (2002) notes that despite a similar period of repose between tsunami and volcanic eruptions, signs of eruption onset at Rabaul were more readily recognised by village elders and were translated into pre-emptive evacuation before any warnings from authorities. However, in this case it should be noted that local volcanologists had carried out extensive public education activities in this area for more than 10 years prior to the eruption.

In another modern example, Crittenden and Rodolfo (2002) showed that a clear correlation between environmental threat and cultural response cannot be assumed. Following the 1991 eruption of Mt Pinatubo (Philippines), Bacalor township experienced years of repeated lahar activity, ultimately burying some buildings under >9 m of deposits. This repeated threat was not enough, however, to make around 2,000 of its residents abandon the site and move elsewhere, largely because they had nowhere else to go. Similarly, traditional knowledge of catastrophic lahars that had covered Armero, Colombia, in the mid-17th and 19th centuries did not stop resettlement of the area and were of no use in preventing the lahar-induced disaster of 1985 (Voight 1990). In fact, there is currently intense pressure on the government to allow people to return to this area. In all of these cases, high population pressures, particularly in valleys that boast good farmland in addition to lahar hazards, clearly reduce the effectiveness of the 'lessons' in the oral traditions, although an additional factor may be the declining role or awareness of oral traditions in modern societies.

Recovery, Migration, and New Beginnings

Some oral traditions extend beyond the event itself to include discussion of the cultural aftermath of destructive eruptions together with issues and problems of resettlement. For example, stories surrounding the 1452 AD Kuwae eruption in Vanuatu (Galipaud 2002) describe how the population returned to the largest island remnant, Tongoa, once the birds had resettled, apparently six years after the eruption. As the island had lost approximately two-thirds of its area, much of the cultural impact was related to land-ownership conflicts, which persisted for many generations.

Oral traditions may also provide specific information that can be used to aid recovery from eruptions. Particularly in the case of far-field effects, oral traditions may include details of how long ash falls lasted, how long the ash covered the ground, and both negative and positive effects of ash inundation (Blong 1982; Gossen 2002). For instance, soil creation is described in Chamula traditions as the consequence of falling hot water that broke up rocks that people could use. Also recorded is the beneficial, fertilising effect of volcanic ash on crop production (Cronin *et al* 1998), which has led some groups in the Papua New Guinea highlands to attempt to invoke *bingi* to replenish crop-depleted soils (Blong 1982).

Often the longest lasting impacts of volcanic disasters are psychosocial (Paton 1996). Oral traditions may provide a strategy for coping with psychological effects of disaster as people discuss the event in an attempt to understand it. The act of casting an overwhelming event into the language of a story, using personification or contextualisation within local belief systems, is likely one of the most important community psychological coping mechanisms for major catastrophes.

Finally, volcanic interpretations of oral traditions may also offer an important context for important archaeological research. In Vanuatu, stories about the Kuwae eruption describe migrations of people to other islands and the conflicts resulting from the new arrivals. These internal migrations may be the trigger for cultural change, or at least a temporary diversion of effort from domestic/artistic activities towards defence. In central areas of Tonga, apparent changes in agricultural practice are evidenced by sudden changes in paleo-vegetation records (Burley 1998; Flenley *et al* 1999). However, rather than being purely cultural change, these modifications may have been forced by the coincident deposition of >30 cm of volcanic ash in places (Cronin, Smith *et al* 2004). As described above, these ash-fall events were momentous enough to be recorded in oral tradition on at least two island subgroups of the archipelago (Taylor 1995). In eastern Mexico, the Chamula legends of darkness and boiling rain record a series of creations, each resulting in people who are more sophisticated than those previous. These include specific reference points that could be investigated in archaeological studies, such as living in houses or caves and the advent of corn cultivation. In Africa, oral traditions related to exploding lakes explain seemingly straightforward migration histories in Cameroon (Shanklin

1989). All of these examples present intriguing possibilities for developing linkages between volcanic and archaeological research.

Place Names

The best long-term mechanism for modern-day hazard mitigation is land use planning. In the absence of formal legislation, this may be achieved in nonliterate societies by establishing forbidden or *tambu* areas (as described above). Defining place names is another way to provide warning about dangerous areas, or at least pointing out locations where momentous events have occurred. Examples abound of place names recording the sites of ancient eruptions. This form of memory may lie at the oldest end of the oral tradition spectrum, with the name remaining at a site of eruption, even when there is no supporting legend or even contemporary language. This longevity may be aided by the distinctiveness of volcanic landforms such as mountains and crater lakes. Additionally, in some cases the origin of a volcanic landform may be recognised and given a 'volcano' name by analogy rather than through an inheritance from oral traditions.

Of particular note is the profusion of place names relating volcanic features to fire, such as the Polynesian/Melanesian cognate Tofua/Tafua/Tavuyaga used to describe 'places of fire'. Tofua is a sporadically active volcano in Tonga, although it has had no widespread eruptions since before c 400 yrs BP (Cronin and Taylor unpublished data). Tafua is an obvious monogenetic vent complex on the eastern coast of Savai'i, Samoa, which is dated at >600 yrs BP (Cronin *et al* 2001). Similarly, Tavuyaga is a monogenetic eruption vent of at least 1,100 yrs BP in age (Cronin and Neall 2001) on Taveuni in Fiji, with the word 'Tavuyaga' an archaic form of Fijian (Geraghty 1996).

Other examples of such associations are plentiful around the world. In New Zealand, the name Tarawera is interpreted as 'burnt-peak', a name that existed prior to its catastrophic 1886 eruption. Likewise 'Rangipo', the name given to the barren tephra plain downwind of Ruapehu and Tongariro/Ngauruhoe volcanoes, means 'place of darkness'. This area is described in several stories of ash fall, including one in which an ash fall came to the aid of local Maori when repelling invading tribes (Cowan 1930). Other Polynesian examples include the 1760 Mauga Afi (mountain of fire) in Samoa (Kear and Wood 1959) and the c 1800 Puhi a Pele (Pele's bonfire) in Hawaii (Maguire 1966). Such 'fire mountains' or 'smoking mountains' are not restricted to Polynesia, but can be found throughout the world, including names associated with Mount St Helens (eg, Lawelatla/Tah-one-lat-chah/ Loo-wit [Williams 1980]) and with Mexico City's towering volcano Popocatépetl (Vitaliano 1973).

Implications for Education and Awareness Programmes

In preliterate communities, oral traditions were one of the most durable ways in which catastrophic events in the history of a society were recorded and passed on to the next generation. They may have also represented an important psychosocial disaster recovery process. Almost all such volcanic traditions have value in community hazard management, even if only to impart awareness that such events are possible. In communities that still display hostility and distrust towards outsiders and scientists, such as rural Vanuatu and Solomon Islands (Cronin, Gaylord *et al* 2004; Cronin, Petterson *et al* 2004), oral traditions have been an excellent starting point to establish dialogue with scientists. Features of legends can be related to geological facts and utilised in hazard mapping, assessments, response planning, education and awareness, and physical hazard mitigation activities through participatory education activities.

With the spread of literacy and development, rural Pacific community leaders report the waning of traditional ways (Cronin, Gaylord *et al* 2004; Cronin, Petterson *et al* 2004), including the active transmission of oral traditions and the trust placed in these traditions by younger generations. Along with the death of this cultural practice goes its utility as an active public education tool. As the demise of these traditions may be inevitable, it is all the more important to recognise the rich store of knowledge within recorded oral traditions and to work to preserve still extant stories. Additionally, analysing the efficacy of such oral traditions in transmitting hazard information within the few remaining oral societies may provide important lessons for improvement of modern hazard forecasting and mitigation activities.

CONCLUSIONS

Despite the multitude of factors that obscure the volcanic information contained within oral traditions, we have demonstrated ways in which their origin as collective or singular eyewitness descriptions of catastrophic eruptions makes them a valuable source of information about the eruption history of a region. If used with care, they can provide much additional information to volcanologists attempting to assess the volcanic hazards of a region, although the weakening role of traditional culture in some Pacific Island countries may now have rendered this less effective. Stories and songs contexturalising volcanic catastrophes were probably also an important aid to psychosocial recovery.

In modern societies, oral traditions are a valuable resource for scientists if the information contained within them is interpreted carefully and within the context of both the culture and the geological evidence from the area. Details gleaned from oral traditions can extend the geological interpretation of hazards in an area by providing careful descriptions of the common hazards, the frequency of recurrence of eruptive episodes, the areas likely to be impacted, the duration and scale of individual events, and common sequences of eruptive activity. Similarly, oral traditions may be an important asset to disaster managers when devising community disaster response and recovery strategies, as oral traditions commonly identify local refuge areas and evacuation routes necessary for response planning and development of land-use planning guidelines.

We further conclude that oral traditions may represent the oldest public awareness campaigns for volcanic hazards and guides to hazard mitigation. The dramatic, but highly sporadic, nature of volcanic activity may promote longevity of these oral traditions for the specific purpose of helping future descendants avoid a similar disaster. Finally, we suggest that oral traditions are an important cultural asset that is fading from world culture. Loss of these traditions means not only loss of culture in general but also loss of what we believe to have been one of their original purposes – that of warning future generations of hazards they may face and helping communities cope with the aftermath of disasters.

ACKNOWLEDGMENTS

SJC acknowledges several years of assistance from the UNESCO Office for the Pacific that allowed exposure to south-west Pacific oral traditions and thanks Douglas Charley (Port Vila), Thomas Toba (Honiara), Patrick Nunn (Suva), and Paul Taylor (Sydney) for discussions on local oral traditions. KVC thanks Jim Kauahikaua for combined years of lava flow studies and Hawaiian stories that have inspired this work. This collaboration was supported by the Royal Society of New Zealand ISAT-linkages programme and FRST grant MAUX0401.

REFERENCES

Beauclerc, GAFW (1909) 'Legend of the elevation of Mount Washington, Kadavu', *Transactions of the Fiji Society* 1909, 22–24

Blong, RJ (1982) *The Time of Darkness, Local Legends and Volcanic Reality in Papua New Guinea*, Seattle: University of Washington Press

Blong, RJ (1984) *Volcanic Hazards. A Source Book on the Effects of Eruptions*, Sydney: Academic Press

Blot, C and Priam, R (1962) *Volcanisme et Séismicité dans l'Archipel des Nouvelles-Hebrides* Noumea, New Caledonia: ORSTOM

Burley, DV (1998) 'Tongan archaeology and the Tongan past, 2850–150 BP', *Journal of World Prehistory* 12, 337–92

Capasso, L (2000) 'Herculaneum victims of the volcanic eruptions of Vesuvius in 79 AD', *Lancet* 356, 1344–46

Cioni, R, Gurioli, L, Sbrana, A and Vougioukalakis, G (2000) 'Precursory phenomena and destructive events related to the Late Bronze Age Minoan (Thera, Greece) and AD 79 (Vesuvius, Italy) Plinian eruptions; inferences from the stratigraphy in the archaeological areas', in McGuire, WG, Griffiths, DR, Hancock, PL and Stewart, IS (eds), *The Archaeology of Geological Catastrophes, Geological Society of London Special Publications* 171, 123–41

Clark, EE (1953) *Indian Legends of the Pacific Northwest*, Berkeley: University of California Press

Cowan, J (1930) *Fairy Folk Tales of the Maori*, Auckland, New Zealand: Whitcombe and Toombs

Crittenden, KS and Rodolfo, KS (2002) 'Bacalor town and Pinatubo volcano, Philippines: coping with recurrent lahar disaster', in Torrence, R and Grattan, JP (eds), *Natural Disasters and Cultural Change*, pp 43–65, London: Routledge

Cronin, SJ, Ferland, MA and Terry, JP (2004) 'Nabukelevu volcano (Mt. Washington), Kadavu – a source of hitherto unknown volcanic hazard in Fiji', *Journal of Volcanology and Geothermal Research* 131, 371–96

Cronin, SJ, Gaylord, DR, Charley, D, Wallez, S, Alloway, B and Esau, J (2004) 'Participatory methods of incorporating scientific with traditional knowledge for volcanic hazard management on Ambae Island, Vanuatu', *Bulletin of Volcanology* 66, 652–68

Cronin, SJ, Hedley, MJ, Neall, VE and Smith, G (1998) 'Agronomic impact of tephra fallout from 1995 and 1996 Ruapehu volcano eruptions, New Zealand', *Environmental Geology* 34, 21–30

Cronin, SJ and Neall, VE (2001) 'Holocene volcanic geology, volcanic hazard and risk on Taveuni, Fiji', *New Zealand Journal of Geology and Geophysics* 44, 417–37

Cronin, SJ, Petterson, MG, Taylor, PW and Biliki, R (2004) 'Maximising multi-stakeholder participation in government and community volcanic hazard management programs, a case study from Savo, Solomon Islands', *Natural Hazards* 33, 105–36

Cronin, SJ, Smith, I, Taylor, P and Platz, T (2004) 'New evidence for widespread, late Holocene, explosive volcanism along the Tongan arc', Abstract S12b_o_05, Proceedings of the International Association of Volcanology and Chemistry of the Earth's Interior (IAVCEI) General Assembly, Pucon, Chile

Cronin, SJ, Taylor, PW and Malele, F (2001) 'Final report on the Savaii volcanic hazards project, Samoa', *SOPAC Technical Report* 343, Suva, Fiji: SOPAC

Davies, H (2002) 'Tsunamis and the coastal communities of Papua New Guinea' in Torrence, R and Grattan, JP (eds) *Natural Disasters and Cultural Change*, pp 28–42, London: Routledge

Deane, W (1909) 'Tanovo – the god of Ono', *Transactions of the Fijian Society* 1909, 39–42

Duffield, WA (2001) 'At least Noah had some warning', *EOS Transactions of the American Geophysical Union* 82, 305, 309

Echo-Hawk, R (2000) 'Ancient history in the New World: integrating oral traditions and the archaeological record in Deep Time', *American Antiquity* 65, 267–90

Endo, ET, Malone, SD, Noson, LL and Weaver, CS (1981) 'Locations, magnitudes, and statistics of the March 20–May 18 earthquake sequence', in Lipman, PW and Mulleneaux, DL (eds), *The 1980 Eruptions of Mount St. Helens Washington*, pp 93–108, Geological Survey Professional Paper 1250, Washington: US Geological Survey

Ewert, JW, Lockhart, AB, Marcial, S and Ambubuyog, G (1996) 'Ground deformation prior to the 1991 eruptions of Mount Pinatubo', in Newhall, CG and Punongbayan, RS (eds), *Fire and Mud. Eruptions and Lahars of Mount Pinatubo, Philippines*, pp 329–38, Seattle: Philippine Institute of Volcanology and Seismology and University of Washington Press

Finnegan, R (1992) *Oral Traditions and the Verbal Arts: A Guide to Research Practices*, London: Routledge

Finnegan, R (1995) 'Introduction; or, why the comparativist should take account of the South Pacific', in Finnegan, R and Orbell, M (eds), *South Pacific Oral Traditions*, pp 1–6, Bloomington: Indiana University Press

Flenley, JR, Hannan, CT and Farelly, MJ (1999) 'Final report on the stratigraphy and palynology of swamps on the islands of Ha'afeva and Foa, Ha'apai, Tonga', *Geography Programme, School of Global Studies Miscellaneous Publication Series 99/3*, Palmerson North, New Zealand: Massey University

Frater, M (1917) 'The volcanic eruption of 1913 on Ambrym volcano, New Hebrides', *Geological Magazine* 6, 496–503

Galipaud, J-C (2002) 'Under the volcano: Ni-Vanuatu and their environment', in Torrence, R and Grattan, JP (eds), *Natural Disasters and Cultural Change*, pp 162–71, London: Routledge

Geraghty, P (1996) 'Fiji's lands of fire: Past, present and future 12', *Fiji Post*, 3 Oct

Gifford, EW (1924) *Tongan Myths and Tales*, Bernice P. Bishop Museum Bulletin 8, Honolulu: Bishop Museum

Glasse, RM (1963) 'Bingi at Tari', *Journal of the Polynesian Society* 72, 270–71

Gossen, GH (1974) *Chamulas in the World of the Sun: Time and Space in a Maya Oral Tradition*, Cambridge, MA: Harvard University Press

Gossen, GH (2002) *Four Creations: An Epic Story of the Chiapas Mayas*, Civilization of the American Indian Series 245, Norman: University of Oklahoma Press

Grattan, JP, Gilbertson, DD and Dill, A (2000) '"A fire-spitting volcano in our dear Germany": documentary evidence for a low-intensity volcanic eruption of the Gleichberg in 1783?', in McGuire, WG, Griffiths, DR, Hancock, PL and Stewart, IS (eds), *The Archaeology of Geological Catastrophes, Geolological Society of London Special Publication* 171, 307–15, London: Geological Society of London

Harlow, DH, Power, JA, Laguerta, EP, Ambubuyog, G, White, RG and Hoblit, RP (1996) 'Precursory seismicity and forecasting of the June 15, 1991 eruption of Mt Pinatubo', in Newhall, CG, Punongbayan, RS (eds) *Fire and Mud: Eruptions and Lahars of Mount Pinatubo, Philippines*, pp 282–306. Seattle: Philippine Institute of Volcanology and Seismology and University of Washington Press

Izett, J (1904) *Maori Lore. The Traditions of the Maori People with the More Important of their Legends*, Wellington: Government Printer

Kauahikaua, J, Cashman, KV, Clague, DA, Champion, D and Hagstrum, JT (2002) 'Emplacement of the most recent lava flows on Hualālai volcano, Hawai'i', *Bulletin of Volcanology* 64, 229–53

Kear, D and Wood, BL (1959) *The Geology and Hydrology of Western Samoa*, New Zealand Geological Survey Bulletin 63, Wellington: Government Printing Office

Kienle, J, Kowalik, Z and Murty, TS (1987) 'Tsunamis generated by eruptions from Mount St. Augustine volcano, Alaska', *Science* 236, 1442–47

Kling, GW, Clark, MA, Compton, HR, Devine, JD, Evans, WC, Humphrey, AM, Koenigsberg, EJ, Lockwood, JP, Tuttle, ML and Wagner, GN (1987) 'The 1986 Lake Nyos gas disaster in Cameroon, West Africa', *Science* 236, 169–97

Lindstrom, L (1995) *Cargo Cult: Strange Stories of Desire from Melanesia and beyond* Honolulu: University of Hawaii Press

Loughlan, SC, Baxter, PJ, Aspinall, WP, Darroux, B, Harford, CL and Miller, AD (2002) 'Eyewitness accounts of the 25 June 1997 pyroclastic flows and surges at Soufrière Hills volcano, Montserrat and implications for disaster management', in Druitt, TH and Kokelaar, BP (eds), *The Eruption of Soufrière Hills Volcano, Montserrat, from 1995 to 1999*, pp 211–30, Geological Society of London Memoirs 21, London: Geological Society of London

Lowe, DJ, Newnham, RM, McFadgen, BG and Higham, TFG (2000) 'Tephras and New Zealand archaeology', *Journal of Archaeological Science* 27, 859–70

Lowie, R (1915) 'Oral tradition and history', *American Anthropologist* 17, 597–99

Maguire, ED (1966) *Kona Legends*, Hilo, HI: Petroglyph Press

Mahony, BG (1915) 'Legends of the Nuia Islands', *Journal of the Polynesian Society* 24, 116–17

Mason, R (2000) 'Archaeology and Native North American oral traditions', *American Antiquity* 65, 239–66

Moodie, DW, Catchpole, AJW and Abel, K (1992) 'Northern Anthapaskan oral traditions and the White River Volcano', *Ethnohistory* 39, 148–71

Paton, D (1996) 'Disasters, communities and mental health: psychological influences on long-term impact', *Community Mental Health in New Zealand* 9, 3–14

Petterson, MG, Cronin, SJ, Taylor, PW, Tolia, D, Papabatu, A, Toba, T and Qopoto, C (2003) 'The eruptive history and volcanic hazards of Savo, Solomon Islands', *Bulletin of Volcanology* 65, 165–81

Purey-Cust, HE (1896) 'The eruption of Ambrym Island, New Hebridies, South-west Pacific, 1894', *Geographical Journal* (Royal Geographical Society) 8, 585–602

Ray, VF (1980) *The Sanpoil and Nesepelem: Salishan Peoples of Northeastern Washington*, New York: AMS Press

Reid, SJ and Penney, AC (1982) *Upper-level Wind Frequencies and Mean Speeds for New Zealand and Pacific Island Stations*, New Zealand Meteorological Service Miscellaneous Publication 174, Wellington: Government Printing Office

Robin, C, Monzier M and Eissen, J-P (1994) 'Formation of the mid-fifteenth century Kuwae caldera (Vanuatu) by an initial hydroclastic and subsequent ignimbrite eruption', *Bulletin of Volcanology* 56, 170–86

Rosenbaum, JG and Waitt, RB Jr (1981) 'Summary of eye-witness accounts of the May 18 eruption', in Lipman, PW and Mulleneaux, DL (eds), *The 1980 Eruptions of Mount St. Helens, Washington*, pp 53–67, US Geological Survey Professional Paper 1250, Washington, DC: US Geological Survey

Schmincke, H-U (2000) *Vulkanismus*, Darmstadt, Germany: Wissenschaftliche Buchgesellschaft

Scott, KM, Vallance, JW and Pringle, PT (1995) 'Sedimentology, behaviour, and hazards of debris flows at Mount Rainier, Washington', *U.S. Geological Survey Professional Paper* 56, 1547

Shanklin, E (1989) 'Exploding lakes and maleficent water in Grassfiel legends and myths', *Journal of Volcanological and Geothermal Research* 39, 233–46

Sigvaldason, GE (1989) 'International conference on Lake Nyos disaster, Yaounde, Cameroon 16–20 March, 1987: conclusions and recommendations', *Journal of Volcanology and Geothermal Research* 39, 97–107

Smith, AL and Roobol, MJ (1990) *Mt Pelée, Martinique: A Study of an Active Island-Arc Volcano*, Geological Society of America Memoir 175, Washington, DC: Geological Society of America

Steingrimsson, J and Kunz, K (1998) *Fires of the Earth: The Laki Eruption 1783–1784*, Reykjavik: University of Iceland Press and the Nordic Volcanological Institute

Tanguy, J-C (1994) 'The 1902–1905 eruptions of Montange Pelée, Martinique: anatomy and retrospection', *Journal of Volcanology and Geothermal Research* 60, 87–107

Taylor, PW (1995) 'Myths, legends and volcanic activity: an example from northern Tonga', *Journal of the Polynesian Society* 104, 323–46

Tierney, KJ (1989) 'The social and community contexts of disaster', in Gist, R and Lubin, B (eds) *Psychological Aspects of Disaster*, pp 11–39, New York: Wiley

Toba, T (1993) *Analysis of Savo Custom Stories on the Eruption of Savo Volcano*, technical report TR4/93, Honiara, Solomon Islands: Water and Mineral Resources Division, Ministry of Energy, Water and Mineral Resources

Vansina, J (1985) *Oral Traditions as History*, Madison: University of Wisconsin Press

Vitaliano, DB (1973) *Legends of the Earth: Their Geologic Origins*, Bloomington: Indiana University Press

Voight, B (1990) 'The 1985 Nevado del Ruiz volcano catastrophe: anatomy and retrospection', *Journal of Volcanological and Geothermal Research* 41, 151–88

Warden, AJ (1970) 'Evolution of Aoba caldera volcano, New Hebrides', *Bulletin of Volcanology* 34, 107–40

Westervelt, W (1963) *Hawaii Legends of Volcanoes*, Ruttland, New Zealand: Charles Tuttle Company

Whiteley, PM (2002) 'Archaeology and oral tradition: the scientific importance of dialogue', *American Antiquity* 67, 405–16

Wiart, P (1995) 'Impact et gestion des risques volcaniques au Vanuatu' 13 Notes Techniques, Sciences de la Terre, Geologie-geophysique, Vanuatu: ORSTOM

Will, J (1903) 'Report on the medical relief expeditions to Martinique and St. Vincent in aid of the sufferers from the volcanic eruptions', in *Further Correspondence Relating to the Volcanic Eruptions in St. Vincent and Martinique in 1902 and 1903*, London: His Majesty's Stationery Office

Williams, C (1980) *Mount St Helens: A Changing Landscape*, Portland, OR: Graphic Arts Centre Publishing

Wohletz, KH and Sheridan, MF (1979) 'A model of pyroclastic surge', *Geological Society of America Special Paper* 180, 177–93

CHAPTER 10

Geomythology, Theodicy, and the Continuing Relevance of Religious Worldviews on Responses to Volcanic Eruptions

David K Chester and Angus M Duncan

GEOMYTHOLOGY

Geomythology is the study of oral traditions that perpetuate memories of prehistoric geological events, whereas *theodicy* is any attempt to reconcile theistic belief with the reality of human suffering. Although legends based on the writings of outside observers should be viewed with caution because cultural presuppositions and prejudices may be read into the reactions of indigenous peoples (eg, see Freeth 1993) – an issue sometimes known as *syncretism* – there are numerous examples in which myths, related to volcanic eruptions and the suffering they have caused, have been analysed by scholars. For societies that have left written records, such as Greece and Rome during the classical era, eye-witness accounts may be consulted, but these frequently contain mythological elements (eg, Duncan *et al* 2005). Interpretation of artefact suites provides a further source of information. Effigies of Popocatépetl have been convincingly interpreted as evidence of deistic propitiation on the part of ancient societies that once populated this region of Mexico (Plunket and Uruñuela 1998). In most archaeological sites, however, such evidence of human reactions to eruptions is at worst lacking and at best equivocal. In the case of the earliest known painting of a volcano on the wall of a shrine (dated at c 6200 BC) from the Neolithic town of Çatal Hüyük in Anatolia (Turkey) (Mellaart 1965, 1967: 59–60, 176–77), the artist's placement of the settlement very close to the volcano may be suggestive of anxiety about volcanic activity (Harris 2000: 1308), but could equally merely reflect the artist's aesthetic sensibility (Figure 10.1).

These notes of caution apart, there can be little doubt that in *pre-industrial societies* (White 1973) studied by historians, anthropologists,

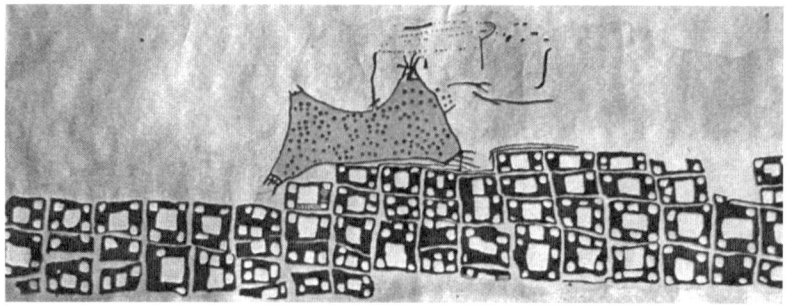

Figure 10.1 Drawing of a wall painting from a shrine at Çatal Hüyük, Anatolia. The painting is from level VII and is dated at c 6,200 BC. The eruption depicted is possibly from Hasan Dağ volcano. (The figure was originally drawn by Mrs Mellaart and is reproduced from *Earliest Civilisations of the Near East* [1965], by James Mellaart, Thames and Hudson, Ltd, London, figure 51/52, p 82, reproduced with permission.)

and archaeologists, responses to volcanic eruptions that are couched in theistic terms have transcended place, time, and culture. Three contrasting examples – Mediterranean Europe, North America, and New Zealand – may be used by way of illustration.

During the 'classical' period in Mediterranean Europe, the Etruscan god Velkhan was primarily perceived as an agent of destruction, but was sometimes portrayed as a god of productive fire and the hearth. Hephaestos (Latin, Vulcan) was normally characterised as a constructive craftsman, and this aspect of volcanism is stressed in both Latin and Greek literature, but volcanoes are also frequently personified in the form of the destructive and angry Titan. Hephaestos, cripple and craftsman, represented both human brilliance and a reminder of human mortality and reliance on divine benevolence. Vulcan was a long-standing source of worship in archaic and classical Rome. Lucilius Junior (1st century AD) notes that on Mt Etna, people offered incense to propitiate the gods who were thought to control the mountain (Chester 2005a; Chester *et al* 2000; Doonan 1997; Hyde 1916), whilst in classical literature few islands are mentioned as frequently as Thera (ie, Santorini). The most well-known myth is that Thera may be the setting of Plato's *Atlantis*, created by Poseidon (Friedrich 2000: 13–19).

Much of the day-to-day practice of Christianity in medieval Europe was permeated by the need to placate what many saw as an arbitrary and vengeful God (Soelle 1975). Propitiation took several forms (Hanska 2002). At its most simple, many individuals carried amulets and lucky charms as a means of protection, but the medieval church also had many official and semi-official rites. Official rites were usually modelled on liturgies of Rogation and were repeated when people felt threatened by disaster. The *Sacri Congregazione dei Riti* was established beginning in 1588. 'After any significant catastrophe such as a plague, an earthquake or a volcanic eruption, numerous requests of canonisation reached the congregation' (Hanska 2002: 97). Saints were frequently viewed as protectors, especially if they had 'miraculously' prevented losses and the same was often true of saintly relics.

In Italy from the classical era until ~1900 AD, the skull and two vials of the blood of St Januarius (San Gennaro), who was martyred in the reign of Diocletian (285–305 AD), were often appealed to by the citizens of Naples during eruptions of Vesuvius. These appeals to God included intercession and displays of the saintly relics at the fronts of lava flows during the 685, 1631, 1707, and 1767 eruptions (Fisher *et al* 1997). In 252 AD, the veil of St Agatha, who had been martyred the previous year, was used by the people of Catania in Sicily to halt a lava flow from Etna, and this and other relics were used on many future occasions (Chester *et al* 2000; Rodwell 1878). In fact, so efficacious was the veil thought to be that, following the Lisbon earthquake in 1755, some influential clergy in Portugal believed that St Agatha should be adopted as their patron saint to ward off any recurrence. In the event, the Iberian saint, Francis Borgia, was afforded this singular honour (Kendrick 1956: 72).

A second set of examples may be drawn from North America. Harris (2000: 1312–13), for instance, has argued that Native Americans may have preserved oral myths about the Mt Mazama eruption (Oregon ~7,500 BP) for around 250 generations. The Klamath tribe interpreted this great eruption as a battle between the god, Llao, whom they believed lived in Mt Mazama, and the sky god, Skell (Harris 1988: 116). These myths were eventually recorded in 1865 by a young soldier, William M Colvig, who was stationed at Fort Klamath. In the gorge of Columbia River, legends about volcanic events of unknown date were transmitted to missionaries in the 19th century, whilst in Sunset Crater volcano (northern Arizona) basaltic lava flows of mid–late 11th century AD date show corn casts. It has been suggested that the casts are indicative of ritual practices, perhaps serving to appease the divine forces responsible for eruptions (Elson *et al* 2002: 119, Chapter 6 in this volume).

A final instance of theistic appeasement comes from New Zealand, though many other examples from around the world are listed in Table 10.1. Before the European settlement, each group within Maori society had its own sacred mountain, and Te Heubeu, chief of the Ngati Tuwharetoa tribe of the Taupo district, explained to an Austrian geologist in 1859 that fire was sent from Hawaiiki (the mythical Maori homeland) in response to a call from Ngatoroirangi the high priest (Lowe *et al* 2002: 149). Maoris took advantage of the benefits of volcanic activity, but were only too aware of its dangers, always taking care to appease the volcano god.

Table 10.1 Selected examples of pre-industrial societies in which eruptions have been interpreted in deistic terms (based on Chester 2005a and the sources cited). Further examples may be obtained from Fisher *et al* (1997:179–98) and Sigurdsson (1999: 11–20)

Example	Nature of response
Northern Europe	a. Hekla, Iceland. In the 12th century AD and according to Cistercian monks, Hekla was the gateway to hell; a symbol to deter heresy. Such views persisted until at least the 16th century (Blong 1984: 175; Thorarinsson 1970: 6).
	b. In Icelandic mythology, the god Surtur was the incarnation of eruptions (Sigurdsson 1999: 17).
	c. One effect of the Laki eruption in 1783 was the fear of Armageddon in several European countries. In France, the reddened sun and smoky air 'alarmed the superstitious part of the people, who had been wrought on by their priests to believe that the end of the world was at hand.' Priests were forced to wear vestments to exorcise the fog (Grattan *et al* 2002: 98).
	In England, the hymn writer and clergyman William Cowper recorded that some people thought that the end of the world was at hand. The clergyman and naturalist Gilbert White describes a 'superstitious kind of dread' amongst many people (Grattan *et al* 2002: 98).
Africa	Societies near the Nyamuragira and Niyragongo volcanoes (Central Africa) annually sacrificed 10 of their warriors (Sigurdsson 1999: 13).
	In ancient Egypt, some natural catastrophes were followed by the execution of the Pharoah (Bell 1971).
	In the large igneous province of Ethiopia, some elder inhabitants still lay offerings at the bottom of the Serpent-God Dyke. Legend holds that Arwe, the serpent-god was a god of terror that was appeased by the annual sacrifice of maidens. The god was killed around 1,000 BC, but worship is still active in some remote areas (Mege and Korme 2004).
Central Mongolia	In the Mongolian language, the name of the little-known Har-Togoo volcano means black pot, and there is a local belief that a dragon lives in the volcano (Anon 2003a: 9–11).
Indonesia	In Java and until recent times, human sacrifice to appease Broma volcano was practised. Chickens are now substituted (Sigurdsson 1999: 13).
Japan	The Oni monster is a horned red giant, whose effigy is still to be found in souvenir shops near to active volcanoes. Many volcanic features in Japan are called *Jigoku*, a term for hell derived from Buddhist notions of an underground prison (Sigurdsson 1999: 17–18).
	In a study of archaeological sites in southern Kyushu, Shimoyama (2002: 336) adduced clear evidence of spiritual activity in response to an eruption. There is an example of a pot being offered to propitiate disaster during the accumulation of tephra around 1,300 BP.

(continued)

Table 10.1   (Continued)

Example	Nature of response
Hawaii	There are numerous legends about the need to propitiate the Goddess Pelé, who controls human fate (de Boer and Sanders 2002: 22–46).
Vanuatu and Fiji	In Vanuatu, volcano-related disasters 'are sometimes described in myths and legends and are usually attributed to demons or spirits who wish to punish a breach of a social or cultural taboo' (Galipaud 2002: 164; see also Cronin, Gaylord *et al* 2004). In Fiji, legends concerning the two gods, Tanovo, who presided over Ono Island, and Tautaumolau, who ruled south-western Kadalvu, may relate to known volcanic events (Cronin, Ferland, and Terry 2004b).
Central and South [America]	a. Settlements on the slopes of Popocatépetl volcano in Mexico were destroyed by a Plinian eruption around 50 BC. Excavations show shrines with effigies dedicated to the 'volcano' god, suggestive of attempts at divine appeasement (Plunket and Uruñuela 1997, 1998). b. Human sacrifices are a feature of Maya, Aztec, and Inca reactions to active volcanoes. Nicaraguans believed Cosequina would stay quiet if a child was sacrificed every 25 years, and virginal sacrifice was a feature of the society living on the flanks of Masaya (Poole and Poole 1962; Stephens 1969; Vitaliano 1973). c. In the years before the Spanish conquest, people living in the vicinity of Huaynaputina in Peru sacrificed sheep, birds, and personal clothing to the volcano. Some inhabitants even claimed that they conversed directly with the demons that supposedly controlled the mountain.

There are several historic instances of pre-industrial beliefs becoming more prominent during times of volcano-related crisis, examples from Ethiopia, Indonesia, and Hawaii being noteworthy (Table 10.1). Sometimes this trend was extremely complex and involved far more than simple reversion. Before conversion to Catholicism, residents in the vicinity of Huaynaputina volcano in Peru, for example, developed a cult that involved sacrifice of sheep, birds, and clothing to the volcano, some people even claiming that they conversed directly with the demons who controlled the mountain. Following the catastrophic eruption of 1600 AD, a number of contemporary accounts record that the inhabitants resumed their former religious practices. Processions to Huaynaputina took place, together with renewed sacrifices. Although Christian priests and pagan leaders differed in their detailed interpretation of the events and the appeasement demanded, both groups were agreed that the eruption should be interpreted, not simply as a natural event, but as a punishment for sins committed in the months and years leading up to the eruption (Jara *el al* 2000).

THE CONTINUING RELEVANCE OF THEISTIC WORLDVIEWS

It is well known from detailed psychological research – much of it involving quantitative studies of large numbers of respondents – that people from many cultures turn to deities for coping during times of stress (Paragament 1997; Paragament and Hahn 1986). From our research in Italy (Chester *et al* 1985; Chester *et al* 1999) and the Azores (Dibben and Chester 1999), it has became clear that geomythology and theodicy are not only features of ancient and present-day *pre-industrial* societies, but are also part of contemporary culture in many economically more developed volcanic regions. In the village of Furnas (São Miguel, Azores), a settlement that would be threatened by any future eruption of either the Fogo or Furnas volcano, the intrinsic fatalism of some interview respondents was clearly framed within a religious worldview, with God being seen as controlling both individual and familial futures (Dibben and Chester 1999: 140–41). On the slopes of Mt Etna in Sicily, the veil of St Agatha and other saintly relics have been placed in front of lava flows during many 20th-century eruptions, and as recently as July 2001, the archbishop of Catania celebrated Mass in the town of Belpasso to halt the progress of a lava flow; but to popular disapproval, however, he rejected the use of the martyr's veil. The archbishop explained, 'I bless this mountain and invoke the mercy of God on these craters so that they close up. … It is not the veil that will stop the lava but our prayer. … The warmer the prayer the cooler the lava'. According to Kennedy (2001: 10), a teacher claimed that 'local people still believe in miracles. If human technology can't keep the lava back, the eternal father is our only salvation.' Interesting features of these events in 2001 are that they occurred in a well-educated community; involved respected community leaders; and represented views shared by much of the population at risk. For instance about 7,000 people attended Mass in Belpasso, which is about one-third of the total population of the town.

To test how widespread these findings from Italy and the Azores are, detailed research has been done looking at human reactions to eruptions that have occurred over the past 150 years. This has involved conflating standard lists of eruptions that have caused loss of life (ie, the catalogues produced by Sigurdsson *et al* 2000; Tanguy *et al* 1998; Tilling 1989, and others) with records of events that have produced economic losses, causing human disruption but fatalities. Information has also been compiled on the ways in which the major religions of the world and local faith communities have dealt with theodicy, both in the abstract and when encountering the reality of human suffering. Although data on the theology of suffering in world religions have been relatively straightforward to compile (Table 10.2),

Table 10.2 The relationships between theodicy and eruption responses since 1852

Religious Tradition, 'Classic' Models of Theodicy and Attitudes to Human Suffering	*Examples from Eruptions over the Past 150 Years*
Judaism and Christianity	
The word 'theodicy' was first coined by Leibniz in 1712 (1952) and there are several models (Chester 1998), the principal ones being:	In virtually all eruptions within predominantly Christian countries, free-will theodicies have been prominent. Suffering has been viewed in terms of divine punishment, the answer being
a. Free-will (*Augustinian*). Humans have freedom. Suffering is not only a consequence of the operation of free-will, which may include sinfulness, but is also contrary to the divine purpose. Punishment is sometimes emphasised.	prayer and intercession. Examples include: Vesuvius 1872 and 1906; Mt Pelée 1902; Taal 1911; Parícutin 1943–52; Lamington 1951; Arenal 1968; Mount St Helens 1980; Nevado del Ruiz 1985; Galeras 1993;
b. Best-possible world (*Irenaean*). The world is governed by natural laws, not by divine providence. Good may come out of suffering (eg, without eruptions there would be no planetary atmospheres). The world facilitates 'soul making', and human ethical growth is expressed in helping disaster victims.	Montserrat 1997; Popocatépetl 1997; Nyiragongo (DRC), plus many eruptions of Etna. In many cases, an *Irenaean* response is also present, which includes relief of victims, whilst in Catholic countries the use of saintly and/or other relics has often been used as a form of divine propitiation.
Islam	
Islam means submission to the will of God. In the *Qur'an*, suffering is a punishment for sin, a test of faith and is *instrumental* to the purposes of God. Suffering must be endured as a test of faithfulness. Although it is an unfortunate component of human life, suffering should be alleviated when possible, and good works to relieve the suffering of others are commendable (Bowker 1970).	Krakatau 1883; many other eruptions in Java and the rest of Indonesia, and the Philippines.
South Asian Religions	
All hold the common belief that a person's behaviour leads irrevocably to an appropriate reward or punishment (Kogen 1987: 261; see also Pilgrim 1999) and may encourage a fatalistic attitude to disasters. *Karman* (nominative – *karma*) – means that current actions produce the seeds of future happiness or suffering. There is no theology of innocent suffering.	Bandai San 1888; Sakura-jima 1914; Agung 1963.

(continued)

Table 10.2 (Continued)

Religious Tradition, 'Classic' Models of Theodicy and Attitudes to Human Suffering	Examples from Eruptions over the Past 150 Years
In **Buddhism**, reincarnation is emphasised, meaning that sins committed in former lives are of importance. There is no notion of 'first cause' (ie, a creator and/or an almighty god). Both happiness and suffering are deserved. In **Hinduism**, there is less certainty about deserved suffering than in Buddhism, and there is a sense that the gods also participate in the conquest of evil. **Shintoism** is a syncretic religion and shows many similar features to Buddhism, plus many unique features. The most important of these is *kami* – the idea that the heart of being is ultimately mysterious and that there is harmony in creation. Humans are children of *kami* and the lives of all humans are sacred. The notion that *kami* resides in shrines, in some cases is permanently present within them, is a feature of Shintoism. Shrine worship to mountains and volcanoes such as Mt Fuji is strongly developed.	
Complex: Syncretic interactions between:	
a. Indigenous religions and world faiths.	a. Tarawera 1886; Soufriere (St Vincent) 1902; Sakura-jima 1916; Rabaul 1937; Lamington 1951; Arenal 1968; Galunggung 1982; Nevado del Ruiz 1985; Lake Nyos 1986; Pinatubo 1991; Galeras 1993; Merapi 1994.
b. Between world faiths and secular humanism	b. Virtually all eruptions to some extent. Very evident in the eruptions of: Vesuvius 1872; Heimaey 1973; Mount St Helens 1980; Unzen 1990–95 and Pinatubo 1991.

information on deistic responses to eruptions has been much more difficult to acquire. Both volcanology and hazard planning have roots in the 18th-century Enlightenment and have paid little attention to mythological and/or religious interpretations of eruptions and their consequences. Scientists and social scientists invariably explain human responses to eruptions in purely scientific and secular humanistic terms, and the notion of the 'Act of God' has long been replaced by a perspective that views natural catastrophes as outcomes of human vulnerability and a demoralised nature (Steinberg 2000; but also see Dynes and Yutzy 1965). The quotations in Table 10.3 show just how deep seated these secular attitudes are within volcanology and hazard studies. Reactions to disastrous events that make use of the 'languages' and modes of thought of faith communities have been virtually eliminated from official records in learned journals and eruption reports, and perspectives

Table 10.3 Examples of attitudes of volcanologists and social scientists to religious and mythological interpretations of eruption losses (based on the references cited; see also Fritscher 1998)

'The grand and striking phenomena displayed by volcanoes are especially calculated to inspire terror and to excite superstition, and such feelings must operate in preventing those close and accurate observations which alone can form the basis of scientific reasoning' (Judd 1881: 2–3).

'That such feelings of superstitious terror in connection with volcanoes are, even at the present day, far from being extinct, will be attested by every traveller who, in carrying on investigations about volcanic centres, has had to avail himself of the assistance of guides and attendants from among the common [sic] people' (Judd 1881: 3–4).

'Among the great writers of antiquity we find several who had so far emancipated their minds from the popular superstitions as to be able to enunciate just and rational views upon the subject of volcanoes' (Judd 1881: 4).

'Theorizing independent of the Bible began in America with the Enlightenment. In the mid-century the danger that geology represented to religious orthodoxy was apparent and, although, there were attempts at reconciliation the mechanistic view quickly became dominant' (Dean 1979: 291).

'Science is the first human craft to treat prophesy without superstition' (Jaggar 1937a: 2).

'Medieval science mingled art, religion and science. We have learned to separate the compartments' (Jaggar 1937a: 3).

'The nether regions of the Earth are inaccessible in the ordinary sense. Before the time of Newton, when evidence about them was nearly totally lacking, it was not necessarily unreasonable to describe the Earth in terms of models involving say a Hell, or a subterranean monster shaking itself to cause earthquakes. The subsequent growth of evidence has lowered the plausibility of such models' (KE Bullen 1975, quoted by Sigurdsson 1999: 11).

James Hutton (1788) made the first attempt 'to explode the myth associated with volcanic eruptions' (Sigurdsson 1999: 13).

'The beginning of the Renaissance marked the transition from centuries of intellectual domination of theology during the Middle Ages to a new era of human reason in the Western World' (Sigurdsson 1999: 84).

'The repetitiveness of impacts and forms of damage, the deliberate or inadvertent creation of vulnerability, and the gross predictability of the consequences of disaster all add up to human, not supernatural, responsibility' (Alexander 2000: 186–67).

'By the beginning of the 1830s the old, predominantly religious, ways of explaining natural disasters had become the prerogative of the decreasing ranks of fundamentalist Christians, and this continues to be the situation today' (Hanska 2002: 178).

from theodicy are frequently part of a 'hidden history'. This deficiency has been remedied recently for some eruptions – works on Parícutin, Mexico, 1943/53 (Luhr *et al* 1993; Scarth 1999), Krakatau, Indonesia, 1883 (Simkin and Fiske 1983), and Mt Pelée (Scarth 2002) are particularly noteworthy. Nevertheless, for many eruptions – even some that have occurred during the past few decades – there is no mention of human appeals to deistic explanations, even when these are well attested in anthropological studies and accounts are published in international newspapers of record, such as the *New York Times* and the London *Times*. For example, standard sociological and policy reviews of the 1980 Mount St Helens eruption (eg, Anderson 1987; Perry and Greene 1983; Saarinen and Sell 1987) do not mention theodicy although local radio preachers warned of God's anger towards blasphemers and drunkards. Other ministers in this region reportedly said 'that God was using Mount St Helens, to tell Kelso and Longview residents to be more religious, charitable and caring towards their families' (Blong 1984: 176; also see Anon 1980, 1983 and Tiedemann 1992: 338).

Additional examples come from the 1991 eruption of Pinatubo in the Philippines. An often quoted account of this eruption strongly supports the work of the Philippines Institute of Volcanology and stresses technology and planning approaches, with responses grounded in theodicy being viewed as signs of backwardness, even shame. 'We have come a long way from the passive extreme, where people viewed volcanic eruptions as expressions of God's displeasure and hence beyond man's Tayag [*sic*] power of understanding and intervention' (Tayag and Punongbayan 1994: 2). Unfortunately for the authors, this statement completely ignores well-attested reports of the highly theistic responses of the Aeta ethnic community and some Christian groups (Anon 1991a, 1991b; England 1993; Leone and Gaillard 1999: 230–31).

In the case of the Aeta community, their religion is animist in character, and the universal creator is believed to reside on Mt Pinatubo. So powerful is their attachment to their mountain and their god that 300 Aeta families refused to evacuate the region when the eruption began (Leone and Gaillard 1999: 230). In referring to these responses, some scientists and social scientists were dismissive. In a detailed survey of responses to the eruption, Tayag *et al* (1996: 94–95) grouped prayer in the same category as 'running about aimlessly and weeping', whilst Bautista (1996) classified religious responses under the heading 'psychological effects' and described 'pandemonium (with) people … screaming and crying as they (called) on their God for help and deliverance' (p 157). Christian responses to the 1991 eruption were no less complex. The Philippines is the only predominantly Christian country in eastern Asia; only an estimated 1% of the population is openly atheist or agnostic. Most people – even the then President Corazon Aquino – thought God

Table 10.4   A selection of major eruptions since 1850. **Bold type** shows eruptions that have a detailed record of human responses cast in deistic terms. *Italic type* indicates eruptions where deistic interpretations are mentioned but not discussed in depth. Only 16 eruptions (plain type) show no significant recorded religious interpretation of events

1856 Awu (Sangiha Island, Indonesia)[1]	1971 Villarrica (Chile)
1871–75 Hibok Hibok (central Philippines)[2]	**1973 Heimaey (Iceland)**[26]
1872 Vesuvius (Italy)[3]	1977 Nyiragongo (Zaire)
1877 Cotopaxi (Equador)[4]	**1980 Mount St Helens (USA)**[27]
1883 Krakatau (Sunda Strait)	1982 El Chichon (Mexico)[30]
1886[5] *Tarawera (New Zealand)*[6]	*1982 Galunggung (Java, Indonesia)*[28]
1888 Bandai San (Japan)[7]	1983 Colo (Sulawesi, Indonesia)
1892 Awu (Sangiha Island, Indonesia)[8]	**1985 Nevado del Ruiz (Colombia)**[29]
1897 Mayon (Philippines) Blong	*1986 Lake Nyos (Cameroon)*[30]
(1984:176–77)	**1990 Kelut (Java, Indonesia)**[31]
1902 Mt Pelée (Martinique)[9]	1990–95 Unzen (Kyushu, Japan)[32]
1902 Santa Maria (Guatemala)[10]	**1991 Pinatubo (Luzon, Philippines)**[33]
1902 Soufriere (St Vincent)[11]	**1993 Galeras (Colombia)**[34]
1906 Vesuvius (Italy)[12]	1994 Nyiragongo (Democratic Republic
1911 Taal (Central Philippines)[13]	of Congo)
1914 Sakura-jima (Japan)[14]	**1994 Merapi (Java)**[35]
1919 Kelud (Java, Indonesia)[15]	1995 Rabaul (Papua New Guinea)
1929 Santa Maria (Santiaguito, Guatemala)[16]	1995–96 Ruapehu (New Zealand)
1930 Merapi (Java, Indonesia)[17]	1996 Manam (Papua New Guinea)
1937 Rabaul (Papua New Guinea)[18]	**1997 Montserrat (Caribbean)**[36]
1943–52 Parícutin[19]	**1997 Popocatépetl (Mexico)**[37]
1944 Vesuvius (Italy)[12]	**2002 Nyiragongo (Democratic**
1951 Lamington (Papua New Guinea) [20]	**Republic of Congo)**[38]
1951 Hibok-Hibok (Philippines)[21]	**Plus: Persistent activity of Etna (Sicily)**[39]
1953 Ruapehu (New Zealand)[22]	**and the Hawaiian volcanoes**[40]
1963 Agung (Bali, Indonesia)[23]	
1963 Surtsey (Iceland)	
1966 Kelud (Java, Indonesia)[24]	
1968 Arenal (Costa Rica)[25]	

The table is based on information in Blong 1984 and: [1]Anon 1856; [2]Anon 1875; Alcaraz *et al* 1952a; [3]Anon 1872a, 1872b, 1872c; Hull 1892: 58; [4]Whymper 1892; [5]Simkin and Fiske 1983: 73, 74, 77, 80, 84, 85, 95, 97, 117, 132, and 134; Schlehe 1996; Scarth 1999: 145–46; [6]Keam 1988; [7]Sekiya and Kikuchi 1890: 100, 105; [8]Anon 1892; [9] Anon 1902a; Heilprin 1903; Scarth 1999: 164, 165, and 171, 2002: 51; [10]Anderson 1908a; Anon 1902b; [11]Anderson and Fleet 1903: 378, 379, 413, and 430; Heilprin 1903: 251–52; Anderson 1908b: 289; Jaggar 1931a; [12]Anon 1906a, 1906b; Perret 1924: 48; Anon 1944; Lewis 1983: 105; Tiedemann 1992: 338; Scarth and Tanguy 2001; 15; [13]Pratt 1911: 63–86; Worcester 1912: 320; [14]Koto 1916: 44, 55; Onishi 1930; Jaggar 1924: 465–66; [15]Van Bemmelen 1949: 222–24; Schlehe 1996; [16]Jaggar 1931b: 2; Mercado and Rose 1988; [17]see[15]; [18]Jaggar 1937b: 2; Johnson and Threlfall 1985; Sigurdsson 1999: 17; [19]Nolan 1979: 306–08, 322–24; Scarth 1999: 193–206; [20]Belshaw 1951; Anon 1952a, 1952b; Taylor 1958: 22; Ingleby 1966: 30–32; Schwimmer 1969: 5, 7, 71–72, 91, 123, 129; [21]Alcarez *et al* 1952a, 1952b; [22]Anon 2003b, 2003c; [23]Mathews 1965; [24]see[15]; [25]Alvararado-Induni 1993: 21, 68–81, 113; [26]Clapperton 1973: 500; [27]Anon 1980, 1983; Perry and Green 1983: 39; Saarinen and Sell 1987: 50–51, Tiedemann 1992: 338; [28]Anon 1982; Sundradjar and Tilling 1984; also see [15]; [29]Bruce

Table 10.4   (Continued)

2001: 29, 33; Voight 1988: 22, 1990: 173–74; [30]Le Guern *et al* 1992: 173; [31]Schlehe 1996; [32]Katsuya and Takahashi 1992: 127; [33]Anon 1991a, 1991b; England 1993: 31; Newhall and Punongbayan 1996: 825; Leone and Gaillard; 1999: 230–31; Scarth 1999: 258; [34]Bruce 2001: 29, 33, 35, 121, 193–94, Williams and Montaigne 2001: 12; [35]Schlehe 1996: 391–93, 404–07; [36]Huggins *et al* 1997; Kennedy 1997; Possekel 1999; Pattullo 2000: 4, 5, 9–10, 18, 75–76, 93–95; [37]Davison 1997: 14; [38]Dummett 2002; [39]Rodwell 1878; Chester *et al* 1985: 345–65; Chester *et al* 2005; Kennedy 2001; [40]Lachman and Bonk 1960; Murton and Shimabukuro 1974; Hodge *et al* 1979.

was either testing or punishing his people, and the churches felt it expedient in some of the worst-affected areas to organise prayer meetings to counter these simple fatalistic theodicies (Bankoff 2004).

In our survey of eruptions over the past 150 years (Table 10.4), beginning with the 1856 eruption of Awu in Indonesia and ending with the eruption in 2002 of Nyiragongo (Democratic Republic of Congo) and including persistent activity at volcanoes such as Etna and those in Hawaii, detailed bibliographic research has revealed that for some eruptions there is a rich record of human responses being couched in deistic terms. These are: Vesuvius (Italy) 1872 and 1944; Krakatau (Indonesia) 1883; Mt Pelée 1902 (Martinique); eruptions of virtually all the Javan volcanoes throughout the review period; Rabaul 1937 (Papua New Guinea); Parícutin 1943–52 (Mexico); Vesuvius 1944 (Italy); Mt Lamington 1951 (Papua New Guinea); Agung 1963 (Bali, Indonesia); Arenal 1968 (Costa Rica); Heimaey 1973 (Iceland); Mount St Helens 1980 (US); Nevado del Ruiz 1985 (Colombia); Pinatubo 1991 (Philippines); Galeras 1993 (Colombia); Montserrat 1997 (Caribbean); Popocatépetl 1997 (Mexico); Nyiragongo 2002 (Democratic Republic of Congo), and many eruptions of Etna and volcanoes in Hawaii.

Although reactions vary between societies because of the differing theodicies of the particular faith communities, there are relatively few eruptions where no religious elements in human responses are recorded. For many other eruptions, there is mention of deistic interpretation but little in-depth discussion, suggesting that local archives could contain much of value if they were to be more fully interrogated. The same may also be true for some of the eruptions for which no record of theistic interpretation of events has been found within the internationally accessible literature.

Whereas until the 1970s information about attitudes came largely from third-person accounts, in recent years these have been supplemented by detailed social surveys. On the slopes of Etna in Sicily – specifically in the village of Trecastagni – Christopher Dibben concluded that 'for many (people) religious beliefs play a significant role in their representation of the volcano' (Dibben 1999: 196), whereas, after the eruption on Montserrat (Caribbean), which began in 1995, 15 out of a representative 70 people surveyed during in-depth interviews ascribed the eruption to an 'Act of God' (Possekel 1999: 161–63). At the time of eruption Montserrat had a well-educated population, excellent health services, close links with the global economy, and established migration flows to the United Kingdom, Canada, and the United States (Possekel 1999: 88–90). Despite these features of modernity the people of Montserrat showed reactions to the eruption that were not dissimilar to those recorded in letters and other accounts from those affected by the destruction of town of Saint-Pierre in Martinique 93 years earlier (Scarth 1999: 164–65, 2002: 51). In both cases, ideas of punishment for personal sinfulness and 'fate' being in God's hands were common features, though there were detailed differences. In predominantly Catholic Martinique, there was a rush of applications for baptism and requests for confession and the last rites, whereas in the more scripturally based and denominationally diverse island of Montserrat, religious imagery such as 'first darkness' and 'Ash Monday' was commonly used to describe volcanic events (Pattullo, 2000: 4, 9–10, 75, 76; Scarth 1999: 164–65, 2002: 51).

THE CONTEMPORARY RELEVANCE OF THEODICY IN HAZARD STUDIES

The emphasis in disaster research changed during the course of the International Decade for Natural Disaster Reduction (IDNDR) (1990–2000). Understanding local culture is now considered crucially important if responses to disasters are to be successful (Chester 2005a; Chester *et al* 2002). This changing emphasis may be seen when papers and reports published at the beginning of the IDNDR (eg, Lechat 1990) are compared with those that have appeared more recently (Eades 1998; United Nations 1995). At present, policy is guided by the United Nations' International Strategy for Disaster Reduction (ISDR), which began at the start of the current decade. Specifically, the reduction of disaster risk is now viewed as 'a complex array of related political, social, economic and environmental challenges' (Hamilton 2005: 31). Concerns that are central to current research are public awareness; moving from cultures of reaction to cultures of prevention; and the necessity of identifying vulnerable groups and maintaining the sustainability of hazard prone areas (Hamilton 1999, 2005; United Nations 1999, 2002). Within this context, religious perspectives on the social construction of hazard are clearly important. There is a need for a dialogue between hazard planners and volcanologists, on the one hand, and members of faith communities, on the other.

There are two ways in which this could take place. First, it is now widely recognised that in many disasters religious organisations and their leaders have the potential to provide community as well as spiritual leadership, aid, solace to victims, and important information on people and places in need (Alexander 2002: 123; see also Mitchell, 2003). One frequently quoted example of where this has occurred was during and following flooding in West Virginia (US) in 1985, with clergy both caring for and shaping the opinions of their congregations (Bradfield *et al* 1989). A more recent example comes from the Philippines following the 1991 Pinatubo eruption. The town of Bacolor, one of the cradles of Kapampangan culture, was amongst the settlements most badly affected, and the role of religion, in general, and the Catholic Church, in particular, was critical in restoring the morale and identity of the community, which is now located on several sites (Gaillard 2003: 81–82). These remain, however, isolated examples and for dialogue to take place, there needs to be a recognition that religious interpretations of suffering are not only widely held but should also be respected.

A second, more significant factor is that models of theodicy are now much more sophisticated than was the case in the past, offering far greater scope for incorporating the views of religious groups in the process of hazard planning (Chester 2005b). For instance, in Christian tradition, what are termed *liberationist theodicies* have emerged (Chester 1998; Soelle 1975). Strongly influenced by liberation theology (Boff and Boff 1987; Gutiérrez 1988), these theodicies hold that human sinfulness and suffering are linked. Sin is, however, collective not personal, and involves such factors as global and national disparities in wealth, poverty, power, and access to strategies of hazard reduction. By interpreting the effects of a disastrous eruption in this manner, common ground has emerged between social scientists, theologians, and members of Christian churches and their leaders.

In Montserrat, for example, a common critique of the alleged ineffectiveness of the colonial government before and during the emergency has been articulated by Christian groups (eg, Huggins *et al* 1997; Kennedy 1997), social scientists (eg, Pattullo 2000), and many journalists (eg, Masood 1998; see Figure 10.2). Some idea of the religious and cultural complexity of the response of the islanders to the eruptions on Montserrat can be appreciated from the following, almost poetic, quotation.

> In many ways the dead (*from the eruption*) came to represent all that was virtuous about Montserrat and its people. They became symbols of all old-fashioned, God-fearing society in which the values of an emancipated peasantry – individualism, independence, devotion to land and home – triumphed over the circumstances of death. ... Some were reaping their crops in the fields ... to feed the island; others were tending livestock (*note the biblical imagery*), and others stayed resolutely at home. In a sense, they became the heroic dead, the victims of a colonial war. (Pattullo 2000: 5, emphasis added)

Figure 10.2 Cartoonist's view on where the responsibility for the Montserrat disaster should be placed (David Brown, *The Independent* [London], 22 August 1997, p 17, reproduced with permission of the *The Independent* [London].)

Other religious traditions are also more heterogeneous in their approaches to disasters than is admitted by many Western commentators. Java (Indonesia) is a land of frequent eruption, and Schlehe (1996) has highlighted the complex, often syncretic, manner in which losses are explained by means of a conflation of Islamic, Hindu, Christian, and pre-Christian spiritual understandings. Islam literally means submission, and many Western commentators have emphasised its strongly 'instrumentalist' view of suffering: God uses disasters to bring followers back to the prophet's teaching (eg, Anon 1997: 968; Bemporad 1987; Bowker 1970: 113). Islam is much more theologically heterogeneous, however, than is commonly supposed, and there is no typical response (Dhaoudi 1992: 41).

Muslim culture has had more than a thousand years of experience to come to terms with differing social and political conditions (Al-Azmeh 1996: 44; Halliday 1994: 96), and this is expressed in considerable diversity. Although we are unaware of any overtly volcanological examples, this 'theological accommodation' may be seen after the 1992 Dahshûr earthquake in Egypt. Shortly after the disaster, the Egyptian government produced a report, *Earthquake Catastrophes and the Role of People in Facing Them*, that successfully reconciled scientific, planning, and Islamic approaches to disaster losses (Degg and Homan 2005; Homan 2001). Such an approach holds out much hope for volcanology in volcanic regions such as Indonesia and the Philippines where Islam and Enlightenment–influenced earth and social sciences have frequently viewed disasters with mutual incomprehension.

Effective response to volcanic hazard to mitigate risk requires engagement with local communities. In most parts of the world, religious institutions are part of the fabric of society and play a prominent role during disasters. In line with ISDR strategy to take into account public awareness and the need to work with and through local community structures, bringing religious networks into play has the potential to improve interaction between scientists, civil defence authorities, and the local population. We contend that the interface between volcanology, social science, and theology is a potentially fruitful research frontier, which deserves more research.

REFERENCES

Al-Azmeh, A (1996) *Islams and Modernities*, London: Verso

Alcaraz, A, Abad, LF and Quema, JC (1952a) 'Hibok-Hibok volcano, Philippines Islands, and its activity since 1948', *The Volcano Letter* 516, 1–6

Alcaraz, A, Abad, LF and Quema, JC (1952b) 'Hibok-Hibok volcano, Philippines Islands, and its activity since 1948', *The Volcano Letter* 517, 1–4

Alexander, D (2000) *Confronting Catastrophe*, Harpenden, UK: Terra Publishing

Alexander, D (2002) *Principles of Emergency Planning and Management*, Harpenden, UK: Terra Publishing

Alvarardo-Induni, G (1993) *Costa Rica: Land of Volcanoes*, San José, Costa Rica: Gallo Pinto Press

Anderson, J (1987) 'Learning from Mount St. Helens – catastrophic events as educational opportunities', *Journal of Geography* 86, 229–33

Anderson, T (1908a) 'The volcanoes of Guatemala', *Geographical Magazine* 31, 473–89

Anderson, T (1908b) 'Report of the eruption of Soufrière in St. Vincent in 1902, and on a visit to Montagne Pelée in Martinique', *Philosophical Transactions of the Royal Society of London* 208, 275–332

Anderson, T and Fleet, JS (1903) 'Report on the eruptions of Soufriere in St Vincent, in 1902 and on a visit to Montagne Pelée in Martinique', *Transactions of the Royal Society of London* 200A, 353–553

Anon (1856) 'Report', *The Times* (London), 19 July, p 10, column f

Anon (1872a) 'Report', *The Times* (London), 6 May, p 6, column d

Anon (1872b) 'Report', *The Times* (London), 9 May, p 6, column d

Anon (1872c) 'Report', *The Times* (London), 5 May, p 6, column c

Anon (1875) 'Report', *The Times* (London), 30 May, p 4, columns d and e

Anon (1892) 'Report', *The Times* (London), 28 May, p 5, column b

Anon (1902a) 'Report', *The Times* (London), 15 May, p 7, column a

Anon (1902b) 'Volcanic disturbances in Guatemala', *National Geographic Magazine* 113, 461–62

Anon (1906a) 'Report', *The Times* (London), 13 April, p 3, column e

Anon (1906b) 'Report', *The Times* (London), 17 April, p 4, columns c and d

Anon (1944) *War Office Manuscript on the 1944 eruption of Vesuvius*, Public Record Office London, file WO220/439

Anon (1952a) 'Report', *The Times* (London), 22 January, p 6, column d

Anon (1952b) 'Report', *The Times* (London), 24 January, p 6, column b

Anon (1980) 'Report', *The Christian Century*, 2–9 July, p 695
Anon (1982) 'Report', *The Washington Press*, 25 November, p G2
Anon (1983) 'Report', *The Columbian* (Clark County, WA), 31 March
Anon (1991a) 'Report', *The Times* (London), 11 June, p 12, column d
Anon (1991b) 'Report', *The Times* (London), 17 June, p 10
Anon (1997) 'Theodicy', in Bowker, J (ed), *The Oxford Dictionary of World Religions*, pp 968–69, New York: Oxford University Press
Anon (2003a) 'Har-Togoo, central Mongolia', *Bulletin of the Global Volcanism Network* 28, 9–11
Anon (2003b) *Police Response to Disaster: Tangiwai*, available online at www.nzhistory.net.nz/Gallery/police/tangi-wai.html (accessed 4 March 2003)
Anon (2003c) *The Tangiwai Railway Disaster*, available online at www.nzhistory.net.nz/Gallery/Tangiwai.html (accessed 4 March 2003)
Bankoff, G (2004) 'In the eye of the storm: the social construction of the forces of nature and the climatic and seismic construction of God in the Philippines', *Journal of Southeast Asian Studies* 35, 91–111
Bautista, CB (1996) 'The Mount Pinatubo disaster and the people of central Luzor', in Newhall, CG and Punongbayan, RS (eds) *Fire and Mud: Eruptions and Lahars of Mount Pinatubo, Philippines*, pp 151–61, Quezon City, Philippines: Philippines Institute of Volcanology and Seismology and Seattle: University of Washington Press
Bell, B (1971) 'The Dark Ages in ancient history 1. The first dark age in Egypt', *American Journal of Archaeology* 75, 1–26
Belshaw, CD (1951) 'Social consequences of the Mount Lamington eruption', *Oceania* 21, 241–52
Bemporad, J (1987) 'Suffering', in Eliade, M (ed), *The Encyclopedia of Religion*, vol 14, pp 99–104, New York: Macmillan
Blong, RA (1984) *Volcanic Hazards*, Sydney: Academic Press
Boff, L and Boff, C (1987) *Introducing Liberation Theology*, Tunbridge Wells, UK: Burns and Oates
Bowker, J (1970) *Problems of Suffering in the Religions of the World*, Cambridge: Cambridge University Press
Bradfield, C, Wylie, ML and Echterling, LG (1989) 'After the flood: the response of ministers to a natural disaster', *Sociological Analysis* 49, 397–407
Bruce, V (2001) *No apparent danger – The true story of volcanic disaster at Galeras and Nevado del Ruiz*, New York: Harper Collins
Chester, DK (1998) 'The theodicy of natural disasters', *Scottish Journal of Theology* 51, 485–505
Chester, DK (2005a) 'Volcanoes, society and culture', in Marti, J and Ernst, GJ (eds), *Volcanoes and the Environment*, pp 404–39, Cambridge: Cambridge University Press
Chester, DK (2005b) 'Theology and disaster studies: The need for dialogue', *Journal of Volcanology and Geothermal Research* 146, 319–28
Chester, DK, Dibben, C and Duncan, AM (2002) 'Volcanic hazard assessment in western Europe', *Journal of Volcanology and Geothermal Research* 115, 411–35
Chester, DK, Duncan, AM, Dibben, C, Guest, JE and Lister, PH (1999) 'Mascali, Mount Etna Region, Sicily: an example of fascist planning during and after the 1928 eruption', *Natural Hazards* 19, 29–46
Chester, DK, Duncan, AM and Guest, JE (2005) 'Responses to eruptions of Etna from the Classical Period to 1900', in Balmuth, MS, Chester, DK and Johnston, P (eds), *The Cultural Response to Volcanic Landscape*, pp 93–108, Boston: Archaeological Institute of America
Chester, DK, Duncan, AM, Guest, JE, Johnston, PA and Smolenaars, JJL (2000) 'Human responses to Etna volcano during the Classical Period', *Geological Society, London, Special Publication* 171, 179–88
Chester, DK, Duncan, AM, Guest, JE and Kilburn, CRJ (1985) *Mount Etna: The Anatomy of a Volcano*, London: Chapman and Hall

Clapperton, CM (1973) 'Thrice threatened Heimaey', *Geographical Magazine* 45, 495–500

Cronin, SJ, Ferland, MA and Terry, JP (2004) 'Nabukelevu volcano (Mt. Washington), Kadavu – a source of hitherto unknown hazard in Fiji', *Journal of Volcanology and Geothermal Research* 131, 371–96

Cronin, SJ, Gaylord, DR, Charley, D, Alloway, BV, Wallez, S and Esau, JW (2004) 'Participatory methods of incorporating scientific and traditional knowledge for volcanic hazard management on Ambae Island Vanuatu', *Bulletin of Volcanology* 66, 652–68

Davison, P (1997) '"El Popo" tips ashtray into Mexico City smog', *The Independent* (UK), 2 July, p 14

de Boer, JZ and Sanders, DT (2002) *Volcanoes in Human History*, Princeton, NJ: Princeton University Press

Dean, DR (1979) 'The influence of geology on American literature and thought', in Schneer, CJ (ed), *Two Hundred Years of Geology in America*, pp 289–303, Lebanon, NH: University Press of New England

Degg, M and Homan, J (2005) 'Earthquake vulnerability in the Middle East', *Geography* 90, 54–66

Dhaoudi, M (1992) 'An operational analysis of the phenomenon of the other underdevelopment in the Arab world and the Third World', in Albrow, M and King, E (eds), *Globalization, Knowledge and Society*, pp 193–208, London: Sage

Dibben, C (1999) 'Looking beyond eruptions for an explanation of volcanic disasters: vulnerability in volcanic environments', unpublished PhD thesis, Faculty of Science, Technology and Design, University of Luton, Luton, UK

Dibben, C and Chester, DK (1999) 'Human vulnerability in volcanic environments: the case of Furnas, Sao Miguel, Acores', *Journal of Volcanology and Geothermal Research* 92, 133–50

Doonan, RCP (1997) 'Vulcanism and the furnace: the social production of the technological metaphor', in *Volcanoes, Earthquakes and Archaeology* (conference abstracts), London: The Geological Society

Dummett, M (2002) 'DR Congo's endless suffering', *BBC News online*, http://news.bbc.couk/1/hi/world/africa/1780540.stm (accessed 26 March 2003)

Duncan, AM, Chester, DK and Guest, JE (2005) 'Eruptive activity of Etna before 1600 CE with particular reference to the classical period', in Balmuth, MS, Chester, DK and Johnston, P (eds) *The Cultural Response to Volcanic Landscape*, pp 57–70, Boston: Archaeological Institute of America

Dynes, RR and Yutzy, D (1965) 'The religious interpretation of disaster', *Topic* (Washington, DC) 5, 34–48

Eades, T (1998) 'The international decade for natural disaster reduction', in Twigg, J (ed), *Development at Risk: Natural Disasters and the Third World*, pp 20–22, Oxford: Oxford Centre for Disaster Studies

Elson, MD, Ort, MH, Hesse, SJ and Duffield, WA (2002) 'Lava, corn and ritual in the northern southwest', *American Antiquity* 67, 119–35

England, V (1993) 'Return to Pinatubo', *The Geographical Magazine* 55, 28–31

Fisher, RV, Heiken, G and Hulen, JB (1997) *Volcanoes: Crucibles of Change*, Princeton, NJ: Princeton University Press

Freeth, SJ (1993) 'On the problems of translation in the investigation of the Lake NCOs disaster', *Journal of Volcanology and Geothermal Research* 54, 353–56

Friedrich, WL (2000) *Fire in the Sea: The Santorini Volcano: Natural History and the Legend of Atlantis*, Cambridge: Cambridge University Press

Fritscher, B (1998) 'Volcanoes and the "wealth of nations". Relations between the emerging sciences of political economy and geology in 18th century Scotland', in Morello, N (ed), *Volcanoes and History*, pp 209–28, Genova, Italy: Brigati

Gaillard, J-C (2003) 'Territorial and ethno-cultural implications of a volcanic crisis: The case of Mount Pinatubo eruption', *Alaya: Kapampangan Research Journal* 1, 83–88

Galipaud, J-C (2002) 'Under the volcano: Ni-Vanuatu and their environment', in Torrance, R and Grattan, J (eds), *Natural Disasters and Cultural Change*, pp 162–71, London: Routledge

Grattan, J, Brayshaw, M and Schüttenhelm, RTE (2002) 'The end is nigh? Social and environmental responses to volcanic gas pollution', in Torrence, R and Grattan, J (eds), *Natural Disasters and Cultural Change*, pp 87–106, London: Routledge

Gutiérrez, G (1988) *A Theology of Liberation*, New York: Orbis Books

Halliday, F (1994) 'The politics of Islamic fundamentalism: Iran, Tunisia and the challenge to the secular state', in Ahmed, AS, Doonan, H (eds), *Islam, Globalization and Postmodernity*, pp 91–113, London: Routledge

Hamilton, RM (1999) 'Natural disaster reduction in the 21st century', in Ingleton, J (ed), *Natural DisasterManagement*, pp 304–07, Leicester, UK: Tudor Rose

Hamilton, RM (2005) 'Evolution in approaches to disaster reduction', in Anon (ed), *At Risk. International Strategy for Disaster Reduction, United Nations*, pp 31–32, Leicester, UK: Tudor Rose

Hanska, J (2002) *Strategies of Sanity and Survival. Religious Responses to Natural Disasters in the Middle Ages*, Studia Fennica Historica 2, Helsinki: Finnish Literature Society

Harris, SL (1988) *Fire Mountains of the West*, Missoula, MT: Mountain Press

Harris, SL (2000) 'Archaeology and volcanism', in Sigurdsson, H, Houghton. B, Mc Nutt, SR, Rymer, H and Stix, J (eds), *The Encyclopedia of Volcanoes*, pp 1301–14, San Diego, CA: Academic Press

Heilprin, A (1903) *Mont Pelee and the Tragedy of Martinique*, Philadelphia: Lippincott

Hodge, D, Sharp, V and Marts, M (1979) 'Contemporary responses to volcanism: Case studies from the Cascades and Hawaii', in Sheets, PD and Grayson, DK (eds), *Volcanic Activity and Human Ecology*, pp 221–47, New York: Academic Press

Homan, J (2001) 'A culturally sensitive approach to risk? "Natural" hazard perception in Egypt and the UK', *Australian Journal of Emergency Management* 16, 14–18

Huggins, J, Cunningham, R and Panton, J (1997) *Submission from Christian Aid and the Montserrat Aid Committee to the Committee on International Development of the House of Commons*, London: Christian Aid and the Montserrat Aid Committee

Hull, E (1892) *Volcanoes, Past and Present*, London: W. Scott

Hyde, WW (1916) 'The volcanic history of Etna', *Geographical Review* 1, 401–18

Ingleby, I (1966) 'Mount Lamington fifteen years later', *Australian Territories* 6, 23–34

Jaggar, TA (1924) 'Sakura-jima, Japan's greatest volcanic eruption', *National Geographic Magazine* 45, 441–70

Jaggar, TA (1931a) 'The crater of Soufriere volcano', *The Volcano Letter* 359, 1–4

Jaggar, TA (1931b) 'Eruption of Santa Maria November 1929', *The Volcano Letter* 356, 1–4

Jaggar, TA (1937a) 'Trends in the philosophy of science', *The Volcano Letter* 447, 2–6

Jaggar, TA (1937b) 'Eruptions at Rabaul, New Guinea', *The Volcano Letter* 448, 2–6

Jara, LA, Thouret, J-C, Siebe, C and Dávila, J (2000) 'The AD 1600 eruption of Huaynaputina as described in early Spanish chronicles', *Boletín de la Sociedad Geológica del Perú* 90, 121–32

Johnson, RW and Threlfall, NA (1985) 'Volcano town. The 1937–43 Rabaul eruptions', Bathurst, Australia: Robert Brown and Associates

Judd, JW (1881) *Volcanoes: What They Are, and What They Teach*, London: Kegan Paul, Trench and Trubner

Katsuya, M and Takahashi, K (1992) 'Decision making process of both the administrative bodies and the inhabitants for evacuation during the eruption of Mt. Fugen in Unzen', in Yanagi, T, Okada, H and Ohta, K (eds), *Unzen Volcano, the 1990–1992 eruption*, pp 120–27, Fukuoka, Japan: Nishinippon and Kyushu University Press

Keam, RF (1988) 'Tarawera: the volcanic eruption of 10 June 1886', Department of Physics, University of Auckland (private publication)

Kendrick, TD (1956) *The Lisbon Earthquake*, London: Methuen

Kennedy, F (2001) 'Sicilians pray as technology fails to stop lava', *The Independent* (London), 30 July, p 10

Kennedy, J (1997) 'Natural disaster versus man made volcanoes', *The Independent* (London), 30 August, p 15

Kogen, M (1987) 'Karman', in Eliade, M (ed), *The Encyclopedia of Religion*, vol 8, pp 261–68, New York: Macmillan

Koto, B (1916) 'The great eruption of Sakuri-jima in 1914', *Journal of the College of Science Imperial University of Tokyo* 38, 1–237

Lachman, R and Bonk, WJ (1960) 'Behavior and beliefs during the recent volcanic eruption at Kapoho, Hawaii', *Science* 131, 1095–96

Lechat, MF (1990) 'The international decade for natural disaster reduction: background and objectives', *Disasters* 14, 1–6

Le Guern, F, Shanklin, E and Tebor, S (1992) 'Witness accounts of the catastrophic event of August 1986 at Lake Nyos (Cameroon)', *Journal of Volcanology and Geothermal Research* 51, 174–84

Leibniz, GW (1952) *Theodicy*, London: Routledge and Kegan Paul

Leone, F and Gaillard, J-C (1999) 'Analysis of the institutional and social responses to the eruption and the lahars of Mount Pinatubo volcano from 1991 to 1998 (Central Luzon, Philippines)', *GeoJournal* 49, 223–38

Lewis, N (1983) *Naples '44: An intelligence Officer in the Italian Labyrinth*, London: Eland

Lowe, DJ, Newnham, RM and McCraw, JD (2002) 'Volcanism and early Maori society in New Zealand', in Torrence, R and Grattan, J (eds), *Natural Disasters and Cultural Change*, pp 127–61, London: Routledge

Luhr, JF, Simkin, T and Cuasay, M (1993) *Parícutin: The Volcano Born in a Mexican Cornfield*, Phoenix: Geoscience Press

Masood, E (1998) 'Montserrat residents "lost faith" in volcanologists' warnings', *Nature* 392, 743–44

Mathews, A (1965) *The Night of Purnama*, London: Jonathan Cape

Mercado, R and Rose, WI (1988) 'November 1929 dome collapse and pyroclastic flow at Santiaguito done, Guatemala', *EOS* 69, 1487

Mege, D and Korme, T (2004) 'Dyke swarm emplacement in the Ethiopian large igneous province: not only matter of stress', *Journal of Volcanology and Geothermal Research* 132, 283–310

Mellaart, J (1965) *Earliest Civilizations of the Near East*, London: Thames and Hudson

Mellaart, J (1967) *Çatal Hüyük: A Neolithic Town in Anatolia*, London: Thames and Hudson

Mitchell, JT (2003) 'Prayer in disaster: case study of Christian clergy', *Natural Hazards Review* 4, 20–26

Murton, BJ and Shimabukuro, S (1974) 'Human adjustment to volcanic hazard in the Puna District, Hawaii', in White, GF (ed), *Natural Hazards: Local, National, Global*, pp 151–61, New York: Oxford University Press

Newhall, CG and Punonbayan, RS (1996) 'The narrow margin of volcanic-risk mitigation', in Scarpa, R and Tilling, RI (eds), *Monitoring and Mitigation of Volcanic Hazards*, pp 807–38, Heidelberg, Germany: Springer-Verlag

Nolan, ML (1979) 'Impact of Parícutin on five communities', in Sheets, PD and Grayson, DK (eds), *Volcanic Activity and Human Ecology*, pp 293–338, New York: Academic Press

Onishi, K (1930) 'Eruption of Sakurajima 1914. Review by K Onishi of F Omori's article in the *Bulletin of the Imperial Earthquake Investigation Committee*, vol 8, no 1, Sept 1914', *Volcano Letter* 308, 1–3

Paragament, KI (1997) *The Psychology of Religion and Coping. Theory, Research, Practice*, New York: The Guilford Press

Paragament, KI and Hahn, J (1986) 'God and a just world: casual and coping attributions to God in health situations', *Journal for the Scientific Study of Religion* 25, 193–207

Pattullo, P (2000) *Fire from the Mountain: The Tragedy of Montserrat and the Betrayal of its People*, London: Constable

Perret, FA (1924) *The Vesuvius Eruption of 1906: Study of a Volcanic Cycle*, Publication 339, Washington, DC: Carnegie Institution of Washington

Perry, RW and Greene, MR (1983). *Citizen Response to Volcanic Eruptions*, New York: Irvington

Pilgrim, NK (1999) 'Landslides, risk and decision-making in Kinnaur District: bridging the gap between science and public opinion', *Disasters* 23, 45–65

Plunket, P and Uruñuela, G (1997) 'Revelations of a Plinian eruption of the Popocatépetl volcano in central Mexico', in *Volcanoes, Earthquakes and Archaeology* conference abstracts), pp 33–34, London: The Geological Society

Plunket, P and Uruñuela G (1998) 'Appeasing the volcano gods', *Archaeology* 51, 36–42

Poole, L and Poole, G (1962) *Volcanoes in Action: Science and Legend*, New York: McGraw Hill

Possekel, AK (1999) *Living with the Unexpected*, Berlin: Springer

Pratt, WE (1911) 'The eruption of Taal volcano, January 30, 1911', *Philippine Journal of Science* 6, 63–86

Rodwell, GF (1878) *Etna: A history of the Mountain and Its Eruptions*, London: CK Paul

Saarinen, TF and Sell, JL (1987) *Warning and Response to the Mount St. Helens eruption*, Albany: State University of New York Press

Scarth, A (1999) *Vulcan's Fury: Man against the Volcano*, New Haven, CT: Yale University Press

Scarth, A (2002) *La Catastrophe: Mount Pelée and the Destruction of Saint-Pierre, Martinique*, Harpenden, UK: Terra Publications

Scarth, A and Tanguy, J-C (2001) *Volcanoes of Europe*, Harpenden, UK: Terra Publishing

Schlehe, J (1996) 'Reinterpretation of mystical traditions – explanations of a volcanic eruption in Java', *Anthropos* 91, 391–409

Schwimmer, EG (1969) *Cultural Consequences of a Volcanic Eruption Experienced by Mount Lamington, Orokaiva*, Report 9, Department of Anthropology, University of Oregon, Eugene

Sekiya, S and Kikuchi, Y (1890) 'The eruption of Bandai-San', *Journal of the College of Science, Imperial University Japan* 3, 91–171

Shimoyama, S (2002) 'Volcanic disasters and archaeological sites in Southern Kyushu, Japan', in Torrence, R and Grattan, J (eds), *Natural Disasters and Cultural Change*, pp 326–41, London: Routledge

Sigurdsson, H (1999) *Melting the Earth: The History of Ideas on Volcanic Eruptions*, Oxford: Oxford University Press

Sigurdsson, H, Houghton. B, Mc Nutt, SR, Rymer, H and Stix, J (2000) *The Encyclopedia of Volcanoes*, San Diego, CA: Academic Press

Simkin, T and Fiske, RS (1983) *Krakatau 1883: The Volcanic Eruption and Its Effects*, Washington, DC: Smithsonian Institution Press

Soelle, D (1975) *Suffering*, London: Darton, Longman and Todd

Steinberg, T (2000) *Acts of God: The Unnatural History of Natural Disaster in America*, New York: Oxford University Press

Stephens, JL (1969) *Incidents of Travel in Central America, Chiapas and Yucatan*, vol 2, New York: Dover

Sundradjar, A and Tilling, R (1984) 'Volcanic hazards in Indonesia', *Episodes* 7, 13–19

Tanguy, J-C, Ribière, Ch, Scarth, A and Tjetjep, WS (1998) 'Victims from volcanic eruptions: a revised database', *Bulletin of Volcanology* 60, 137–44

Tayag, JC and Punongbayan, RS (1994) 'Volcanic disaster mitigation in the Philippines: experience from Mt Pinatubo', *Disasters* 18, 1–15

Tayag, J, Insauriga, S, Ringor, A and Belo, M (1996) 'People's response to eruption warning: the Pinatubo experience, 1991–92', in Newhall, CG and Punongbayan, RS (eds), *Fire and Mud: Eruptions and Lahars of Mount Pinatubo, Philippines*, pp 87–99, Quezon City, Philippines: Philippines Institute of Volcanology and Seismology and Seattle: University of Washington Press

Taylor, GA (1958) 'The 1951 eruption of Mount Lamington, Papua', *Australian Bureau of Mineral Resources, Geological and Geophysical Bulletin*, 38, entire issue

Thorarinsson, S (1970) *Hekla, a Notorious Volcano*, Reykjavik: Almenna Bókafélagid

Tiedemann, H (1992) *Earthquakes and Volcanic Eruptions*, Zurich: Swiss Reinsurance Company

Tilling, RI (1989) 'Volcanic hazards and their mitigation: progress and problems', *Reviews of Geophysics* 27, 237–69

United Nations (1995) *Yokohama Strategy and Plan of Action for a Safer World: Guidelines for Natural Disaster Prevention, Preparedness and Mitigation*, Geneva: United Nations Press

United Nations (1999) *International Decade for Natural Disaster Reduction: Successor Arrangement*, New York: United Nations Press

United Nations (2002) *Living with Risk*, Geneva: United Nations Press

Van Bemmelen, RW (1949) *The Geology of Indonesia*, vol 1A, The Hague: Government Printing Office

Vitaliano, DB (1973) *Legends of the Earth: Their Geologic Origins*, Bloomington: Indiana University Press

Voight, B (1988) 'Countdown to catastrophe', *Earth and Mineral Sciences* 57, 17–30

Voight, B (1990) 'The 1985 Nevado del Ruiz volcano catastrophe: anatomy and retrospection', *Journal of Volcanology and Geothermal Research* 42, 151–88

White, GF (1973) 'Natural hazards research', in Chorley, RJ (ed), *Directions in Geography*, pp 193–212, London: Methuen

Whymper, E (1892) *Travels amongst the Great Andes of the Equador*, Salt Lake City: Peregrine Smith Books

Williams, S and Montaigne, F (2001) *Surviving Galeras*, New York: Houghton-Mifflin

Worcester, DC (1912) 'Taal volcano and its recent destructive eruption', *National Geographic Magazine* 23, 313–67

CHAPTER 11

Planning for the Future: A Multidisciplinary Approach to Reconstructing the Buag Episode of Mt Pinatubo, Philippines

J-C Gaillard, FG Delfin, Jr, EZ Dizon, VJ Paz,
EG Ramos, CT Remotigue, KS Rodolfo,
FP Siringan, JLA Soria, and JV Umbal

INTRODUCTION

Mt Pinatubo, a stratovolcano on Luzon Island in the Philippine archipelago (Figure 11.1), was the site of the greatest volcanic eruption of the 20th century, an event that reduced its height from at 1,745 m to 1,485 m ASL and replaced the summit with a 2.5 km wide caldera, now occupied by a 100 m deep lake.

Within the Philippines, the 1991 eruption and its aftermath caused economic losses estimated at a billion US dollars, and wreaked havoc in the lives of two million people. Part of the reason for the extent of the damage was the lack of knowledge about the past history of Pinatubo eruptions and consequent human responses. By addressing these deficiencies, our interdisciplinary study provides information useful for future hazard reduction. Newhall et al (1996) have presented good evidence indicating that Pinatubo's past eruptive episodes have followed a cycle. Centuries of repose are terminated by a powerful, caldera-forming eruption of tephra and pyroclastic flows with large contemporaneous lahars. This phase is followed by a period of post-eruption lahars, with minor eruptions and the filling in of the caldera by dome building. Lahar activity is gradually replaced by the distal deposition of enhanced river sediment loads.

To improve understanding of the cyclic process, we undertook a multidisciplinary study of the so-called Buag episode of Mt Pinatubo's previous eruptions, which lasted from about 800 to 500 years ago. We used geological data and satellite imagery to reconstruct how the

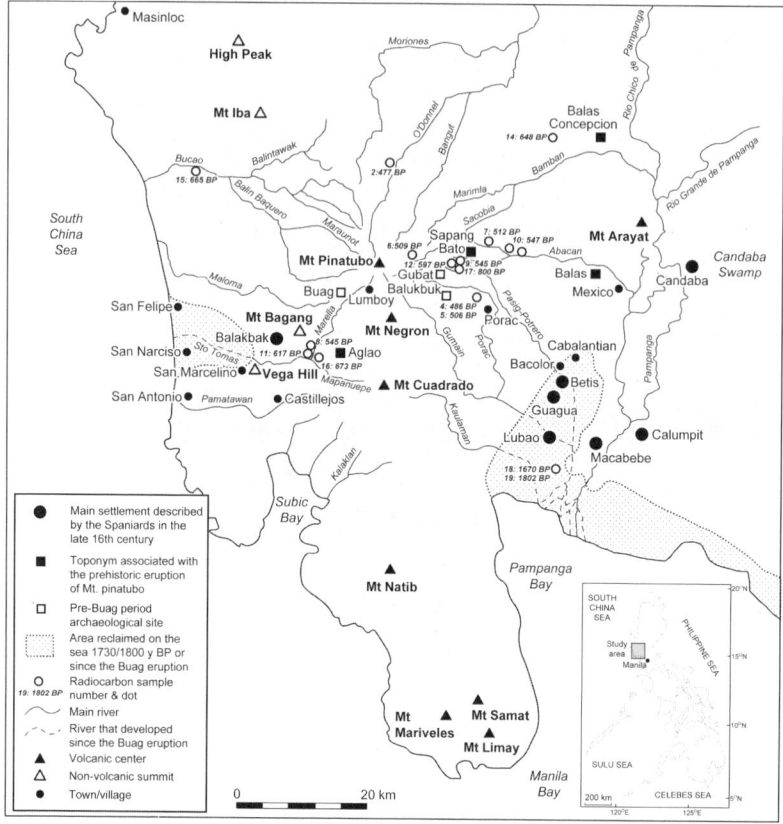

Figure 11.1 Location of the study area showing location of sites, Buag-age radiocarbon samples listed in Table 11.1, and significant changes to coastal topography.

landscape was modified, whilst archaeological and historical data and oral accounts provide an insight into how the eruption and its aftermath affected the people on and around Pinatubo.

BUAG ERUPTION HISTORY

Active volcanism in the vicinity of Pinatubo began at least 1.6 ma and the eruptions that built Pinatubo proper began 1.1 ma or earlier (Delfin 1983, 1984; Delfin *et al* 1996; Newhall *et al* 1996). Since then, the volcano and its surroundings have experienced frequent cataclysmic eruptions of which those of the Buag episode were among the least explosive.

Newhall *et al* (1996) have named the most recent prehistoric eruption period after Buag, a village of San Marcelino, Zambales in the Marella River valley. Charcoal and wood from lahar and fluvial deposits suggest that the Buag eruptive period begun as early as 600 or even 800 BP (Table 11.1 and Figure 11.1). Debris from eruptions of the Buag period buried all the major watersheds of the volcano, except those of the Gumain and Porac rivers. The composition and distribution of these deposits indicate a Plinian-type eruption that produces avertical eruption columns spreading tephra far and wide and numerous smaller pyroclastic flows. This Plinian eruption probably left a caldera comparable in size to that produced in 1991, but the depression was filled and overtopped when glassy andesite lavas of the last Buag eruptions built up Pinatubo's steep-sided, 2 km wide, pre-1991 summit dome.

The topography of Mt Pinatubo prior to the 1991 eruption also hints at the nature and sequence of events associated with the Buag eruption. Commanding the terrain in Figure 11.2, the dome is marked by a prominent summit hill with its north-eastern side apparently breached by a short lava flow. This dome partly filled the large caldera that existed before the dome-forming events, leaving only a circular trend of discontinuous ridges to mark the rim of the former caldera. Further down the western flanks, the topography displays large pyroclastic fans deposited during the Buag event, in the same areas reached by the 1991 pyroclastic flows and filling the valleys in a similar manner. Smaller pyroclastic fans were created on the northern and south-eastern sections of the volcano. Before the 1991 eruption, the pyroclastic fans had been deeply incised with river valleys, just as the recent deposits continue to be dissected.

LAHARIC AFTERMATHS

Each major plinian Pinatubo eruption overloaded the watersheds on the volcano with great volumes of pyroclastic materials, thereby initiating a period of accelerated sedimentation (Major *et al* 1996; Newhall *et al* 1996; Pierson *et al* 1992; Pierson *et al* 1996; Rodolfo *et al* 1996; Scott *et al* 1996; Umbal and Rodolfo 1996). These periods of rapid aggradation by lahars, which lasted for several years or decades after each main eruptive phase, built up and laterally expanded the alluvial fan systems that drape the footslopes of Mt Pinatubo. Streams draining the watersheds outside the volcano were less affected by pyroclastic deposition. Where these were tributaries of a lahar-aggrading river, they would often be dammed by lahar sediments and transformed into ephemeral ponds or longer lasting lakes. Failures of such natural dams cause dangerous lake-breakout floods that pose a significant downstream hazard (Rodolfo *et al* 1996; Umbal 1994; Umbal and Rodolfo 1996). Over millennia, fluvial processes have transported the debris from all the major eruptions downslope, building deltaic coastal plains and pushing the shores seawards.

Table 11.1 Radiocarbon data of the Buag eruptive event (modified from Newhall *et al* 1996). C[14] ages have been defined using the Libby half life of 5,568 years. Calibration ages have been calculated by use of calibration curves developed by Stuiver and Reimer (1993). The calibrated ages are the statistically most-likely equivalent in calendar years before 1950 (BP) using 1-sigma range. Sample locations are shown in Figure 11.2. (Note: * samples from Newhall *et al* 1996; ** new data)

Lab. Number	Latitude (°N)	Longitude (°E)	Material	Occurrence	Drainage	C[14] Age (BP)	Calibrated Age (BP)	Significance
WW-111*	15°04'	120°16'	Charcoal	Cultural layer	Marella	397 ± 70	476	Dates latest occupation; settlement on old lahar deposits.
WW-26*	15°16.8'	120°22.7'	Wood	Fluvial deposit	O'Donnell	400 ± 80	477	Latest major filling of Crow Valley.
W-6478*	15°09.2'	120°34.4'	Wood	Soil above gravel	Abacan	410 ± 55	485	After high-energy stream; before latest pre-1991 lahars.
WW-4684	15°05.4'	120°31.5'	Charcoal	Cultural layer	Porac	415 ± 40	486	Dates latest occupation.
WW-4683**	15°05.4'	120°31.5'	Charcoal	Cultural layer	Porac	455 ± 40	506	Dates latest occupation.
W-6314*	15°09'	Ca 120°24'	Charcoal	Pyroclastic-flow deposit	Upper Sacobia	460 ± 30	509	Youngest pre-1991 pyroclastic flow in Sacobia.
Beta*	15°10.5'	120°29.5'	Wood	Post-pyroclastic-flow soil	Abacan	470 ± 50	512	Start of latest pre-1991 fluvial deposition in Abacan.
W-6509*	14°59.9'	120°15.5'	Charcoal	Lahar deposit	Marella	560 ± 60	545	Covered by 1992 lahars.

Sample	Latitude	Longitude	Material	Deposit type	River	Age	Cal.	Remarks
WW-25*	15°8.2'	120°28.8'	Wood	Lacustrine deposit	Pasig-Potrero	560 ± 70	545	Latest fill in the north fork, Pasig-Potrero.
WW-21*	15°09.7'	120°33.4'	Charcoal	Silt over lahar deposit	Abacan	570 ± 70	547	Age of terrace by Friendship Bridge, south edge, Clark Air Base.
WW-6505*	14°59.9'	120°15.6'	Charcoal	Lahar deposits	Marella	600 ± 60	617	Covered by 1992 lahars.
WW-29*	15°08.2'	120°28.7'	In situ root	Fluvial deposits	Pasig-Potrero	630 ± 70	597	After latest pre-1991 downcutting; before major siltation.
Ebasco*	15°03.3'	120°17.0'	Wood	Lahar deposits	Marella	635 ± 80	593	Lahar just below toe of morphologically youngest pre-1991 pyroclastic-flow deposit.
WW-28*	15°18.5'	120°35.7'	Wood	Flood-plain deposit	Bamban	660 ± 80	648	Sedimentation across broad Bamban-Capas flood plain.
WW-107*	15°16.03'	120°05.3'	Charcoal	Hyper-concentrated flow deposit	Bucao	730 ± 80	665	Latest major sedimentation in Bucao; over-topped in 1993.
WW-6504*	14°59.9'	120°16.6'	Charcoal	Fluvial deposits	Marella	760 ± 60	673	Covered by 1993 lahars.
WW-31*	15°08'	120°29'	Wood	Lahar deposits	Pasig-Potrero	950 ± 70	800	

(continued)

Table 11.1 (Continued)

Lab. Number	Latitude (°N)	Longitude (°E)	Material	Occurrence	Drainage	C¹⁴ Age (BP)	Calibrated Age (BP)	Significance
WW-4686**	14°52.008'	120°32.090'	Wood	Flood-plain deposit	Guagua-Pasac	1730 ± 40	1670	Reworked? Wood fragments in Buag age deposit or undocumented pre-Buag event.
WW-4685**	14°52.008'	120°32.090'	Wood	Flood-plain deposit	Guagua-Pasac	1800 ± 40	1802	Reworked? Wood fragments in Buag age deposit or undocumented pre-Buag event.

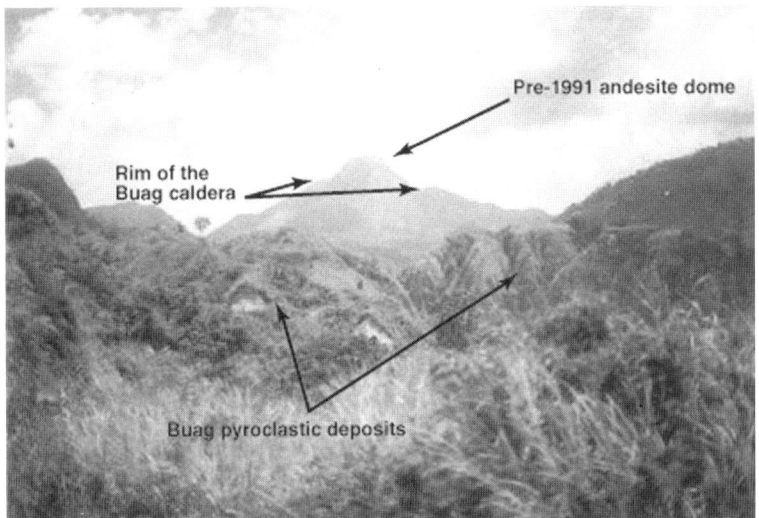

Figure 11.2 View of the east side of Mt Pinatubo in 1990 (photograph by Luminoso H. Dizon).

Lahars of the Buag eruptive period were no exception. They were mostly confined within the kilometre-wide channel banks of major rivers. At the Marella-Mapanuepe river confluence, lahars were built up slightly higher than the 120 m elevation, suggesting that a lake similar in size and shape to the present-day Mapanuepe Lake was formed; however, the lake had ceased to exist even before the Spaniards explored Zambales in the 16th century.

The lahars created during and after the 1991 eruption have had very similar patterns and channel routes as those of the Buag period; however, they rapidly filled channels and overflowed to inundate adjacent alluvial plains that the Buag lahars had not been affected (Umbal 1997). This may be because the repose interval between the 1991 and last Buag eruption was too short to allow the drainages to fully adjust to the influx of Buag sediment before the 1991 debris arrived.

MAJOR ENVIRONMENTAL CONSEQUENCES

The huge lahar events of the prehistoric eruptions of Mt Pinatubo shaped the surrounding environment in major ways. At the mouths of the Pampanga and Santo Tomas rivers, respectively east and west of the volcano, lahars infilled shallow coastal bays. The process has been reconstructed by identifying old beach ridges that appear as lighter toned linear features in both aerial photographs and in 1991 synthetic aperture radar images of the coastal areas (Figure 11.3). To verify and

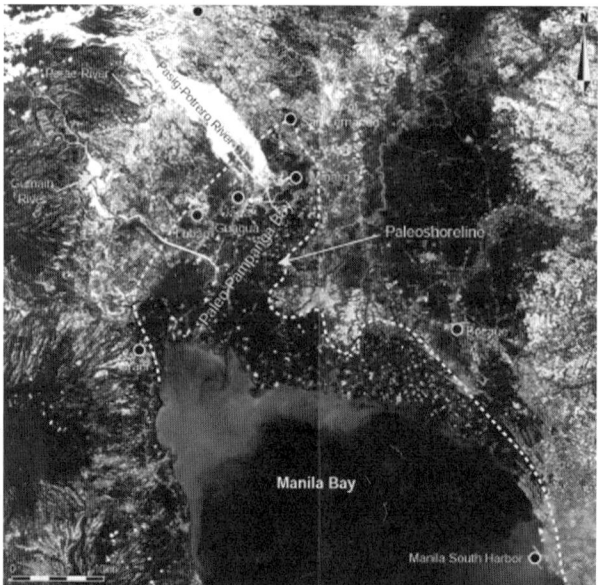

Figure 11.3   Landsat TM satellite image (19 February 1995) of the southern part of the Central Luzon Basin with the tentative alignment of the paleo-shoreline on the northern edge of the Manila Bay. The paleo-shoreline was delineated after the current limit of the wetlands and core data.

evaluate the old environments on the ground, sediment cores up to 10 m long were acquired in the Pampanga delta plain and logged according to differences in colour, grain size, gross composition, and sedimentary structures (Figure 11.4).

Filling of Pampanga Bay

Beach ridges along the north-east coast of Manila Bay diverge north-westward. In the Bulacan River area, the oldest and most landward ridge is situated 7 km north of the present shoreline. Exactly when these paleo-shoreline positions were occupied has not yet been established.

An area of roughly 360 km² has supposedly been reclaimed from the sea. Given an average sea depth of 3–4 m, this suggests a sediment volume of 1.1–1.4 km³ is required to fill the paleo–Pampanga Bay. The volume of pyroclastic materials deposited in the watersheds of the Pasig-Potrero, Abacan, Sacobia, and Bamban rivers that empty into Manila Bay during the 1991 eruption has been estimated at 1.0–1.6 km³ by Pierson *et al* (1992). These data indicate that a 1991 size–like eruption – like the Buag event – may be sufficient to fill the paleo–Pampanga Bay.

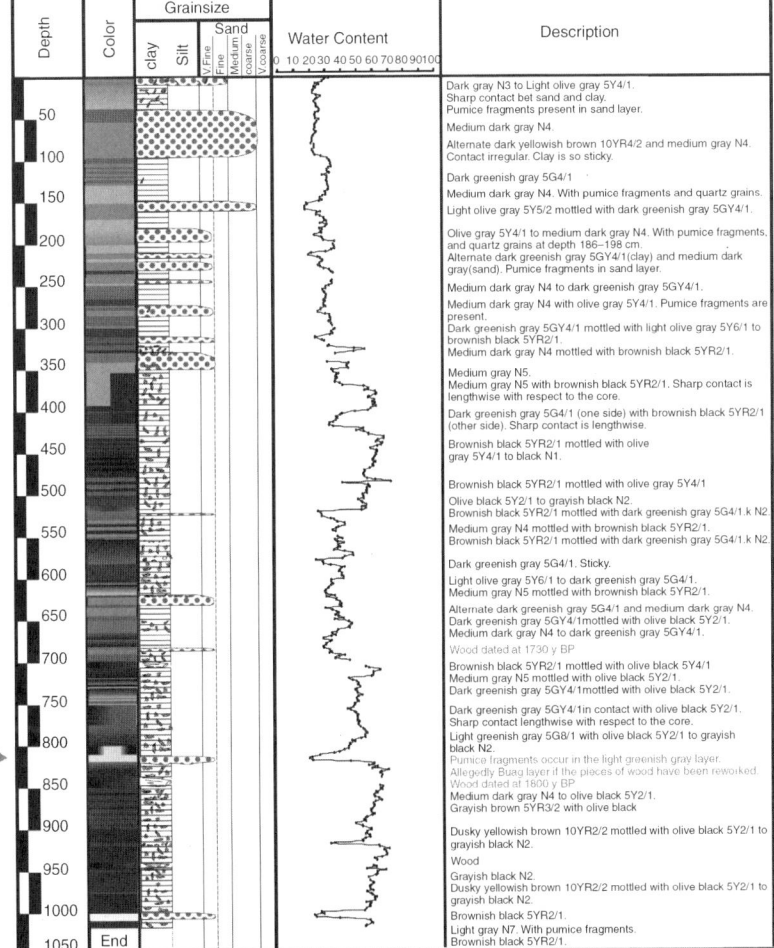

Location: 14°52.008' N, 120°32.090' E
San Rafael Baruya, Lubao, Pampanga
Total Length: 10.20 m
Scale: 1.50 cm

Core ID: PAM 4
Described by: JL A. Soria
Date cored: 29 August 2002
Date logged: 27 September 2002

Figure 11.4   Log of the 10 m push core in San Rafael Baruya, Lubao, Pampanga.

The San Rafael core data suggest that the shoreline once followed the general present-day landward limit of the wetlands (Figure 11.1). Either around 1.73–1.8 ka or soon after the start of the Buag event, the ancient Pampanga Bay may have begun to be filled, thereby forming this low-lying and virtually flat part of the delta plain and cause the shoreline to regress. This may be reflected in the toponym of the town of Guagua around 15 km inland from the current shoreline. Guagua is a corruption of '*Uaua*', meaning 'river mouth', indicating a former coastal location.

Moreover, the name of the town of Lubao, also 12 km from the shore, refers to a floating community.

Evolution of the Zambales Coastal Plain

Siringan and Ringor (1996, 1997) used synthetic-aperture radar satellite images and aerial photographs to trace beach ridges along the southern coastal plain of Zambales. At the mouth of the Pamatawan River, which has no delta at present, they reconstructed a paleo-delta that could have existed approximately 1 km seawards of the present shoreline. They hypothesised that this paleo-delta was formed when Buag eruptions enhanced the Pamatawan sediment load. The return of the sediment load of the Pamatawan River to pre-Buag eruption levels may have led to its eventual erosion. Presently, the Pamatawan River has no delta because dikes now constrain Pinatubo sediment within the Santo Tomas River channel.

Similar phenomena may have occurred at the mouth of the Bucao River. According to anecdotal accounts, the Bucao River mouth was formally an estuary in which barges or galleons docked during the Spanish era and up to the early 1900s. Even by 1680, the writings of the Spanish friar Domingo Pérez indicate that relatively recent major lahars had left unconsolidated, still unvegetated debris on the Bucao plain:

> sandy ground which is very large and full of rocks left by the river which flows from the mountain of Pinatuba [sic]; and in those places where there are no rocks, but only the sand, the road is also very wearisome because the sand has no cohesion, and the least wind that blows lifts the dust. (Blair and Robertson 1903–09, vol 47: 296)

SOCIOECONOMIC CONSEQUENCES

Such large eruptions and their environmental consequences must have seriously disturbed the socioeconomic milieu. Archaeological excavations in both Pampanga and Zambales provinces, respectively on the eastern and western slopes of Mt Pinatubo, have unearthed evidence that prehistoric economic activities and settlement patterns were dislocated following the Buag eruptions.

Evidence of pre-Buag human habitation was uncovered in Babo Balukbuk, Porac, Pampanga, on the western bank of the Porac River (Figure 11.1). This area has been the focus of archaeological research since the 1930s (Beyer 1947; Fox 1960a, 1960b). Recent excavations of a 32 × 40 m site called Dizon-1 unearthed 22,721 artefacts, including locally made earthenware potteries, Chinese and Vietnamese ceramic shards, bronze rings, and iron implements (de la Torre 1999; Dizon 2002; Paz 2003).

A stratigraphic sequence of at least four cultural horizons has been distinguished at Babo Balukbuk (Figure 11.5). Pottery recovered from a 4 m deep trench documents human presence before 2.3 ka, but it is unclear if this habitation was a settlement. The Maraunot eruptive period, starting around 2.3 ka, covered the area with pyroclastic-flow deposits more than 3 m thick on which, some time during the 13th century, a settlement was established by people who grew rice, built sturdy structures, used metal implements, traded for earthenware vessels, and may have also made their own pottery and stoneware. They wore glass beads and metal bangles and they buried their dead with tradeware ceramics near their houses.

It is not known if this site was abandoned because of early Buag eruptions, but the absence of volcanic deposits on the top of this cultural layer does not exclude the possibility that the people fled from volcanic activity. Later, habitation and burial areas were replaced by ploughed fields for rice agriculture that disrupted the surface on top of 13th–14th-century burials. This indicates a discontinuity of settlement and the arrival around 200 years later of a different group of rice farmers who also interred their dead with tradeware ceramics among their houses. Another possible explanation is that the time lapse of over a century was sufficient for a community to have lost lineage ties with or forgotten the existence of the previous community.

The last Buag eruption to leave deposits at Babo Balukbuk covered the fields with airborne tephra (<1 m thick deposit), which buried pottery vessels and other artefacts lying on the land surface. Recent radiocarbon dating of charcoal samples recovered from inside a buried hearth yielded mean calibrated ages of 506 (WW-4683) and 486 BP (WW-4684) (Table 11.1 and Figure 11.1). These dates are consistent with the age of the latest tradeware ceramics recovered from this area (early 16th century) and the age from latest occupation of the Buag site on the western flank of the volcano. This area was not recultivated until the early 20th century. The 1991 Pinatubo eruption left a new layer of light-coloured sandy tephra, around 10 cm thick, but the land continues to be dedicated to agriculture, mostly for sugarcane and cassava.

Inland, near Mt Bagang on the north bank of the upper Marella River (Figure 11.1), systematic archaeological surveys and excavations were conducted in 1990 in the San Marcelino settlements of Buag, Manggahan, and Kakilingan, around 10 km south-west of the pre-1991 Pinatubo summit (Dizon 1990). The pre-1991 stratigraphy of the sites provides the following historical reconstruction (Figure 11.6).

People who used earthenware and stoneware pottery established a settlement, marked by two successive cultural layers, on several metres of sterile ash and pumice deposits from a probable pre-Buag lahar or

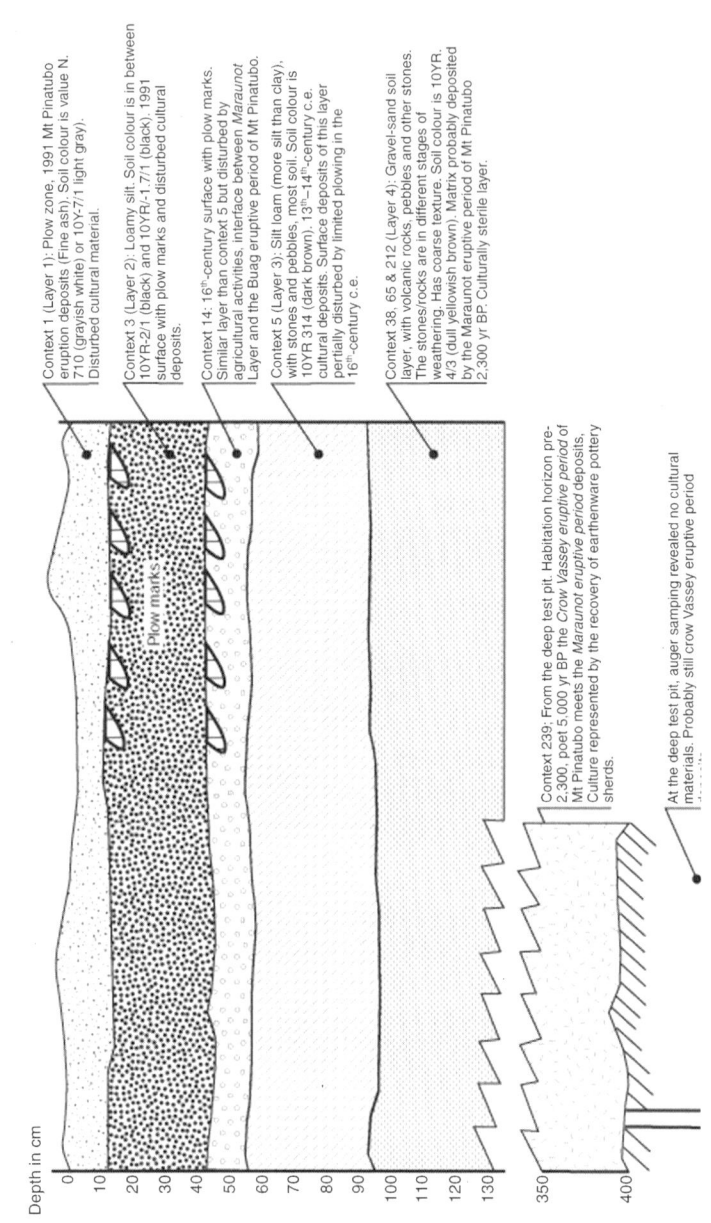

Figure 11.5 Composite stratigraphic profile of the Balukbuk archaeological site in Porac, Pampanga.

Depth in cm

Context 1 (Layer 1): Plow zone, 1991 Mt Pinatubo eruption deposits (Fine ash). Soil colour is value N. 710 (grayish white) or 10Y-7/1 light gray). Disturbed cultural material.

Context 3 (Layer 2): Loamy silt. Soil colour is in between 10YR-2/1 (black) and 10YR/-1.7/1 (black). 1991 surface with plow marks and disturbed cultural deposits.

Context 14: 16th-century surface with plow marks. Similar layer than context 5 but disturbed by agricultural activities, interface between Maraunot Layer and the Buag eruptive period of Mt Pinatubo.

Context 5 (Layer 3): Silt loam (more silt than clay), with stones and pebbles, most soil. Soil colour is 10YR 314 (dark brown). 13th–14th-century c.e. cultural deposits. Surface deposits of this layer partially disturbed by limited plowing in the 16th-century c.e.

Context 38, 65 & 212 (Layer 4): Gravel-sand soil layer, with volcanic rocks, pebbles and other stones. The stones/rocks are in different stages of weathering. Has coarse texture. Soil colour is 10YR 4/3 (dull yellowish brown). Matrix probably deposited by the Maraunot eruptive period of Mt Pinatubo 2,300 yr BP. Culturally sterile layer.

Context 239: From the deep test pit. Habitation horizon pre-2,300, poet 5,000 yr BP the Crow Vassey eruptive period of Mt Pinatubo meets the Maraunot eruptive period deposits. Culture represented by the recovery of earthenware pottery sherds.

At the deep test pit, auger sampling revealed no cultural materials. Probably still crow Vassey eruptive period

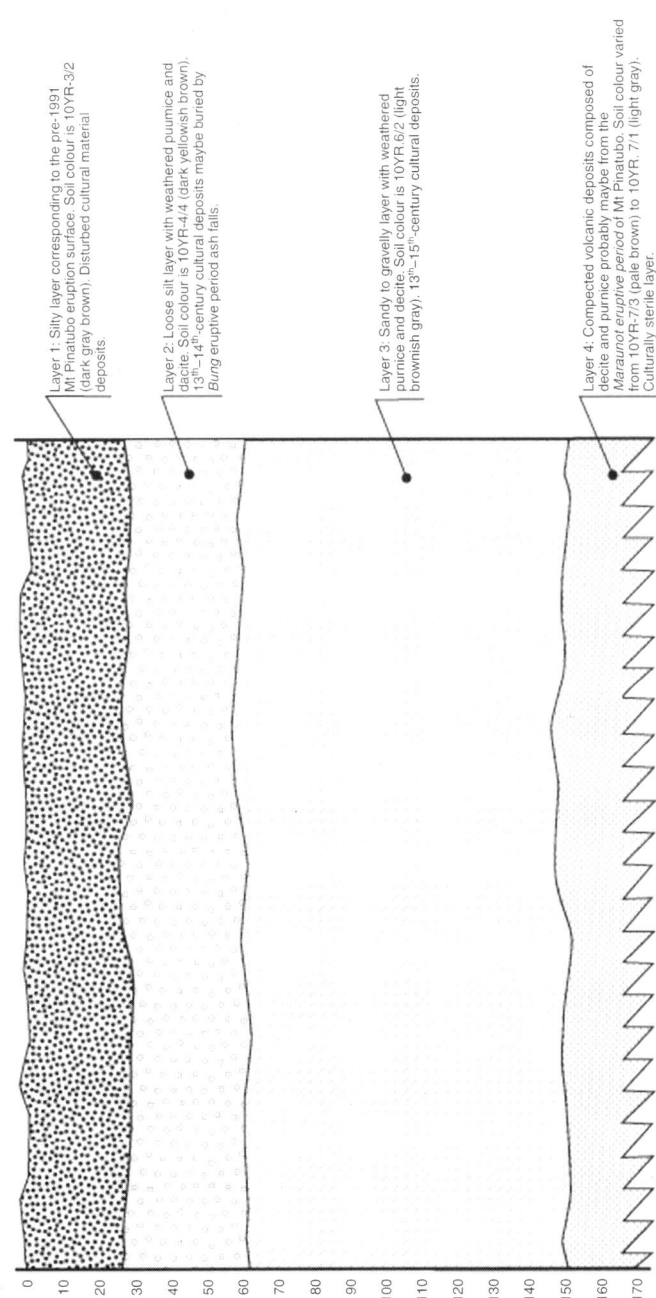

Depth in cm

Layer 1: Silty layer corresponding to the pre-1991 Mt Pinatubo eruption surface. Soil colour is 10YR-3/2 (dark gray brown). Disturbed cultural material deposits.

Layer 2: Loose silt layer with weathered puumice and dacite. Soil colour is 10YR-4/4 (dark yellowish brown). 13th–14th-century cultural deposits maybe buried by *Bung* eruptive period ash falls.

Layer 3: Sandy to gravelly layer with weathered purnice and decite. Soil colour is 10YR-6/2 (light brownish gray). 13th–15th-century cultural deposits.

Layer 4: Compacted volcanic deposits composed of decite and pumice probably maybe from the *Maraunot eruptive period* of Mt Pinatubo. Soil colour varied from 10YR-7/3 (pale brown) to 10YR-7/1 (light gray). Culturally sterile layer.

Figure 11.6 Composite stratigraphic profile of the Buag archaeological site in San Marcelino, Zambales.

pyroclastic flow (Newhall et al 1996). Chinese and other Southeast Asian ceramic tradeware show that this site was occupied at least from 1250–1450 AD. Smelting slag and metal implements demonstrate that the people worked metals, and rice husks within a whole pot indicate that they consumed rice and may have tilled it themselves. Congruence of geological and archaeological records around the mid-15th century suggests that this settlement was abandoned during the last Buag eruption. The ashy, pumice-bearing soils that contained the artefacts indicate that this settlement was lightly buried by a few decimetres of tephra. In 1990, these two cultural layers were overlain by a contemporary ploughed agricultural layer of dark brown loam soils containing remains of root crops. The 1991 Mt Pinatubo eruption added 30 cm of tephra on the top of this layer.

Chinese and Southeast Asian trade ceramics from excavations on both the eastern and western slopes of the volcano indicate that from the 13th to some time between the late 15th and early 16th centuries AD, both the Buag and Babo Balukbuk sites were occupied; Buag would have been easy to reach from the San Marcelino coast. At the time, the interior areas of central Zambales may have been accessible by small boats following river systems from the coastline between San Narciso and San Antonio. Upstream, this would have been eased by the possible existence of a lake along the Mapanuepe River, as suggested by the geological records. Chinese Ming Dynasty annals mention different place names that may have been located near Mt Pinatubo. According to Wang (1964), Li-Gun-Tiong-Pang may have been a trading post situated near the present town of San Narciso. Ming annals also cite a 'very high mountain' called Piao San or Pio San in the Zambales Mountains that may have been Mt Pinatubo. No archaeological remains have yet been excavated from the paleo-shores of the Pampanga Bay, but this does not exclude the possibility that this area was another trading port for Southeast Asian traders in prehistoric times, where artefacts that eventually reached Porac landed.

The last Buag eruptions appear to have disrupted these commercial activities, burying or forcing abandonment of villages and trading markets such as the Buag and Babo Balukbuk settlements. Possible infilling of the Santo Tomas and Pampanga estuaries also would have eliminated the inland anchorages of merchant ships. Moreover, Mapanuepe Lake, which previously may have facilitated the upstream incursion of boats, disappeared following the Buag eruption, as suggested by the historical record (Wang 1964) and the lack of mention of the lake in the early Spanish accounts. New pyroclastic deposits on the flanks of the volcano would have disrupted travel and trade for years.

POPULATION REDISTRIBUTION

The population occupying the region around Mt Pinatubo at the time of the Buag eruption probably differed greatly from that encountered by the Spanish conquistadors led by Martín de Goiti in 1571. This may have been particularly true on the eastern side of the volcano. The Spaniards described many densely populated riverbanks in the Pampanga River delta, within a triangle extending from Macabebe and Lubao to Betis or Vitis, now part of Guagua (Blair and Robertson 1903–09; de San Agustin 1998). Some fortified settlements were quite large, notably Betis (7,000) and Macabebe (2,600) (Jocano 1975). If this area had been reclaimed by pre-Buag or Buag lahars, these settlements must have been established well before the Spaniards arrived but after the last Buag eruptions that affected the south-east sector of the volcano.

The Buag eruptions must have drastically redistributed any population southwards away from the volcano. Indeed, Spanish chroniclers reported at the end of the 16th century that populations near Mt Pinatubo were very sparse. Some local place names in this upper region, however, may reflect oral memory of the Buag eruption. The three main lahar channels of the east side of the volcano are the Pasig-Potrero, Abacan, and Sacobia rivers, respectively in the municipalities of Bacolor and Mexico in Pampanga and Concepción in Tarlac. Each of these towns has villages named Balas, which means 'sand' in Kapampangan. This might suggest that these communities had been affected by lahar onslaughts. The name of Sapang Bato village in Angeles City literally means 'the river of stones', possibly referring to the coarse material transported down the Abacan River by lahars.

A somewhat similar, but more eastern, redistribution of the population also occurred after the 1991 eruption, because southern Pampanga experienced heavy flooding. The municipalities of the eastern half of Pampanga have indeed displayed the highest rates of population increase between 1990 and 2000 (Gaillard 2002). If the lahar threat may have guided the population towards safer grounds, the resettlement policy of the government definitely played a great role in this redistribution.

South-west of Mt Pinatubo, the plain of the Santo Tomas River is one of the few large agriculture flatlands in Zambales province. The towns of San Antonio, San Narciso, San Marcelino, San Felipe, Casillejos on the plain, together with most of their villages, have Spanish names. This is because Castillejos, the oldest town was settled by Tagalogs from Subic Bay in 1743, and those bearing the names of saints were settled by Ilokanos from the north who followed the coast of the South China Sea to migrate there in the mid-19th century, more than two centuries into the Spanish colonial period (Apostol 1956; De Jesus 1990).

One reason why the plain remained unsettled for so long is because the Sambals along the coast were fierce, and the Aytas of the surrounding mountains were 'the cruelest of that scattered nation'; in 1648, they killed Fray Pedro de Valenzuela and took his head (Pérez 1680 in Blair and Robertson 1903–09, vol 37: 170). Even an 18th-century history (De Santa Teresa 1743 in Blair and Robertson 1903–09, vol 36: 174) refers to the 'great risk' posed by Aytas in revolt.

The eruptive history may also have played a role in the settlement of the Santo Tomas River region. First, the prehistoric Santo Tomas lahars were destructive enough to bury or discourage all human settlements. Second, the Santo Tomas plain formerly was a bay of the South China Sea that was filled up by Buag lahars not too long before the Spaniards arrived. The absence of settlements on the Santo Tomas plain soon after the Buag eruptive period is consistent with the 17th-century Spanish chronicles, which state that the people of this area lived only in the mountains and the Spaniards had a hard time making them resettle in the lowlands (Scott 1986). This would also indicate that on the western side of Mt Pinatubo, unlike on the eastern side, the Buag eruptions had little effect on the population, which remained concentrated on the volcano flanks.

ORAL TRADITIONS

During the Buag eruption period, the slopes and surroundings of Mt Pinatubo were probably inhabited by three major ethnolinguistic groups. On the east side, the Kapampangans occupied the southwestern part of the Luzon central plain. Early Spanish accounts suggest that the narrow coastal plain of Zambales, and possibly also inland Pampanga were inhabited by Sambals (Blair and Robertson 1903–09), which are nowadays divided into Sambal Tina and Sambal Botolan. Finally, the Aytas inhabited the east and west flanks of the volcano. These three groups preserve traces of the prehistoric eruption in their local legends.

The toponym Pinatubo itself, literally 'made to grow' in the Sambal language, implies that people may have witnessed the growth of the dome that followed the eruption. The Aytas oral tradition best describes what could have been a prehistoric eruption (Rodríguez 1918). In this legend, two supernatural beings called Blit and Aglao fight a mischievous turtle. The most realistic part describes the eruption as follows:

> Finding the lake a useless place of refuge, he climbed the Mount Pinatubo in exactly twenty-one tremendous leaps. When he had reached the top, he at once began to dig a big hole into the mountain. Big pieces of rock, mud, dust, and other things began to fall in showers all around the mountain. During all the while, he howled and howled so loudly

that the earth shook under the foot of Blit, Aglao, and his hosts. The fire that escaped from his mouth became so thick and so hot that the pursuing party had to run. (Rodríguez 1918)

This description evokes all the destructive, frightening phenomena experienced during the 1991 eruption: falling rock debris, mud, dusty tephra, noise, and sustained earthquakes. Once the eruption stopped, a great hole is said to have replaced the former summit of the mountain and a lake with clear water had been filled with rocks and mud. The geological evidence indicate that this lake may have been located along the Mapanuepe River, an idea reinforced by the toponym Aglao of a village that was drowned in 1991 when lahars descending the volcano along the Marella River dammed the Mapanuepe River to form a lake.

A few other authors have also referred to behaviour that could be viewed as traces of oral memory among the Ayta people. Speaking about hunting practices, John Garvan (1963: 72) recalled that jesting remarks about the quarry irritate the spirits of Mt 'Pinyatubu' and may cause petrification. Rodolfo (1995: 88) reported that Ayta parents would warn children who will not go to sleep by saying '*Apo Namalyari*' will wake up and start throwing stones if you don't behave'.

Among the Kapampangan community that occupies the eastern foothills of the volcano, blurred traces of a prehistoric eruption appear in a couple of legends, transmitted from one generation to another. In all the versions of the different legends, Mt Pinatubo and Mt Arayat, sometimes referred to as giants or brothers, fight by throwing boulders at each other. One of the missiles supposedly decapitates Mt Arayat, forever destroying its perfect conical shape. The war lasted for a few days before calm is restored for several centuries (Castillo 1918; David 1918; Galang 1940; Hilario-Lacson 1984; Nicdao 1918a, 1918b; Quizon 1918). The boulders spewed out by Mt Pinatubo could refer to volcanic blocks. It is also noteworthy that the Kapampangan people named Mt Pinatubo '*Apung Punsalan*', literally the 'lord of enmity', possibly referring to past and destructive eruptions of the volcano.

More puzzling are the words '*silakbu*' and '*margaha*' in the lexicon of the village of Cabalantian, in the Pampanga municipality of Bacolor. According to Forman (1971), '*silakbu*' stands for 'volcanic eruption', whereas '*martgaha*' means 'volcanic ash' and in rare cases 'lava'. This last term is not Kapampangan because there is no letter 'H' in this language. '*Margaha*', however, is a word used by the Bukidnon indigenous group that inhabits the flanks of Mt Kanlaon, an active volcano on Negros island in the central Philippines, where it designates the old crater. It is difficult to understand how the word reached the Mt Pinatubo area. Spanish friar D Bergaño (1732) further provides a translation for the word 'volcano' or '*sibul*' (which also means 'source' in the Kapampangan language).

The Sambal Tina ethnolinguistic group may preserve a memory of the eruption in its local folklore. A local story recorded by Santos (1979: 258) says that a mountain rose from the place where a magician planted a stone as if it was a seed. But 'the magician neglected to check the rapid growth of the Pinatubo (planted) mountain that grew to such proportion that it destroyed the regions surrounding it. The whole plain was soon covered with rocky outcroppings.'

THREE PHASE SYNTHESIS

Bringing the geological, geomorphic, archaeological, and oral records together, we suggest the Buag eruptive period of Mt Pinatubo can be divided into three phases, as synthesised in Table 11.2. First, an initial Plinian eruption generated widespread tephra falls and accompanying pyroclastic flows and lahars close to the volcano; next came the extrusion and growth of a dome; and, finally, the collapse of part of the growing dome generated the youngest, distinctive ash flow sheet filling the Marella valley. The lahars of the second phase gradually became the more normal, rain-induced sedimentation that continued through and beyond the third phase, slowly filling in the shallow bays in Pampanga and Zambales. On a geologic time scale, all three phases can be treated as contemporaneous, but the opening Plinian eruption may have preceded the second and third phases by years, decades, or even centuries.

Had the main caldera-forming Buag eruptions happened only at 500 BP, it is surprising that the Spaniards, who arrived in Pampanga and conquered it only a few generations later in 1571, did not hear about such cataclysmic events from inhabitants of the many large, well-settled towns there. Either the eruptions occurred so long ago that they were lost to memory or the people the Spaniards encountered had settled the delta region and Manila after the major eruptions. People in the Papua New Guinea highlands have a strong oral tradition about the Long Island eruption that occurred more than three centuries ago (Blong 1982), and inhabitants of Mt Parker in Mindanao have a sketchy oral history of its 1641 eruption (Delfin et al 1997). This reinforces the radiocarbon dates that indicate main plinian eruptions of the Buag period occurred significantly earlier than the ca 500 BP pyroclastic deposits at the Buag-type locality.

The best temporally constrained event is in the south-west sector of Mt Pinatubo. Here the emplacement of the 'Lumboy ash flow sheet' apparently forced settlers to abandon the lower Marella valley and provided the sediment for the final filling of the Pamatawan and Santo Tomas river valleys. At least two dates from the Marella valley, one from the Buag archaeological site and one farther downstream, suggest calibrated calendar dates ranging from 476–545 cal BP (WW-111 and W-6509) (Table 11.1 and Figure 11.1). This would put the event some-time in the early-to-mid-15th century, which is consistent with the ages of Chinese and Anamese ceramics found at Buag. The destruction of the old Buag settlement following the partial collapse of the dome may also be retained in the toponym Buag, which means 'to collapse over' or 'to pull over' in local languages.

If the final aggradation above sea level of the pre-1991 Santo Tomas and Pamatawan river valleys were consequences of this event, then it is indeed likely that around 500 BP, a shallow embayment of the South China Sea separated what are now the towns of San Felipe in the north and San Antonio and Castillejos in the south. Such an embayment would have allowed Chinese or Anamese trading junks to penetrate further eastwards, closer to the Pinatubo foothills. This is supported by reference to trading in this area within Chinese Ming Dynasty annals. Old lahar deposits at the confluence of the Marella and Mapanuepe rivers, together with the Ayta myth on the origin of Mt Pinatubo, sug-gest that a lake occupied the present location of the Mapanuepe Lake before the 500 BP eruption, which no longer existed when the Spaniards arrived in the late 16th century.

The early Plinian phase of the Buag period that showered tephra over a wide area and formed a caldera is well retained in both oral his-tory and toponymy. The best account is the Ayta legend about how the supernatural spirits Blit and Aglao fought a turtle that scaled the sum-mit, dug a hole, howled loudly and breathed fire, shaking the earth. In Bacolor, Mexico, and Concepción, villages named Balas, which means 'sand' in Kapampangan, indicate memories of lahars or pyroclastic flows. That the Porac river valley was not inundated by a major pyro-clastic flow during is consistent with the tephra fall recorded in the Babo Balukbuk archaeological site. Despite the uncertainty regarding the wood samples recently dated at 1670–1802 BP (samples WW-4685 and WW-4686) (Table 11.1 and Figure 11.1), the Lubao-Bacolor area very likely was a shallow northward extension of Manila Bay until filled in with pre-Buag or early Buag-period sediment. This is further indicated by the toponym Uaua, meaning 'river mouth', 15 km inland from the current shore.

Finally, the Buag eruptions would have triggered a general redistri-bution of the population from the foothills of the volcano towards the delta plains if this had been reclaimed downstream by lahar depos-ition. This is underlined by the toponymic records on the eastern side of Mt Pinatubo, and the early Spanish chronicles in the late 16th century. Oral accounts also confirm that the Pinatubo environs at the time of the Buag eruptive period were predominantly inhabited by three ethnolin-guistic groups, the Aytas, Kapampangans, and Sambals.

Table 11.2 Compilation of geological, geomorphic, archaeological, and oral records of the 2,300 y BP Maraunot and 500 y BP Buag eruptions of Mt Pinatubo

Eruptive Period	Mt Pinatubo Geology	Sedimentological-Geomorphic Records	Lahar Records	Archaeological-Cultural Evidence	Geographic Toponymy and Oral and Historical Records
Buag (~800–500 y BP)	Dome-collapse generated 'Lumboy' ash flow sheet' in Marella River valley.	Sediment loading of Pamatawan River and consequent filling of Pamatawan River paleo-delta (San Marcelino–Castillejos flood plain).	C^{14} dating indicates that lahar deposits comprising the upper layers of the Pamatawan Plain are from the 2,500-year-old Maraunot eruption period; the in-channel terraces along the Santo Tomas River date back to the Buag eruptive period.	Buag site reveals Chinese ceramics from mid-Ming Dynasty, and Anamese ceramics from 14th–16th century.	Several alleged mentions in the Ming Annals of trading posts and other land marks along the shore of the Santo Tomas Bay or in the Zambales Mountains. Village named 'Buag', meaning 'to pile over', buried by the collapse of the dome. Towns with Spanish names in the Santo Tomas valley were settled only in the 18th and 19th centuries. Town named Uaua ('mouth of the river') 15 km inland indicates seaward displacement of the shoreline following the Buag eruptions.

	Extrusion and growth of Pinatubo dome, probably filling or destroying the summit caldera.			'Pinatubo' name suggests that dome growth was witnessed by Ayta people.
	Caldera-forming Plinian eruption with related pyroclastic flows and lahars in major watersheds (except Gumain and Porac).	500-year-old lake deposits indicate the occurrence of a lake upstream of Mapanuepe River during the Buag-period eruptions. Absence of a lake before the 1991 eruption and from historical records suggest that it was filled or drained before the Spanish explorations. 2 cm light olive grey fine-grained ash layer of marine environment; 8 m from surface in Baruya, Lubao. Wood samples from immediately below and above the ashy layer dated at 1,800 and 1,730 y BP.	Babo Balukubuk site (Porac) with Chinese and Anamese ceramic shards suggestive of 13th–16th-century human occupation. Charcoal samples recovered inside a buried hearth yielded ages dated at 506 and 486 y BP.	Ayta legend of supernatural spirits Blit and Aglao fighting a turtle that scaled the summit, dug a hole, howled loudly, and, breathing with fire, shook the earth. Villages named Balas (sand in Kapampangan) in Bacolor, Mexico, and Concepción suggest memory of lahars or pyroclastic-affected lands.
Maraunot (3,900?–2,300 y BP)	Pyroclastic flows, lahars, and streamflows.	Areas of the highest relief on the plain and alluvial fan of the Pasig-Potrero and Santo Tomas rivers are underlain by Inararo deposits. C^{14} dates indicating that the uppermost layers of the Pamatawan Plain comprise 2,500-year-old Maraunot-period lahars. Possible alignment of the Marella-Santo Tomas with the Pamatawan River.	Babo Balukubuk site (Porac) revealed pottery in layer of Maraunot eruptive period.	

(continued)

Table 11.2 (Continued)

Eruptive Period	Mt Pinatubo Geology	Sedimentological-Geomorphic Records	Lahar Records	Archaeological-Cultural Evidence	Geographic Toponymy and Oral and Historical Records
	Widespread distribution of pyroclastic-flow and lahar deposits.	Pre-1991 eruption channels of most of the major drainages and final configuration of their watershed areas were probably developed during the waning phase of lahar activity.	Rapid aggradation at the confluence of the Marella-Mapanuepe rivers formed a lahar-dammed lake that probably experienced repeated damming and lake-breakouts.		

IMPLICATIONS FOR MT PINATUBO ERUPTIONS

This study of the Buag eruptive period is useful for anticipating further consequences of the 1991 and future eruptions. The 500-year gap between the Buag eruptive period and the 1991 eruption is one of the shortest in the history of Mt Pinatubo. Repose periods have tended to decrease with time, according to Newhall *et al* (1996), who concluded that the 1991 eruption was the biggest that could be expected from the magma body, but the same authors, however, warned that moderate-size explosive events are still possible. Indeed, if Mt Pinatubo's eruptive style is truly cyclical and always ends when the caldera left by the initial Plinian eruptions is entirely filled by a dome, then the eruptive period that commenced so spectacularly in 1991 is not finished, so we cannot rule out the possibility that dome-building eruptions are yet to come in the coming decades or centuries.

Our study identifies environmental changes that could occur as a consequence of the 1991 and future eruptions. All deltas, including those of the Pampanga and Santo Tomas rivers, subside naturally from isostatic sinking. Due to increasing use of groundwater by growing populations in the Pampanga delta, the rate of subsidence has been increasing in recent decades (Siringan and Rodolfo 2003). However, since the 1991 eruption, lahars have tended to fill these flood plains, temporarily counteracting the subsidence. Since 1991, dikes have been manufactured for all the lahar pathways on Mt Pinatubo to channel volcanic sediments away from the most densely populated areas. These measures have prevented the natural filling of the deltas, thus preventing a natural counteraction to subsidence. The study of the Buag eruptions has shown, however, that volcaniclastic sediment accumulation in shallow coastal environments has reclaimed large tracts of land from the sea. Knowing which areas were covered with lahar deposits of the pre-Buag and Buag eruptive periods may, therefore, help predict the extent of landform transformations that may still take place because of the 1991 and future eruptions. It may also help orient resettlement of flood victims or threatened communities towards safer grounds. Archaeological excavations in Porac have also confirmed the attachment of affected communities to their land, and this should be a major factor to consider in disaster management planning for future eruptions.

Finally, despite the presence of clues in the local cultures, the oral memory of the Buag eruptions had no visible impact in terms of disaster management during the 1991 eruption and subsequent lahar crises. It is, however, very important that the 1991 eruption experience be remembered by the people so that they can prepare properly for the next Mt Pinatubo eruption. Indeed, it has been clearly demonstrated that prior experience of the victims is amongst the most important factors influencing how they cope with disaster, especially in the case of Mt Pinatubo lahars.

CONCLUSIONS AND OUTLOOK

This multidisciplinary study has compared geological and archaeo-logical data with oral tradition records of the last prehistoric eruption of Mt Pinatubo. The results are important for future risk management because the Buag episode is a good model for future volcanic activity at Pinatubo, even if the magnitude of the eruptions appears to be less-ening through time. Cultural artefacts such as tradeware ceramics from both San Marcelino and Porac are consistent with dates of ca 500 ± 50 BP for the last eruption of the Buag period. Because the Chinese were trading in the Mt Pinatubo area, more precise dating may be found through mention in the historical records. The socioeco-nomic environment at the time of the Buag eruption period still requires detailed study, particularly in terms of the nature and form of connections with China and other Southeast Asian kingdoms.

The present study suggests that victims of volcanic eruptions in ancient times may turn to different alternatives to cope with changes in their immediate environment. Hints from the archaeological and geographical records as well as clues from early written accounts show that people from the eastern and western flanks of Mt Pinatubo did not exploit the creation of large tracts of land in the coastal low-lands. It appears there was a massive redistribution of the population within the plain east of the volcano, whereas the western side of the volcano may have undergone less change. The reasons for such differ-entiated coping strategies may be diverse and rooted in a large spec-trum of factors, as noted in both ancient (Shimoyama 2002) and contemporary cases (Gaillard forthcoming). Our data set on the Buag event is still very sketchy, but we propose that human responses were shaped by the magnitude of the event, the socioeconomic context at the time of the disaster, and the ethnicity of the place.

Our study of the last prehistoric eruptions of Mt Pinatubo should be useful for disaster managers assessing the environmental and socio-economic consequences of the 1991 and later eruptions. Clear lessons for the future can be learned. Pyroclastic flows may inundate all the river basins draining the volcano. The Mapanuepe River may be dammed and form a lake. Massive lahar flows and subsequent enhanced fluvial sedimentation will infill of shallow coastal embayments and extended the deltaic plains. As these factors are now shown to be typical of volcanic activity from Mt Pinatubo, they require more attention from disaster managers and land-use planners.

The area around Mt Pinatubo was already experiencing extensive trade with China and other foreign kingdoms 500 years ago, and is now one of the leading economic regions of the Philippines. However, the current development plans for the region only take into account the reoccurrence of lahars and ignore potential dome-building leading to further explosive eruptions. The evidence we present indicates that the last Buag eruption may have occurred several centuries after the major plinian eruptions. A dome-building eruption of the magnitudes of the ca 500 BP event at Buag would have serious consequences for modern residents and the Philippines economy.

Our study of the Buag episode shows that the current phase of volcanic activity at Pinatubo is unlikely to be over and that local populations are still at risk; the population is increasing rapidly and the government has concentrated economic interests, such as huge special economic zones, within the 40 km perimeter containing the most devastation in 1991. There is a pressing need to seek more precise timing and delineation of areas affected by the Buag plinian and dome-building eruptions as a model to be used in current risk management. As we have demonstrated here, collaborative studies of volcanic history that integrate findings from geology, geography, archaeology, anthropology, and history are essential to the future of the Phillipines and other regions in highly active volcanic regions.

ACKNOWLEDGMENTS

The authors extend their gratitude to Jack McGeehin (US Geological Survey, [14]C Laboratory, US) for the recent radiocarbon dates presented in this study and thank John Grattan (University of Wales), Christopher G Newhall (US Geological Survey and University of Washington, US), David K Chester (University of Liverpool, UK), and Frédéric Leone (Université Paul Valéry, Montpellier III, France) for their valuable reviews of this chapter.

REFERENCES

Apostol, JP (1956) 'The Ilocanos in Zambales' *Journal of History* 4, 3–15

Bergaño, D (1732) *Vocabulario de la Lengua Pampanga en Romance*, reprinted in 1860, Manila: Imprenta Ramírez y Giraudier

Beyer, HO (1947) 'Outline review of Philippine archaeology by islands and provinces', *The Philippine Journal of Science* 77, 205–390

Blair, EH and Robertson, JA (eds) (1903–09) *The Philippine Islands 1493–1898*, Cleveland: Arthur H Clark, 55 volumes

Blong, RJ (1982) *The Time of Darkness: Local Legends and Volcanic Reality in Papua New Guinea*, Canberra: Australian National University Press

Castillo, JS (1918) 'The story of Mt. Arayat', in Beyer, HO (ed), *Philippine Folklore, Social Customs and Beliefs (A Collection of Original Sources)*, vol 19, From the Pampangan Peoples, Microfilm, University of the Philippines Diliman Library, Quezon City, Philippines

David, DG (1918) 'Folklore stories', in Beyer, HO (ed), *Philippine Folklore, Social Customs and Beliefs (A Collection of Original Sources)*, vol 19, From the Pampangan Peoples, Microfilm, University of the Philippines Diliman Library, Quezon City, Philippines

De Jesus, RV (1990) *Zambales*, Manila: Union Zambaleña

de la Torre, AA (1999) 'Archaeological excavation of Dizon I site (III-1999-N), Sitio Babo Balukbuk, Barangay Hacienda Dolores, Municipality of Porac, Province of Pampanga: a preliminary report of the first phase (April 17–May 6, 1999)', unpublished manuscript, Archaeological Studies Program Library, University of the Philippines Diliman Library, Quezon City, Philippines

Delfin Jr, FG (1983) 'Geology of the Mt. Pinatubo geothermal project', unpublished internal report, Philippine National Oil Company Energy Development Corporation

Delfin Jr, FG (1984) 'Geology and geothermal potential of Mt. Pinatubo', unpublished internal report, Philippine National Oil Company, Energy Development Corporation

Delfin Jr, FG, Newhall, CG, Martinez, ML, Salonga, ND, Bayon, FEB, Trimble, D and Solidum, R (1997) 'Geological, ^{14}C, and historical evidence for a 17th century eruption of Parker volcano, Mindanao, Philippines', *Journal of the Geological Society of the Philippines* 52, 25–42

Delfin Jr, FG, Villarosa, HG, Laguyan, DB, Clemente, VC, Candelaria, MR and Ruaya, JR (1996) 'Geothermal exploration of the pre-1991 Mount Pinatubo hydrothermal system', in Newhall, CG and Punongbayan, RS (eds), *Fire and Mud: Eruption and Lahars of Mount Pinatubo, Philippines*, pp 197–212, Seattle and Quezon City: University of Washington Press and Phivolcs Press

de San Agustin, G (1998) *Conquistas de las Islas Filipinas (1565–1615)*, Manila: San Agustin Museum

Dizon, EZ (1990) 'Zambales archaeology project: archaeological report on Manggahan 2, Kakilingan, Buag, San Marcelino, Zambales', unpublished manuscript, National Museum of the Philippines, Manila

Dizon, EZ (2002) 'Southeast Asian protohistoric archaeology at Porac, Pampanga, central Luzon, Philippines: an integrated report of the systematic archaeological excavation of Dizon I site (III-99-N) at sitio Babo Balukbuk, barangay Hacienda Dolores, Porac, Pampanga, manuscript', unpublished manuscript, Archaeological Studies Program Library, University of the Philippines Diliman Library, Quezon City, Philippines

Forman, ML (1971) *Kapampangan Dictionary*, Honolulu: University of Hawaii Press

Fox, RB (1960a) 'Report on the first month of excavation at Porac, Pampanga', unpublished manuscript, National Museum of the Philippines, Archaeology Division, Manila

Fox, RB (1960b) 'Report on the second month of excavation at Porac, Pampanga', unpublished manuscript, National Museum of the Philippines, Archaeology Division, Manila

Gaillard, J-C (2002) 'Implications territoriales et ethno-culturelles d'une crise volcanique: le cas de l'éruption du Mont Pinatubo aux Philippines', *Annales de Géographie* 627–28, 574–91

Gaillard, J-C (forthcoming) 'Catastrophes naturelles et changement culturel au sein des sociétés traditionnelles', Proceedings of *Rencontre Géo-risque 2005 – La réduction de la vulnérabilité de l'existant face aux menaces naturelles*, 8 February 2005, Université Paul Valéry – Montpellier III, France

Galang, RE (1940) *Ethnographic Study of the Pampangans*, Manila: The National History Museum Division

Garvan, JM (1963) in Hochegger, H (ed), *The Negritos of the Philippines*, Institut für Volkerkunde der Univesität Wien, Wiener Beitrage zur Kulturgeschichte und Linguistik 14, Vienna, Austria

Hilario-Lacson, E (1984) *Kapampangan Writing: A Selected Compendium and Critique*, Manila: National Historical Institute

Jocano, FL (1975) *The Philippines at the Spanish Contact*, Manila: MCS Enterprises Inc

Major, JJ, Janda, RJ and Daag, AS (1996) 'Watershed disturbance on the east side of Mount Pinatubo during the mid-June 1991 eruptions', in Newhall, CG and Punongbayan, RS (eds), *Fire and Mud: Eruption and Lahars of Mount Pinatubo, Philippines*, pp 895–920, Seattle and Quezon City: University of Washington Press and Phivolcs Press

Newhall, CG, Daag, AS, Delfin Jr, FJ, Hoblitt, RP, McGeehin, J, Pallister, JS, Regalado, MTM, Rubin, M, Tubianosa, BS, Tamayo Jr, RA and Umbal, JV (1996) 'Eruptive history of Mount Pinatubo' in Newhall, CG and Punongbayan, RS (eds), *Fire and Mud: Eruption and Lahars of Mount Pinatubo, Philippines*, pp 165–95, Seattle and Quezon City: University of Washington Press and Phivolcs Press

Nicdao, A (1918a) 'Sinukuan', in Beyer, HO (ed), *Philippine Folklore, Social Customs and Beliefs (A Collection of Original Sources)*, vol 19, From the Pampangan Peoples, Microfilm, University of the Philippines Diliman Library, Quezon City, Philippines

Nicdao, A (1918b) 'Sinukuan of Mount Arayat', in Beyer, HO (ed), *Philippine Folklore, Social Customs and Beliefs (A Collection of Original Sources)*, vol 19, From the Pampangan Peoples, Microfilm, University of the Philippines Diliman Library, Quezon City, Philippines

Paz, VJ (2003) 'Advancing settlement archaeology study through archaeobotany', unpublished preliminary report, Archaeological Studies Program, University of the Philippines Diliman Library, Quezon City, Philippines

Pierson, TC, Daag, AS, Delos Reyes, PJ, Regalado, MTM, Solidum, RU and Tubianosa, BS (1996) 'Flow and deposition of post-eruption hot lahars on the east side of Mount Pinatubo, July–October 1991', in Newhall, CG and Punongbayan, RS (eds), *Fire and Mud: Eruption and Lahars of Mount Pinatubo, Philippines*, pp 921–50, Seattle and Quezon City: University of Washington Press and Phivolcs Press

Pierson, TC, Janda, RJ, Daag, AA and Umbal, JV (1992) 'Immediate and long-term hazards from lahars and excess sedimentation in rivers draining Mount Pinatubo, Philippines', United States Geological Survey Water Resources Investigations Report 92–4039, US Geological Survey

Quizon, PG (1918) 'A legend of Mt. Arayat of Pampanga', in Beyer, HO (ed), *Philippine Folklore, Social Customs and Beliefs (A Collection of Original Sources)*, vol 19, From the Pampangan Peoples, Microfilm, University of the Philippines Diliman Library, Quezon City, Philippines

Rodolfo, KS (1995) *Pinatubo and the Politics of Lahar*, Quezon City: University of the Philippines Press

Rodolfo, KS, Umbal, JV, Alonso, RA, Remotigue, CT, Paladio-Melosantos, ML, Salvador, JHG, Evangelista, D and Miller, Y (1996) 'Two years of lahars on the western flank of Mount Pinatubo: initiation, flow processes, deposits, and attendant geomorphic and hydraulic changes', in Newhall, CG and Punongbayan, RS (eds), *Fire and Mud: Eruption and Lahars of Mount Pinatubo, Philippines*, pp 989–1014, Seattle and Quezon City: University of Washington Press and Phivolcs Press

Rodriguez, JN (1918) 'The origin of Pinatubu volcano (a negrito myth)', in Beyer, HO (ed), *Philippine Folklore, Social Customs and Beliefs (A Collection of Original Sources)*, vol 22, Ethnography of the Negrito-Ayta peoples, Microfilm, University of the Philippines Diliman Library, Quezon City, Philippines

Santos, AP (1979) *Romance in Philippine Names: Mythical Origins of Philippine Names and Objects*, Manila: National Book Store

Scott, KM, Janda, RJ, de la Cruz, EG, Gabinete, E, Eto, I, Isada, M, Sexon, M and Hadley, KC (1996) 'Channel and sedimentation responses to large volumes of 1991 volcanic deposits on the east side of Mount Pinatubo', in Newhall, CG and Punongbayan, RS (eds), *Fire and Mud: Eruption and Lahars of Mount Pinatubo, Philippines*, pp 970–88, Seattle and Quezon City: University of Washington Press and Phivolcs Press

Scott, WH (1986) 'Life, religion and customs of the 17th century Zambals, as reflected in the missionary labors of Father Domingo Perez, O.P.', *Philippiniana Sacra* 21, 117–61

Shimoyama, S (2002) 'Volcanic disasters and archaeological sites in southern Kyushu, Japan', in Torrence, R and Grattan, J (eds), *Natural Disasters and Cultural Change*, pp 326–42, London: Routledge

Siringan, FP and Ringor, CL (1996) 'Changes in the position of the Zambales shoreline before and after the 1991 Mt. Pinatubo eruption: controls of shoreline change', *Science Diliman* 7–8, 1–13

Siringan, FP and Ringor, CL (1997) 'Influence of rapid massive sediment input in the evolution of the Southern Zambales coast', unpublished terminal report No 09404 NS, University of the Philippines-ORC-NSRC, Quezon City, Philippines

Siringan, FP and Rodolfo, KS (2003) 'Relative sea level changes and worsening floods in the western Pampanga delta: causes and some possible mitigation measures', *Science Diliman* 15, 1–12

Stuiver, M and Reimer, PJ (1993) 'Extended ^{14}C database and revised CALIB radiocarbon calibration program', *Radiocarbon* 35, 215–30

Umbal, JV (1994) 'Lahar-dammed Mapanuepe Lake: two-year evolution after the 1991 eruption of Mount Pinatubo, Philippines', unpublished MS thesis, Department of Geology, University of Illinois at Chicago

Umbal, JV (1997) 'Five years of lahars at Pinatubo volcano: declining but still potentially lethal hazards', *Journal of the Geological Society of the Philippines* 52, 1–19

Umbal, JV and Rodolfo, KS (1996) 'The 1991 lahars of southwestern Mount Pinatubo and evolution of the lahar-dammed Mapanuepe Lake', in Newhall, CG and Punongbayan, RS (eds), *Fire and Mud: Eruption and Lahars of Mount Pinatubo, Philippines*, pp 951–70, Seattle and Quezon City: University of Washington Press and Phivolcs Press

Wang, T-M (1964) 'Sino-Filipino historico-cultural relations', *Philippine Social Sciences and Humanities Review* 29, 277–471

CHAPTER 12

Archaeology of Fire and Glass: Cultural Adoption of Glass Mountain Obsidian

Carolyn D Dillian

INTRODUCTION

In 885 BP, the Glass Mountain volcano erupted (Donnelly-Nolan *et al* 1990: 699). It was a locally massive event with glowing ash and pumice ejected hundreds of metres into the sky, followed by an extensive rhyolitic lava flow that capped preceding pyroclastic deposits with a steep sided dome of glassy obsidian (Anderson 1933). The newly formed obsidian was of a high quality that was ideal for the manufacture of chipped-stone tools, and the obsidian flow was so massive that centuries of quarrying could not significantly deplete the available material. The eruption created a tool-quality obsidian source consisting of over 4 km^2 of large obsidian nodules, which commonly measure as much as 1 m in diameter. Not only was the eruptive event significant in the lives of Native Americans living near Glass Mountain, but the obsidian flow represented an important available natural resource. Based on archaeological case studies of obsidian sources in other areas of the world (Bettinger 1982; Ericson 1982; Spence *et al* 1984; Torrence 1986), widespread use and trade of this material was expected; however, as will be discussed below, archaeological investigations revealed a surprising deviation from predictions. Although the obsidian was exploited, its use was restricted to ritual regalia. This case study illustrates how a particular volcanic event can be culturally adopted and its products deliberately incorporated into ritual practices.

THE PHENOMENOLOGY OF OBSIDIAN FORMATION

As I demonstrate in this chapter, the Glass Mountain eruption had a significant impact on the communities who witnessed it. To understand why this might have been the case, it is useful to reconstruct what people might have experienced during the eruption, based on current knowledge of vulcanian-type events. Glass Mountain, located at the southern tip of the Cascade Range in north-eastern California, United States, in the Medicine Lake Highland is a large shield volcano (US Geological Survey 2000; Figure 12.1). Of the 17 different eruptive events, the obsidian-forming eruption of Glass Mountain in 885 BP was the most recent (Donnelly-Nolan *et al* 1990). First and foremost, earthquakes of varying intensity may have preceded the eruption. Then, days or weeks later, bursts of pumice and ash erupted from Glass Mountain (Chesterman 1955; Donnelly-Nolan *et al* 1990), darkening the sky and raining down heavily to the north-east and east.

Following the pumice and ash fall, lava spewed forth from the main Glass Mountain vent and flowed down the eastern side of the Medicine

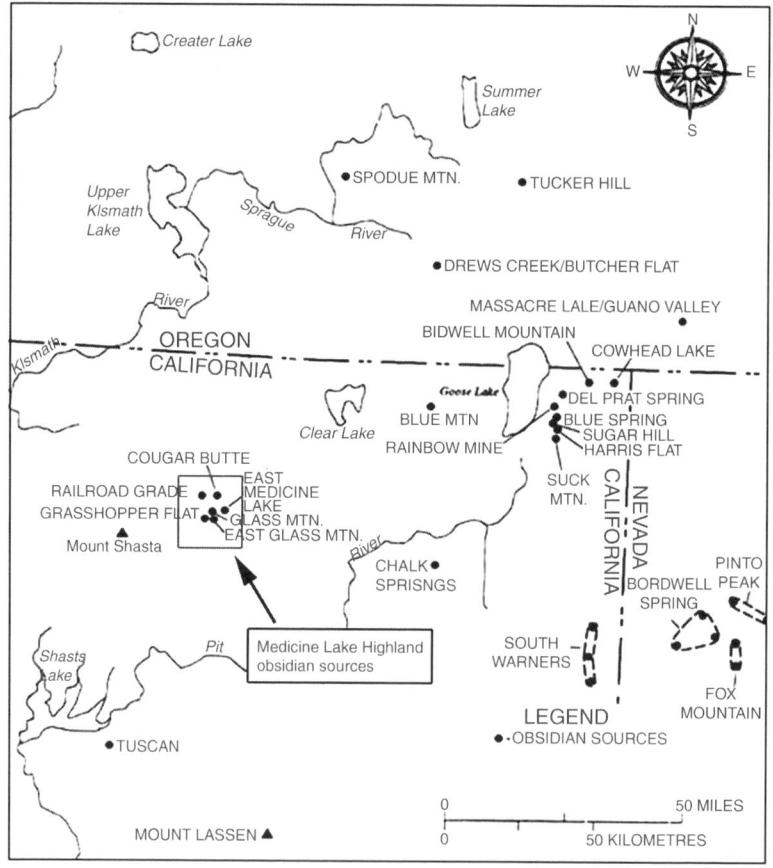

Figure 12.1   Location of Glass Mountain and other Medicine Lake Highland obsidian sources.

Lake Highland. The slowly moving initial lava flow cooled to form high silica dacite and rhyodacite. Fractures in the upper crust of the lava flow exposed molten material, which may have glowed a fiery red. Meanwhile, sulfurous gasses and steam escaped through vents in the crust, creating an inhospitable environment so that a turbulent atmosphere surrounded the volcano. Rapidly melting snow may also have contributed to a massive steam cloud. The eruption was accompanied by bursts of lightning over Glass Mountain and extensive burning of adjacent forests. People living in the region would have had to take shelter from falling ash, and the area may have experienced a perpetual darkness throughout the day. At night, if the air was clear, molten lava glowing atop Glass Mountain would have been visible for more than 100 km, creating an impressive and terrifying sight.

During the final stage of the Glass Mountain eruption, extremely viscous rhyolitic lava plugged many of the secondary vents. Rhyolitic lava squeezed upwards from the central vent and slowly inched down the eastern slope of the Medicine Lake Highland. Because it was so viscous, the lava cooled into an extremely steep-sided flow with a central dome capping the top of the mountain (Anderson 1933, 1941). As the top of the flow cooled, it fractured into angular fragments whilst the interior of the lava flow continued to move. The lava cooled rapidly, preventing crystallisation within the lava matrix, resulting in the formation of a high-quality, glassy black, banded obsidian (Donnelly-Nolan et al 1990). Glass Mountain remained hot and likely contained active fumaroles for centuries after the glass-forming eruption in 885 BP, thereby preventing access to the obsidian immediately following the eruption. The location would have continued to be a memorable, probably frightening, and certainly dangerous place for human visitation. Possibly as in contemporary Pacific communities, access was prohibited by making it sacred (Cronin and Cashman, Chapter 9 this volume).

TRACING GLASS MOUNTAIN OBSIDIAN

As an eruption with a well-known date that produced distinctive volcanic obsidian, tracing the appearance and use of Glass Mountain obsidian in the surrounding settlements offers an unparalleled opportunity to test established paradigms relating to the trade of obsidian (eg, Bettinger 1982; Ericson 1982; Spence et al 1984; Torrence 1986). Initial background research hinted at an unusual pattern of raw material procurement and use from the Glass Mountain source. Examination of obsidian artefact–sourcing studies for sites in the vicinity revealed that Glass Mountain obsidian was relatively rare in the archaeological assemblage from sites within approximately 80 km of Glass Mountain (Gates, personal communication). Instead, Medicine Lake Highland obsidian sources were largely represented by specimens from Grasshopper Flat/ Lost Iron Well, Cougar Butte, and East Medicine Lake (Figure 12.1). This differential pattern of obsidian use in the region prompted a significant question: if not visible in the archaeological assemblage, how had Glass Mountain obsidian been used?

In north-eastern California, at least 25 geochemically distinguishable obsidian sources were used during prehistory, of which six major sources are located within the Medicine Lake Highland: Glass Mountain, East Glass Mountain, East Medicine Lake, Cougar Butte, Railroad Spring, and Grasshopper Flat/Lost Iron Well (Figure 12.1). Of these, Glass Mountain is the only obsidian source to have been formed in the relatively recent past, and it is easily distinguishable from other obsidian sources in the Medicine Lake Highland region using x-ray fluorescence. Fortunately for this study, a large database of x-ray fluorescence analyses of Medicine Lake Highland obsidian sources exists (Hughes 1986). Useful ethnographic data are available concerning obsidian tools, and their use both as practical artefacts and objects of ritual significance was available in several anthropological studies of California Native Americans (Goldschmidt and Driver 1940; Kroeber 1925, 1957; Rust 1905). Our field research was supplemented by published archaeological studies from the local region that provided extensive information on the patterns of obsidian use.

Fieldwork was conducted to identify and record archaeological quarrying sites along the glass flow margins and to document lithic production or retooling at Glass Mountain (Figure 12.2). In addition, previously collected archaeological specimens and geochemical characterisation data were obtained from prior studies (Busby *et al* 1990; Gates 1991; Gates *et al* 2000; Kelly *et al* 1987; McAlister 1988; Shackley 1987; site records on file at Modoc National Forest, Alturas, California) to provide comparative data for analysis of objects recovered from the quarry. Surface sampling combined with *in situ* and field analyses were the preferred methodologies for field investigations because of Native American concerns regarding archaeological collection (Dillian 2002). Archaeological sampling at Glass Mountain consisted of simple random sampling of surface deposits made up almost exclusively of obsidian debitage.

Highly stylised biface fragments recovered during field survey (Figure 12.3) are similar in size and shape to ceremonial artefacts, such as those illustrated in Figure 12.4, which are known from high-status burial contexts in northern California and southern Oregon (Hughes 1978; Loud 1918). Bifaces were classified by stage of reduction (see Dillian 2002) based on stages outlined by Callahan (1979) and Andrefsky (1998).

Figure 12.2 Final stage obsidian bifaces recovered during archaeological survey of Glass Mountain (Dillian 2002).

Figure 12.3 Ceremonial bifaces. Artefact on right measures 33 cm in length; artefact on left measures 23 cm in length (from Kroeber 1925).

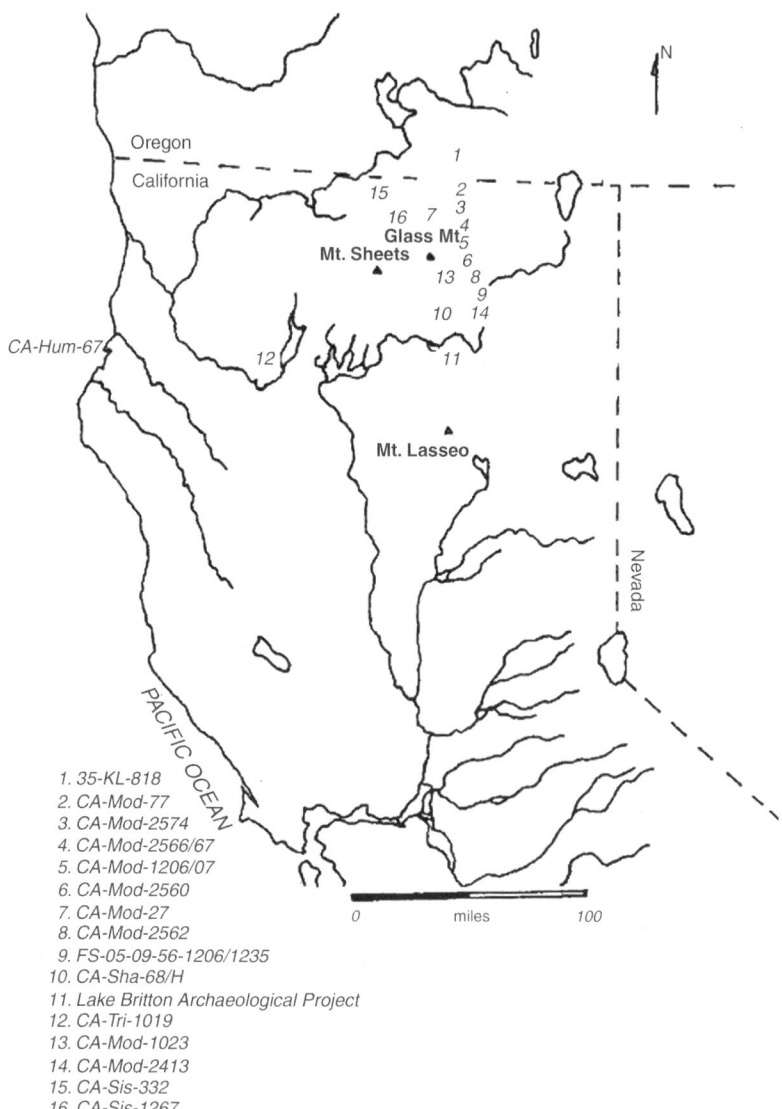

1. 35-KL-818
2. CA-Mod-77
3. CA-Mod-2574
4. CA-Mod-2566/67
5. CA-Mod-1206/07
6. CA-Mod-2560
7. CA-Mod-27
8. CA-Mod-2562
9. FS-05-09-56-1206/1235
10. CA-Sha-68/H
11. Lake Britton Archaeological Project
12. CA-Tri-1019
13. CA-Mod-1023
14. CA-Mod-2413
15. CA-Sis-332
16. CA-Sis-1267

Figure 12.4 Mapped locations of Cultural Resource Management sites presented in Table 12.2.

Results from surface sampling presented in Table 12.1 show that biface-thinning flakes comprise an unusually large percentage of the overall debitage assemblages for each of the sites identified along the Glass Mountain obsidian flow. Furthermore, biface fragments were also extremely common. One can therefore conclude that biface production

Table 12.1 Debitage and biface fragments recovered during the Glass Mountain archaeological survey (Dillian 2002)

FS Site Number	Flakes/ Metre (mean)	Biface Thinning (mean %)	Core Reduction (mean %)	Flake Fragments (mean %)	Biface Fragments (sum in loci)					Total bifaces in loci
					Stage 1	Stage 2	Stage 3	Stage 4	Stage 5	
05-09-56-3000	100.00	3.00	45.00	52.00	0	0	0	0	0	0
05-09-56-3001	158.93	8.93	46.50	44.50	8	2	5	1	0	16
05-09-56-3002	107.14	10.57	41.43	48.00	2	1	0	0	0	3
05-09-56-3003	150.00	1.00	39.50	59.50	0	0	2	0	0	2
05-09-56-3004	500.00	30.00	20.00	50.00	0	0	0	0	0	0
05-09-56-3005	1000.00	5.00	45.00	50.00	3	0	2	1	1	7
05-09-56-3038	375.00	13.00	20.00	67.00	0	0	1	1	0	2
05-09-56-3039	1615.00	7.20	49.00	43.80	0	4	5	1	0	10
05-09-56-3040	453.57	23.14	26.86	50.00	0	2	0	0	0	2
05-09-56-3041	561.36	14.91	29.82	55.27	14	6	0	5	0	25
05-09-56-3042	614.53	6.39	38.39	55.22	34	26	13	11	0	84
05-09-56-3043	510.00	6.33	36.47	57.20	13	14	8	7	0	42
05-09-56-3044	616.25	4.15	32.70	63.05	4	5	3	1	0	13
05-09-56-3045	327.50	5.20	41.45	53.35	6	8	5	0	0	19
05-09-56-3046	512.50	13.50	37.25	49.25	2	2	0	0	0	4
05-09-56-3047	426.00	12.56	42.04	45.40	5	18	6	0	0	29
05-09-56-3048	675.00	11.00	48.00	41.00	0	0	0	0	0	0
05-09-56-3049	325.00	12.00	30.00	58.00	0	0	0	0	0	0
median	476.79	9.75	38.95	51.00						
mean	501.54	10.44	37.19	52.36						
SD	359.02	7.17	8.93	6.90						
							Total Recovered Bifaces:			258

was the primary activity performed at Glass Mountain. In addition, retooling activities were conspicuously absent in the debitage and tool assemblages of sites recorded along the base of the Glass Mountain obsidian flow. Although a handful of utilised flakes were found, no projectile points, projectile point fragments, knives, formed scrapers, drills, or other retouched tools apart from large bifaces and biface fragments were observed.

The exclusive production of large bifaces at Glass Mountain was unexpected, as was the conspicuous absence of typical utilitarian tool types in the archaeological assemblage. Quarry assemblages commonly contain exhausted or broken tools of varying types, suggesting retooling activities at the source (Gilreath and Hildebrandt 1997). Alternatively, production of transportable cores or preforms may indicate manufacture for exchange or long-term planning for resharpening and reduction during the use-life of the object (Ozbun 1991).

In studies of biface manufacture, it is usually assumed that reduction at the quarry was designed to minimise both material bulk and time, and that knappers would remain only long enough to produce easily transportable preforms that could then be completed at another time and place (Bamforth 1986, Binford 1979; Kelly 1988; Ozbun 1991; Roth 1998). This implies that artefactual material found away from the quarry would include retouch or biface reduction debitage from larger bifaces initially prepared at quarry sites. However, it appears that at Glass Mountain, bifaces were knapped far beyond the preform stage at which mass is sufficiently reduced for easy transport. Evidence in the form of tertiary biface-thinning flakes and final-stage biface fragments at the Glass Mountain quarry itself suggests that at least some bifaces were knapped to completion, or nearly so. In contrast, as discussed below, obsidian debitage at sites away from the immediate quarry locale do not contain evidence for retouch or biface thinning of Glass Mountain obsidian objects.

SOURCING STUDIES

To understand the role of the bifaces made at Glass Mountain, it was important to investigate if there was evidence for further working of these artefacts, flakes, or nodules at other sites. Small lithic reduction stations are common archaeological features within the north-eastern California landscape and could represent sequential biface reduction away from quarry locales. Also, besides the bifaces, projectile points and other utilitarian tools might have been made at Glass Mountain as they are with at other quarries in this region. These activities might be invisible in the overwhelming biface production debitage in the assemblages at Glass Mountain. To solve these problems, unpublished sourcing data based on x-ray fluorescence analysis of artefacts studied within archaeological cultural resource projects were examined (Table 12.2)

The sites included in the study were chosen based on the availability of x-ray fluorescence data, were securely dated to the right time period and were located within 80 km of Glass Mountain. Because the Glass Mountain eruption occurred very recently, it was imperative that the sites belong to the appropriate chronological period. Many sites in northern California have been dated using diagnostic projectile point types, which is insufficient to achieve the level of temporal control needed here. Therefore, only securely dated, post–885 BP components, as determined by obsidian hydration or radiocarbon dating of associated material, were included in the study. No components dated to the historic period were included. Because of the paucity of such sites in the region, all relevant available and accessible data were used.

Site records and lithic analysis reports for 16 sites were examined (Busby *et al* 1990; Gates 1991; Gates *et al* 2000; Kelly *et al* 1987; McAlister 1988; Shackley 1987; and site records on file at Modoc National Forest, Alturas, California), yielding data for 1,421 obsidian artefacts. Sites were selected to cluster around Glass Mountain (Figure 12.5), and all except for one are within 80 km of the source. These sites represented a variety of prehistoric activities, including small campsites and larger village sites. None of the selected sites were quarry locales.

Contrary to what might be expected if the use of Glass Mountain obsidian was typical of most quarries, the data presented in Table 12.2 show that raw material from Glass Mountain is extremely rare in the archaeological assemblages of sites that date from the time that Glass Mountain obsidian was available for use by Native Americans. Furthermore, the obsidian sourcing data are consistent with the production patterns observed at the source. They confirm that Glass Mountain obsidian was not used for utilitarian tools. X-ray fluorescence data of the 1,421 obsidian artefacts from within 80 km of Glass Mountain reveal that Glass Mountain obsidian is extremely rare in late prehistoric components, comprising only approximately 5.8% of the debitage and retouched tools. Interestingly, other obsidian sources from the Medicine Lake Highland located only a few kilometres from Glass Mountain including Grasshopper Flat/Lost Iron Well, Cougar Butte, and East Medicine Lake (Figure 12.1), are common and make up over 40% of the same assemblages. Any prohibition against utilising Glass Mountain obsidian for utilitarian objects obviously did not extend to other obsidian sources in the immediate vicinity. Such patterning suggests that cultural and ideological, rather than purely economic, factors influenced the use of Glass Mountain obsidian in prehistory.

Table 12.2 Summary of characterisation results from x-ray fluorescence analysis. Location of the sites is shown in Figure 12.5

Map Number	Site	Reference	Glass Mountain	Total Obsidian	Per cent Glass Mt	Other Medicine Lake	Per cent Other Medicine Lake	Object Type
1	35-KL-818	Site record on file at Modoc NF	1	52	1.9	0	0.0	Flake Tool
2	CA-Mod-77	Site record on file at Modoc NF	4	73	5.5	0	0.0	Debitage
3	CA-Mod-2574	Site record on file at Modoc NF	18	44	40.9	15	34.1	Debitage
4	CA-Mod-2566/2567	Site record on file at Modoc NF	3	129	2.3	0	0.0	Projectile Point/ Debitage
5	CA-Mod-1206/1207	Site record on file at Modoc NF	29	90	32.2	52	57.8	Debitage
6	CA-Mod-2560	Site record on file at Modoc NF	19	155	12.3	126	81.3	Debitage
7	Nightfire Island (CA-Mod-27)	Hughes 1986, Sampson 1985	0	34	0.0	27	79.4	Projectile Points
8	CA-Mod-2562	Site record on file at Modoc NF	1	179	0.6	0	0.0	Multiple
9	OTH-B (05-09-56-1206, 05-09-56-1235)	Busby et al 1990	2	45	4.4	31	68.9	Projectile Point

	Site	Reference						Type
10	CA-Sha-68/H	Site record on file at Modoc NF	1	169	0.6	0	0.0	Debitage
11	Lake Britton	Kelly et al 1987	2	387	0.5	271	70.0	Projectile Point/Debitage
12	CA-Tri-1019	Nilsson 1990	0	15	0.0	9	60.0	Debitage
13	CA-Mod-1023	Gates 1991	1	1	100.0	0	0.0	Projectile Point
14	05-09-56-2413	Gates et al 2000	0	16	0.0	16	100.0	Debitage
15	CA-Sis-332	Shackley 1987	1	7	14.3	6	85.7	Multiple
16	CA-Sis-1267	McAlister 1988	0	25	0.0	25	100.0	Multiple
		Total:	82	1,421	5.8	578	46.1	
		Mean:	*5.1*	*88.8*	*13.5*	*36.1*	*46.1*	
		Median:	*1.0*	*48.5*	*2.1*	*12.0*	*58.9*	
		Std. Dev.:	*8.7*	*99.1*	*26.1*	*70.3*	*40.0*	

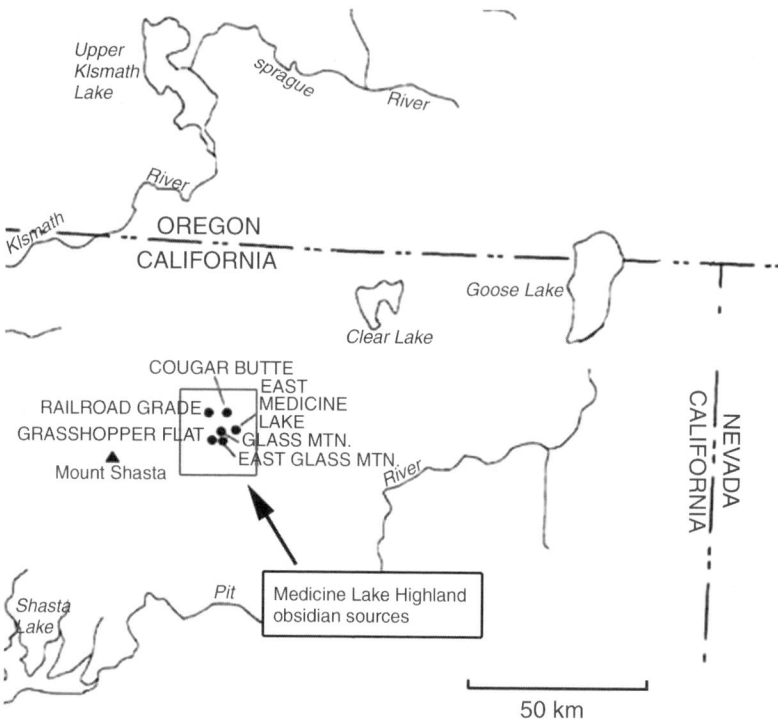

Figure 12.5 Location of sites providing characterisation data in Table 12.2.

NORTHERN CALIFORNIA BIFACES

Based on the presence of other Medicine Lake Highland obsidian sources in the archaeological assemblage, it is obvious that prehistoric peoples were commonly entering the Medicine Lake Highland to obtain obsidian. Despite the quality and quantity of Glass Mountain obsidian that was available during this time, people were specifically selecting other Medicine Lake Highland obsidian and *not* using Glass Mountain for utilitarian things. Although other Medicine Lake Highland obsidian sources were preferred for everyday things, the only procurement activities at Glass Mountain appear to have been exclusively for the manufacture of large, non-utilitarian bifaces.

Large bifaces similar to those found on production sites at the Glass Mountain quarry have been found in high-status and ceremonial contexts in late prehistoric sites along the California coast (Figure 12.4). In the late 1970s, large ceremonial bifaces from burial contexts were geochemically characterised (Hughes 1978) and have more recently been macroscopically analysed (Dillian 2002) to determine the geologic source for the black and red obsidian because the descendant Native American community has requested that no destructive analyses be conducted at this time. If the small sample of black bifaces that was analysed (Hughes 1978) is representative, then 40% of the black bifaces may have originated from Glass Mountain. Though other black obsidian sources were also exploited for large ceremonial bifaces, Glass Mountain obsidian comprised a highly significant portion of the total assemblage. What is interesting is not just that Glass Mountain was used for bifaces, but that it was *only* used for bifaces and not for other types of artefacts.

Large obsidian bifaces from ethnographic and archaeological contexts in northern California and coastal Oregon have long been objects of interest for archaeologists and antiquarians. Their unique beauty and superlative craftsmanship have contributed to their widespread presence in the collections of major museums. However, these objects are much more than merely beautiful things. Bifaces have and continue to serve essential functions in the ceremonial and wealth traditions of the north-west California coast. For some Native American peoples today, these artefacts are still important family heirlooms whose ownership lineage is traceable far back into the past (Yurok Nation NAGPRA Committee, personal communication).

The morphology of ceremonial obsidian bifaces from northern California is somewhat variable, perhaps as a result of raw material quality, nodule size, and knapping ability. Hughes (1978: 53) notes that bifaces 'varied markedly in length and shape, but were usually either bipointed or straight based'. Length was one reflection of value; some biface specimens were up to 75 cm long. In fact, one historic specimen measured 118 cm in length (Heflin 1982: 124). Most of the obsidian bifaces recorded archaeologically and ethnographically are bipointed with parallel or slightly concave margins, otherwise known as 'waisted' types (Heflin 1982: 126). However, some straight-based bifaces have also been found; these probably represent regalia reserved for the Brush Dance (Rust 1905: 694).

Obsidian bifaces were traditionally part of a matched pair consisting of one black and one red biface. Red obsidian was often mined from sources in the Warner Mountains of north-eastern California, where nodules of red obsidian are large enough to be made into sizeable bifaces. Black obsidian was obtained from a variety of sources, including Glass Mountain, Glass Buttes, Vya, Spodue Mountain, and Silver Lake/Sycan Marsh, among others (Hughes 1990: 51). The presence of nodules large enough to be manufactured into ceremonial bifaces limited potential obsidian sources.

Stylised obsidian bifaces were used by the Yurok, Hupa, Tolowa, and Karok of the north-western California coast, located approximately

322 km west of Glass Mountain, as important regalia displayed during the White Deerskin Dance (Figure 12.6). Additionally, ethnographic references mention the use of bifaces by the Shasta, Wiyot, and Wintu, though not necessarily as part of the White Deerskin Dance (Kroeber 1925). This ceremony was important for world renewal and maintenance, but also provided opportunity for wealthy individuals to display valuables, such as bifaces (Heflin 1982: 123; Kroeber 1925: 54). During the dance, bifaces were displayed by special 'flint-carriers' who held a matched pair of red and black obsidian bifaces (Goldschmidt and Driver 1940: 109). This dance was one means to reinforce social status, but it was also important for the cohesion and continuation of the community (Kroeber 1957: 405).

At site CA-Hum-67, obsidian bifaces were found in direct association with human remains. This archaeological site was excavated in 1913 by Llewellyn Loud and is located within the ethnographic territory of Wiyot along the coast of northern California, near the present city of Eureka (Hughes 1978: 56). The site contained 22 burials, with 13 associated obsidian bifaces (Loud 1918: 357–58). Five black obsidian bifaces were found ranging in length from 27.2 cm to 41.0 cm (Loud 1918: 357). All of the black obsidian bifaces were bipointed and all were associated with burned human remains. Loud (1918: 357–58) refers to these as cremations, but Hughes (1978: 56) proposes that this

Figure 12.6 White deerskin dance (from Kroeber 1925).

actually represented grave-pit burning. Hughes's geochemical char-acterisations of many of the obsidian bifaces recovered in Loud's excavations indicated that two of the black bifaces were made of obsidian from the Medicine Lake Highland region (Glass Mountain), two of Vya obsidian, and one of Glass Buttes obsidian (Hughes 1978: 58). In terms of distance, Medicine Lake is the closest to CA-Hum-67, at about 280 km from the site, Glass Buttes is about 480 km distant, and Vya is about 400 km away (Hughes 1978: 60–61).

Loud (1918) also excavated eight red obsidian bifaces found in direct association with human remains. Hughes's (1978: 58) geochem-ical characterisations of the red bifaces determined that all eight origin-ated from the Warner Mountains, located 350 km from the site. The bifaces range in length from 17.1 cm to 30.5 cm (Loud 1918: 358). All exhibited the same bipointed morphology as the black bifaces, though neither Loud nor Hughes discusses the presence or absence of paired specimens. The bifaces from this site have not been dated directly using obsidian hydration. However, radiocarbon dates from a basal peat layer at the site yield a date of 1050 ± 200 yrs BP (Hughes 1978: 56) (lab number not reported in original), which provides an earliest date for the site. Hughes further speculates that the burials themselves and the associated bifaces are approximately 600 years old.

Large obsidian bifaces have also been found in high-status burials along the north-western coast of California and southern Oregon (Loud 1918). These objects were long-distance exchange goods, which consistently appear in non-utilitarian value and wealth contexts. Glass Mountain was one source for black obsidian bifaces, and its repu-tation as a special or significant obsidian source may have further contributed to the valued status of these objects.

OBSIDIAN AND COSMOLOGY

The unique pattern of lithic material utilisation observed at Glass Mountain and in the surrounding area reflects a divergence from trad-itional archaeological views of raw material procurement and use. It was expected that Glass Mountain obsidian, a source containing both extremely high quantities of raw material and high-quality toolstone, would have been commonly used for all types of stone tools. Instead, archaeological evidence indicates that obsidian from Glass Mountain was used more or less exclusively for the manufacture of stylised obsidian bifaces typical of those found in high-value and ceremonial contexts on the California coast. I argue that as witnesses of the large, explosive, pyroclastic eruptions that occurred on Glass Mountain, local populations were instrumental in the creation of an aura of value and 'sacredness' that surrounds this particular quarry. By designating the Glass Mountain quarry to be used exclusively for the manufacture of value objects, people actively created and reinforced a continuing belief system that upheld appropriate behaviours and uses for the Glass Mountain obsidian source. As such, the Glass Mountain source was imbued with a special status, which it bestowed on the objects created from this material.

The phenomenon of the Glass Mountain eruption of 885 BP was an important element in the placement of this obsidian source in the world cosmology and belief systems of prehistoric peoples in northern California. Archaeological sites within 80 km of the volcano do not show abandonment or massive population decline at 885 BP (Baker *et al* 1990; Bevill and Nilsson 1996; Busby *et al* 1990; Delacorte *et al* 1995; Gates 1991; Gates *et al* 2000; Hughes 1986; McAlister 1988; Mikkelsen and Bryson 1997; Moratto 1995; Nilsson 1990; Sampson 1985; Shackley 1987), suggesting minimal population dislocation as a result of volcanic activity. Despite the size of the eruption, much of the environmental impact was limited to the Medicine Lake Highland area, with ash falls to the north-east and east (Anderson 1933, 1941). Archaeological surveys at non-quarry locales within the Medicine Lake Highland do not indicate that this region was densely populated at any point during prehistory (Gates, personal communication). Instead, Native American archaeological sites are generally located at lower elevations, where environmental impact from the Glass Mountain eruption was not as severe. Therefore, the assertion that ancestors of ethnographically documented Native American populations witnessed the Glass Mountain eruption has some validity. Furthermore, a large and explosive eruption such as that which occurred at Glass Mountain was a rare and spectacular event, which could have directly influenced the ideological significance of Glass Mountain obsidian.

The second most recent volcanic eruption in the Medicine Lake Highland in 1050 BP was that of Little Glass Mountain, located approximately 13 km west of Glass Mountain. Stories of the eruption of Little Glass Mountain may have already been part of the local oral histories, but it is highly unlikely that any individuals alive during the Glass Mountain eruption of 885 BP had witnessed such an event before. Furthermore, the Little Glass Mountain eruption did not produce toolquality obsidian, and its lava flow was not a significant raw material used by Native Americans in this region.

Because Glass Mountain remained hot and probably contained active fumaroles for centuries after the eruption, access to the source could have been dangerous for many years. It is also possible to speculate that prehistoric peoples were hesitant to collect and use Glass Mountain obsidian soon after the eruption, even once the danger of heat, earthquakes, and poisonous gasses dissipated. Given the impact of its massive eruption, people may have been fearful of this mountain and therefore avoided the source for several generations. Over time, however, fear may have given rise to reverence of Glass Mountain as a special place, ultimately culminating in its use for high value and ceremonial objects. Through traditions, oral histories, and legends, the Glass Mountain obsidian source may have maintained a reputation as a special source, and consequently was used for ceremonial and high-value objects. Furthermore, the use of Glass Mountain obsidian for special things reinforced the importance of this obsidian, creating a continuing cycle of ideology and value.

Interestingly, the north-western California cultures that used obsidian bifaces in the White Deerskin Dance were geographically quite removed from the experience and phenomenon of the Glass Mountain eruption. Despite this, the obsidian source still figured prominently in the ideological system in the form of large bifaces. Large bifaces, such as those made from Glass Mountain obsidian, were high-status wealth objects as well as important regalia used in the White Deerskin Dance. Obsidian was not locally available to north-western California populations, with the closest sources located at least 200 km distant. The value of a biface may have differed for coastal and inland peoples, yet there is little doubt that these stylised objects retained an important status within the cultures of all northern California peoples. Biface value was intertwined with numerous factors of production, exchange, use, and cultural context.

It is possible to speculate that bifaces, as large, concentrated sources of otherwise relatively rare raw material, were a type of wealth, regardless of the associated ceremonial system. Obsidian, particularly material from distant sources, was a very rare and valuable commodity that could serve utilitarian functions, such as effective cutting tools. In that sense, a biface could conceivably be knapped into points, knives, or flake tools. Perhaps the innate value of a biface is in some degree a measure of the *possible* other tools it could be made into. A large biface was effectively a large piece of raw material. However, these bifaces were not made into utilitarian tools. Instead, they were retained as personal property and wealth. In this way, a biface can be viewed as a form of conspicuous consumption. It was a large cache of potentially useful raw material that was, in effect, *not* used. Instead, it just *was*. It existed as wealth and was incorporated into the ceremonial world renewal system and thus took on a more important significance, verging on sacred (Kroeber 1905: 691).

The value of large bifaces was closely linked with their role as ceremonial objects for the White Deerskin Dance. As ceremonial regalia, bifaces were prominently displayed throughout the dance, and biface owners increased both personal status and fortuitous social connections by including their bifaces and other regalia in the ceremony. The almost sacred significance of bifaces and white deerskins was further reinforced through the context of ceremony and ritual. The value of these objects was transformed through the social context of ceremony and the wealth culture of north-west California Nations.

Translation of value extended across both geographic and cultural boundaries via exchange and interaction of disparate groups. The significance of Glass Mountain in the belief system and worldview of peoples living in the immediate vicinity of the volcano was vastly different from that of peoples along the California coast. However, Glass Mountain and its products retained a valued status in both regions. This translation and transformation of value and ideology further illustrates how both objects and places gain value and how the valued status can be transmitted or altered through interactions between people and groups over large areas.

The Glass Mountain obsidian source retained a special place in the worldview of local peoples. It was perceived as a location for the production of ceremonial and high-value objects, yet cultural prohibitions prevented its use for utilitarian tools. Given the recent date for the eruption of Glass Mountain obsidian, it is highly likely that local peoples witnessed this eruption and the formation of the obsidian flow. Stories of the eruption could have entered into oral histories and legends and contributed to the special status of this obsidian source in the local cosmology. Glass Mountain obsidian was therefore used exclusively for ceremonial and high status objects, such as large bifaces, and neglected for utilitarian purposes. Thus, the cultural memory of the Glass Mountain eruption results in selective procurement and use of this obsidian source. The memory of the Glass Mountain eruption played an active role in the creation and transformation of value of this source and for the large bifaces made from it.

CONCLUSIONS

Volcanic eruptions were rare, spectacular, and destructive events that often had profound impacts on local populations. However, not all volcanic events resulted in negative effects. As demonstrated for the Glass Mountain eruption of 885 BP, volcanoes can produce valuable natural resources, such as the obsidian described in this case study. Furthermore, the volcanic phenomena observed by local populations can enter cultural memory and be the fodder for legends, stories, and sacred places. The archaeological signature of volcanic events exceeds the radius of eruptive deposits and may be visible in the use and distribution of other artefacts as secondary evidence of the cultural and ideological effects of volcanic phenomena.

REFERENCES

Anderson, C (1933) 'Volcanic history of Glass Mountain, Northern California', *American Journal of Science* 26, 485–506

Anderson, C (1941) 'Volcanoes of the Medicine Lake Highland, California', *University of California Publications, Bulletin of the Department of Geological Sciences* 25, 347–422

Andrefsky, W (1998) *Lithics: Macroscopic Approaches to Analysis*, Cambridge: Cambridge University Press

Baker, S, Wagner, H and Simons, D (1990) 'Archaeological excavations at CA-Sha-479 and CA-Sha-195, Whiskeytown unit, Whiskeytown-Shasta-Trinity National Recreation Area, Shasta County, California', unpublished report prepared by Archaeological/Historical Consultants for the National Park Service, Western Regional Office, San Francisco, California

Bamforth, D (1986) 'Technological efficiency and tool curation', *American Antiquity* 51, 38–50

Bettinger, R (1982) 'Aboriginal exchange and territoriality in Owens valley, California', in Ericson, J and Earle, T (eds), *Contexts for Prehistoric Exchange*, pp 103–27, New York: Academic Press

Bevill, R and Nilsson, E (1996) 'Archaeological investigations at CA-Sha-559, Whiskeytown-Shasta-Trinity National Recreation Area, Shasta County, California', unpublished report prepared by Mountain Anthropological Research and Dames and Moore, on file at Modoc National Forest, Alturas, California

Binford, L (1979) 'Organization and formation processes: looking at curated technologies', *Journal of Anthropological Research* 35, 255–72

Busby, C, Bard, J, Dezzani, R, Nissen, K, Findlay, J, Harmon, R and Fong, M (1990) 'OTH-B Cultural resource program site testing: Modoc National Forest', unpublished report prepared by Basin Research Associates, on file at Modoc National Forest, Alturas, California

Callahan, E (1979) 'The basics of biface knapping in the Eastern Fluted Point Tradition: a manual for flintknappers and lithic analysts', *Archaeology of Eastern North America* 7, 1–180

Chesterman, C (1955) 'Age of the obsidian flow at Glass Mountain, Siskiyou County, California', *American Journal of Science* 253, 418–24

Delacorte, M, Reno, R, Burke, T, Mikesell, S, and McGuire, K (1995) 'Report on archaeological test investigations at 209 sites along the proposed Tuscarora pipeline, from Malin, Oregon to Tracy, Nevada', unpublished report prepared by Far Western Anthropological Research Group, Inc, on file at Modoc National Forest, Alturas, California

Dillian, C (2002) 'More than toolstone: differential utilization of Glass Mountain obsidian, Siskiyou County, California', unpublished PhD dissertation, Department of Anthropology, University of California at Berkeley

Donnelly-Nolan, J, Champion, D, Miller, C, Grove, T and Trimble D (1990) 'Post-11,000-year volcanism at Medicine Lake Volcano, Cascade Range, Northern California', *Journal of Geophysical Research* 95, B1219, 693–704

Ericson, J (1982) 'Production for obsidian exchange in California', in Ericson, J and Earle, T (eds), *Contexts for Prehistoric Exchange*, pp 129–48, New York: Academic Press

Gates, G (1991) 'Fairchild Rip-Rap quarry site damage assessment [Supplement to ASR 05-09-700]', unpublished report on file at Modoc National Forest, Alturas, California

Gates, G, Bevill, R and Dillian, C (2000) 'Archaeological investigations at FS-05-09-56-2413: a sparse lithic scatter on the Modoc Plateau of northeastern California', unpublished report prepared for US Forest Service, Alturas, California

Gilreath, A and Hildebrandt, W (1997) *Prehistoric use of the Coso Volcanic Field*, Contributions of the University of California Archaeological Research Facility, No 56, Berkeley: University of California

Goldschmidt, W and Driver, H (1940) 'The Hupa White Deerskin Dance', *University of California Publications in American Archaeology and Ethnology* 35, 103–42

Heflin, E (1982) 'The huge obsidian ceremonial blades of the Pacific northwest', *Central States Archaeological Journal* 29, 122–29

Hughes R (1978) 'Aspects of prehistoric Wiyot exchange and social ranking', *Journal of California Anthropology* 5, 53–66

Hughes, R (1986) *Diachronic Variability in Obsidian Procurement Patterns in Northeastern California and Southcentral Oregon*, University of California Publications in Anthropology, vol 17, Berkeley: University of California Press

Hughes, R (1990) 'The Gold Hill site: evidence for a prehistoric socioceremonial system in Southwestern Oregon', in Hannon, N and Olmo, R (eds), *Living with the Land: The Indians of Southwest Oregon*, pp 48–55, Medford: Southern Oregon Historical Society

Kelly, M, Nilsson, E, and Cleland, J (1987) 'Archaeological investigations at Lake Britton, California', unpublished report prepared by WIRTH Environmental Services, on file at Modoc National Forest, Alturas, California

Kelly, R (1988) 'The three sides of a biface', *American Antiquity* 53, 717–34

Kroeber, A (1905) 'Notes', *American Anthropologist* 7, 690–95

Kroeber, A (1925) *Handbook of the Indians of California*, vol 1, Bureau of American Ethnology Bulletin 78, Washington, DC: Bureau of American Ethnology

Kroeber, A (1957) 'World renewal cult of Northwest California', in Heizer, R and Whipple, M (eds), *The California Indian*, pp 404–11, Berkeley: University of California Press

Loud, L (1918) 'Ethnogeography and archaeology of the Wiyot Territory', *University of California Publications in American Archaeology and Ethnology* 14, 221–436

McAlister, J (1988) 'Kwatuk', unpublished paper prepared for Winema National Forest, on file at Modoc National Forest, Alturas, California

Mikkelsen, P and Bryson, R (1997) 'Culture change along the eastern Sierra Nevada/Cascade front, vol 2, Modoc uplands', unpublished report prepared by Far Western Anthropological Research Group, Inc, on file at Modoc National Forest, Alturas, California

Moratto, M (1995) 'Archaeological investigations PGT-PG & E pipeline expansion project: Idaho, Washington, Oregon, and California, vol 4, Synthesis of findings', unpublished report prepared by INFOTEC Research, Inc and Far Western Anthropological Research Group, Inc, on file at Modoc National Forest, Alturas, California

Nilsson, E (1990) 'Archaeological test excavations at CA-Tri-1019: a late prehistoric site in the Upper Trinity River region of northern California', unpublished report prepared by Mountain Anthropological Research, on file at Modoc National Forest, Alturas, California

Ozbun, T (1991) 'Boulders to bifaces: initial reduction of obsidian at Newberry Crater, Oregon', *Journal of California and Great Basin Anthropology* 13, 147–59

Roth, B (1998) 'Mobility, technology, and Archaic lithic procurement strategies in the Tucson Basin', *Kiva* 63, 241–62

Rust, H (1905) 'The obsidian blades of California', *American Anthropologist* 7, 688–95

Sampson, C (1985) *Nightfire Island: Later Holocene Lakemarsh Adaptation on the Western Edge of the Great Basin*, University of Oregon Anthropological Papers, vol 33, Eugene: University of Oregon Press

Shackley, M (1987) 'U.S. Sprint fiber optic cable project Oroville, California to Eugene, Oregon. Archaeological testing of four sites in California: CA-But-5, Teh-1468, Sha-1684, Sis-332', unpublished report prepared by Dames and Moore, LLP, manuscript on file with the author, Department of Anthropology, Berkeley, California

Spence, M, Kimberlin, J, and Harbottle, G (1984) 'State-controlled procurement and the obsidian workshops of Teotihuacan, Mexico', in Ericson, J and Purdy, B (eds), *Prehistoric Quarries and Lithic Production*, pp 97–105, Cambridge: Cambridge University Press

Torrence, R (1986) *Production and Exchange of Stone Tools*, Cambridge: Cambridge University Press

US Geological Survey (2000) 'Description: Medicine Lake Shield volcano, Medicine Lake caldera', available online at http://vulcan.wr.usgs.gov/Volcanoes/MedicineLake/description_medlake.html

CHAPTER 13

Beyond the Catastrophe: The Volcanic Landscape of Barú, Western Panama

Karen Holmberg

THE VOLCANO IN PERCEPTION AND TIME

A volcano has unquestionable impacts during periods of eruption, but it is also a vibrant presence within the daily experience and memory of people in dormant periods. Archaeologists, however, generally focus solely on the event of eruption. Archaeological assessments frequently note the volcano's catastrophic potential to transform the landscape, contribute to soil fertility, or create useful stratigraphic *termini post quem*. Although these are valid and important considerations in the holistic examination of a volcanic region, they often preclude considerations of the volcano's social context and the great complexity that exists in a volcanic landscape over the long term. The archaeological examination of past volcanic contexts can benefit from a broadening of scope from that of event-based disasters and catastrophism-tinged cultural impacts. This more holistic investigation is framed by conceptions of material landscapes as persistently rich perceptual components of human life (Basso 1996: xvi).

The anthropological divisions between archaeology and sociocultural anthropology unhelpfully divide spheres of information that can fruitfully inform one another. Archaeology is certainly incapable of providing the multi-layered data available to the ethnographer, and it is not my intention to imply that it can. What I do propose is that archaeological investigations of volcanic contexts can be richly informed by the incorporation of ethnography, ethnohistory, and the examination of landscape relationships and volcanic artefacts to get at more textured conceptions of the past volcanic landscape that do not fall back on facile 'prime mover' interpretations of the volcano's role.

This merging of methods takes as its starting point the belief that modernity cannot be seen as a condition that began at a specific post-Enlightenment moment in Europe, encompassed the rest of the globe, and created a complete break with all that came before. The deep past is perfectly modern (Brandt 2004) and 'modernity is deeply prehistoric'

(Matsuda 1996: 12). In a true sense, 'we have never been modern' (Latour 1991). Past and present interimplicate one another and are not separate spheres (Benjamin 1999). The flow of time, rather than being a unilinear movement towards progress, is recursive and carries with it fragments and traces of the past, such as stories or artefacts, that are linked to the physical landscape and the lived relationship that people have with it. Ethnohistoric stories hint at the perceptions of the volcano in the past, archaeology provides material indications of these, and ethnography shows that the intimate incorporation of the volcano into social life is not linked to a pre-modern animism or primitive relationship, but one that is very much alive in the contemporary volcanic landscape as well. The intention of this linkage of methods is not to gloss over the differences between pre- and post-Contact experiences in the New World, but instead to look at the larger duration of time in the volcanic region.

This chapter discusses my ongoing project in the Chiriqui Province of western Panama. The Volcán Barú is prominently discussed in the most systematic archaeological investigations of this region, which were completed in the early 1970s (Linares and Ranere 1980; Linares *et al* 1975). The previous work focuses on the volcano's eruptions and their role in soil fertility and stratigraphic layers and identifies the volcano as the cause for settlement changes interpreted via the artefactual record. My petrographic and stratigraphic analysis of tephra layers as well as recent study of pollen and tephra samples (Behling 2000) indicate, however, that the prior interpretation of an eruption in 600 AD as the last eruption of the volcano should be seriously questioned. Instead, the last eruption of Barú appears to have occurred nearly a thousand years later. As no eruption at 600 AD seems represented in the archaeological record, the volcano cannot be used as an easy catastrophism scapegoat to explain settlement changes interpreted from the material culture. Ethnohistory, artefacts, and markings of the landscape indicate a more social role for the Volcán Barú over the long term that extends to a rich ethnographic presence. The volcano is deeply imbricated with elements of memory, perception, and experience in times of both eruption and acquiescence in both prehistory and modernity, making it less of a purely 'natural' entity than catastrophism interpretations would make it and more of a hybrid form between the natural and social worlds.

IN THE EVENT OF DISASTER

Inability to divorce natural and cultural elements from one another in the investigation of natural disasters or hazards studies is becoming prevalent in academic discussion. Natural disasters provide the focus of a number of important recent edited volumes (eg, Bawden and Reycraft 2000; Hoffman and Oliver-Smith 2002; Torrence and Grattan 2002a). One of the most notable shifts in the recent literature that addresses natural disasters, including volcanism, is an increasing emphasis on the human element of natural disasters and a growing willingness to see social and environmental variables as equally important (Torrence and Grattan 2002b). This strong intermixing of memory, social practices, and natural disaster was vividly expressed in the press coverage from the recent tsunami in Southeast Asia.

The tsunami triggered by a massive earthquake under the Indian Ocean on 26 December 2004 killed over a quarter of a million people in nearly a dozen countries. In terms of loss of life, the tsunami joins the list of epic natural disasters such as Tangshan (1976, north-east China), Krakatoa (1883, Rakata Island, Indonesia), and Tambora (1815, Sumbawa Island, Indonesia). The unprecedented level of press coverage from the affected countries, however, made the disaster a much more 'human' one and provided a far larger and more imme-diate global awareness of the lives and families destroyed than has ever occurred in human history. Newspaper, the Internet, and news magazine reporters brought the varied stories of the victims to the forefront in the weeks following the event, and though patterns did exist in the experiences and reactions of the affected persons, it became clear that those experiences and reactions were highly varied, individual, and tempered by numerous mitigating factors such as age, gender, economic level, and nationality. The experience of sur-vivors regarding speed of rescue and doctors' decisions to amputate limbs rather than use expensive antibiotics to save them, for example, varied considerably depending on whether one was an Indonesian or a foreign tourist, wealthy or poor, an adult or a child.

Many tsunami survivors noted a drastic shift in perception of the sea from a source of tourist revenue, recreation, and food to a source of fear and death. A month after the tsunami, an Acehnese couple originally scheduled to have their wedding on 26 December 2004 held their rescheduled ceremony. The traditional fish curry usually present at weddings was not served at the reception in Banda Aceh, one of the hardest hit areas of Indonesia, because of the connection of fish to the sea and the tsunami and the possibility that fish fed on corpses of loved ones (Soetjipto 2005). The earthquake-generated tsunami was not just a 'natural' event, but one that was intimately intertwined with human life, cultural practices, and perceptions. Politics and press coverage play an intensely strong role in the perception of a disaster's impacts both from within a culture and outside of it as a spectator (Beatty 2005).

The Cartesian separation of culture from nature, particularly in eco-nomically or environmentally deterministic academic perspectives, overlooks human engagement with the environment that was strongly implicated in the tsunami experience. It was not simply the event of the tsunami, however, that was important. The ocean, like the volcano, was a major part of the perceptual life of those who lived near it before the disaster as well as after. Similarly, the volcano not only gains significance through the event of eruption, but is also drawn into the web of social meanings both before and after eruptions.

THE LIVED VOLCANIC LANDSCAPE

Ethnographic Examples from Goma and Vanuatu

The rich interweaving of natural and social elements within volcanic landscapes is readily apparent in the modern context. When Mt Nyiragongo erupted in Congo in January 2002, it killed roughly 100 people and covered the city of Goma in molten rock. Months later, frequent tremors, plumes of white gas pouring from the crater, and a red glow at night continued to remind residents of the volcano's capacity to erupt again. Despite this, life returned somewhat to normal, though with a strong connection to the recent eruption. The central business district was rebuilt in the same place it was buried. An entrepreneur built a Volcano Internet Café. A cellular phone company built an abstract interpretation of Nyiragongo out of lava in the centre of the city. The roads actually improved, as lava was used to fill potholes and build speed bumps. Neatly chopped bricks of lava became available for purchase from enterprising locals along the road. A traditional king, Jean Paul Butsitsi Bigirwa, recommended a return to the practices of the ancestors, who were able to quiet the spirits of the volcano by forcing the first-born daughters of chiefs to remain unmarried and by offering 10% of the harvest and some of their animals to the spirits as tribute (Lacey 2002). Memories of past and extant practices regarding the volcano, symbolic representations of it, experiential elements of eruption, and entrepreneurship mediated the interaction between the people of Goma and the volcano. Such varied responses and strong associations with the volcano are not surprising in a context of recent eruption, although these experiential and perceptual elements are rarely incorporated into hazards studies.

A fascinating hazards study by Cronin *et al* (2004) highlights the importance of accounting for significantly varied cultural perceptions on the volcanic island of Vanuatu. In that case, the clash between scientific knowledge and traditional *kastom* knowledge of the volcano required a closer study of how to mesh the two to create a feasible evacuation plan. Vanuatu islanders noted a number of perceptual elements associated with the volcano before and after eruption. Warning signs of eruption were listed as dreams, strange sounds, unusual animal behaviour, swarms of ants from the ground, and earthquakes. Elements noted after eruptions were the death of trees around Lake Vui, water colour changes and bubbling water, rumblings or booming from the crater, and the rotting of taro in the ground. The crater area was taboo and was avoided to respect the spirits that lived there so as to mitigate future chances of eruption. The islanders noted myriad perceptual aspects of the eruption, although these were not consistent throughout the local population. Cronin and colleagues (2004) discovered that traditional perceptions of the volcano and the volcanic landscape differ widely amongst even a small population and are cross-cut by gender and hierarchy lines. By navigating these differences, the researchers were able to develop a ground-breaking and more useful hazards plan that meshed traditional knowledge and beliefs with scientific ones to form a completely new and customised understanding of the human element in that particular volcanic context. Although the Vanuatu case is a warning against the assumption of a uniform perception of the volcano in the archaeological past, both the Vanuatu and Goma examples remind us that a rich engagement between natural and social worlds occurs in volcanic regions.

The Archaeological Conception of Volcanic Regions

A varied array of archaeological studies concentrate specifically on volcanic contexts in the Americas (eg, Connolly 1999; Cordova *et al* 1994; Isaacson and Zeidler 1998; Mothes 1998; Sanders *et al* 1979; Santley *et al* 2000; Sheets 1983, 1992; Sheets and Grayson 1979; Sheets and McKee 1994). These studies investigate physical impacts on past cultures via eruption; what is outside of their research scope is the social interplay between volcanoes and past peoples. The ethnographic examples of Goma and Vanuatu, discussed above, highlight some of the richly variegated experiential elements present in volcanic regions. In archaeological conception, however, there is a tendency to portray a more single-faceted conception of volcanism focused solely on physical and environmental impacts of eruption. These eruption impacts are often linked in interpretations to discernible culture or settlement changes or use of erupted material for dating purposes. Although this is attributable both to the comparative ease of delineating tephra layers and to research frameworks that do not account for more experiential elements, the lack of focus on more socialised elements of volcanic regions provides an impoverished and catastrophism-based understanding of past life within them.

Artefactual indications do exist in volcanic contexts to prove the intertwining of experience and memory. Plunket and Uruñuela

(1998a, 1998b), for example, detail two small, smoke-producing volcano effigies found at the Terminal Preclassic site of Tetimpa. The site was buried under eruptions of Popocatépetl around the 1st-century AD and again between 700 and 850 AD. The volcano models, which they interpret as shrines, were capable of imitating Popo's puffing of ash and vapour through the use of chimneys and chambers for burning wood underneath them. The researchers propose this as an example of the potential incorporation of volcanic activity into domestic ritual, as the effigies are found within sight of the volcano in houses that were subsequently covered by lava from Popocatépetl.

In a second example, Elson *et al* (2002 and Chapter 6) discuss corn cobs placed within the lava flow of Sunset Crater in the 11th-century AD. Fifty-five pieces of lava with impressions of prehistoric corn were discovered in northern Arizona, approximately 6 km from the volcanic crater at a site (NA 860) first identified in 1928 by Colton (1932). These are interpreted as potentially representing offerings of husked corn ears that were placed in the path of encroaching lava during the eruption, then transported 4 km away where the impressions were removed from the hardened basalt molds. The resulting casts, dubbed 'corn rocks', are posited as apotropaic offerings to the forces responsible for the eruption.

In a third example, Steffian *et al* (1996) describe a painted Kodiak Island box panel recovered by archaeologists in 1987 from the site of Karluk One (see also Beget 2000). The painting, from the Koniag-era (~1550 AD) Alutiiq culture of Alaska, portrays a prehistoric volcanic eruption of the nearby Augustine volcano. The authors interpret the images through analogy with ethnographically recorded practices amongst the Yup'ik Eskimo, the closest cultural group to the Alutiit, who use such paintings to illustrate stories based on events often experienced by family members. These paintings both marked personal property and served as a form of information storage (Himmelheber 1993). The painted boxes were used to augment Alutiiq myths, which are known to make frequent mention of volcanic eruptions (Lantis 1938).

The remnants of volcanic interaction with cultural life found at the Tetimpa, Sunset Crater, and Karluk-One sites hint at the existence of social engagements with the prehistoric volcanic landscape that are far easier to elaborate in modern cases like the post-eruptive Goma context discussed. It has not yet been determined academically how to best conceive of the incorporation of volcanoes into social life in the archaeological past. If one wishes to argue that volcanoes (or mountains in general) are natural monuments, and as such can be referenced over long spans of time to create tradition and a sense of continuity, a growing body of recent literature can be assembled to bolster the argument (eg, Bradley 1984, 1998, 2000; Edmonds 1993: 107; Hobsbawm 1983). Paradoxically, however, it is a well-established dialogue within landscape archaeology literature that landscapes are not static back-drops for human life, but are dynamic and continually changing (eg, Bender 1993; Fullager and Head 1999; Soja 1989: 41). This inherent dynamism obviously speaks against any form of stasis in interpretation between past and present contexts, or even across contiguous archaeological contexts. Furthermore, the traditional conception of monumentality, although helpful in plumbing a more socialised role for the volcano, seems a crude way to address the volcano because it is not 'built', but 'natural' and drawn into the social world.

The most flexible means of conceptualising the volcano's landscape role is perhaps simply by accepting that the volcano's presence in the myriad perceptions and experiences of the past can be inferred via its durable *materiality* (see Holmberg 2005). The sheer scale of substantial landscape forms means that even if meanings or memories attached to them change, their presence in human consciousness can be seen as a constant (Bradley 2002: 111). Hence, we can envision the volcano as an active, natural element that – through its material presence – should be interpreted as an important component of the past social landscape in both eruptive and non-eruptive phases. This repositioning of the volcano within its landscape and the questioning of its nature are particularly apropos in the context of western Panama. The material culture of this cultural region is described by words such as 'unspectacular' (Linares and Ranere 1980: 4), whilst remnants of the past landscape – such as the Volcán Barú and ethnohistoric stories that invoke it – are overlooked as non-data.

BARÚ'S ERUPTION HISTORY: QUESTIONS AND LACUNAE

Chiriqui Province in western Panama is located at the southern end of the Middle American volcanic arc-trench system. This area represents the tectonic conjunction of four major lithospheric plates: the Caribbean, Cocos, Nazca, and South America. Three of these plates – the Cocos, Nazca, and Caribbean – interact beneath the Pacific side of western Panama, creating frequent seismic activity. The Volcán Barú (3,474 m) is the westernmost volcano in Panama, and is a complex andesitic strato-volcano in the Talamanca Range. Its 6 km wide caldera is breached widely to the west, which has created a massive debris-avalanche that extends onto the Pacific coastal plain (Simkin and Siebert 2005).

The current archaeological understanding of the Barú region fore-fronts the role of the volcano during the era of human occupation, although it does so firmly within a catastrophism-tinged focus on eruption as a cause of settlement changes. Volcanic eruption is thought to have first facilitated the original settlement of the highland Chiriqui area through modifications of the local environment (Stewart 1978: 36). These modifications allowed the exploitation of fertile volcanic soils by maize agriculturalists proposed to have entered the highlands sometime prior to 200 AD (Linares and Ranere 1980; Linares *et al* 1975). The decision to move to the area is thought to be specifically related to the volcano both for its agriculturally rich soils and the easy accessibility of volcanic stone for tools (Linares and Ranere 1980: 15, 79, 235, 242).

A series of antipodal dates for eruptions during the span of human occupation are suggested by various studies potentially provide multiple dates for the same eruption (Table 13.1) (Behling 2000; de Boer *et al* 1988; IHRE 1987; Linares and Ranere 1980; Linares *et al* 1975; Montessus de Ballore 1884, 1888; Sapper 1913; Stewart 1978). Alternatively, the inconsistent dates for proposed volcanic events could indicate multiple eruptions that affected different and restricted areas surrounding the Volcán Barú because of factors such as varying wind directions or eruption column heights (Lee Siebert, personal communication).

Despite more recent work and historical records that indicate an eruption occurred during the last millennium, archaeological understanding and popular belief within the Barú area follow the lead of interpretations made in the 1970s that place the most recent eruption

Table 13.1 Proposed dates for eruptions of Barú

Proposed Date of Eruption	*Source*	*Supporting Evidence*
~1550 AD	Montessus de Ballore (1884, 1888)	Unknown; could be from a 17th-century Spanish record showing the volcano in eruption.
~1550 AD	Sapper (1913)	Cited by Montessus de Ballore (1888); Sapper states the date is of uncertain validity.
1315–45/ 1390–45 AD	Behling (2000)	Major eruption. AMS date directly below a tephra layer in a lake core.
1060–1360 AD	Stewart (1978)	Date listed in source as 1210 AD ± 150, but no description of the source material was provided.
1060–1360 AD	de Boer *et al* (1988)	Date listed in source as 740 ± 150 BP from charred wood covered by an ash layer at the site of Barriles.
895–1170 AD	Behling (2000)	Minor eruption. AMS date directly below a tephra layer in a lake core.
~600 AD	Linares and Ranere (1980); Linares *et al* (1975)	Based on radiocarbon dates and artefactual assemblages; no one radiocarbon date or specific stratigraphy is indicated as the source for the date.
120–390 AD	Behling (2000)	Minor eruption. AMS date directly below a tephra layer in a lake core.

of the volcano at roughly 600 AD (Linares and Ranere 1980: 55, 93, 115, 116, 245, 268, 275, 288). The seminal work completed in the 1970s and published as Linares *et al* (1975) and Linares and Ranere (1980) represents the most extensive archaeological study done in the area to date. No systematic archaeological work has been carried out since to corroborate or question the date, prior to my own, and the date has become entrenched in interpretation of the region. The 600 AD eruption is posited as a key factor in the abandonment of prehistoric settlements and the relocation of people to lower plains and the Caribbean coast (Linares and Ranere 1980; Linares *et al* 1975). A 10 cm tephra layer recorded during archaeological survey in the early 1970s is seen as evidence of this indirectly dated eruption (Dahlin 1980; Rosenthal 1980). The 600 AD date is contradicted by a more recent study by Behling (2000), which utilises pollen and charcoal samples from a lake core to propose two minor Barú eruptions at 1,800 and 1,000 C^{14} yr BP and a major eruption at 500 C^{14} yr BP (Table 13.2). The more recent and larger eruption deposited a 20-cm tephra layer and had catastrophic impact on vegetative life and agriculture, whilst the two earlier eruptions did not disrupt human settlement in Behling's assessment.

Description of the Tephra Samples

I sampled tephra from three locations – two lake cores and the archaeological site of Barriles (Figures 13.1 and 13.2). The samples were exposed to a sonic bath of de-ionised water for 15 minutes, dried with a heat lamp, and given an acetone bath to clean and separate the particles for examination with a fibre optic light and binocular scope at 1.5× magnification.

Sample set (A) is from a lake coring from the Laguna Volcán, roughly 16 km south-west of the Barú crater (see Behling 2000). Sample set (B) is from a core sample from an unnamed lake roughly 10 km north-east of the Barú crater; the core was drilled in March 1996

Table 13.2 Proposed dates of Barú eruptions from organic material directly under the three tephra layers found in the Laguna Volcán core, as reported by Behling (2000: 391)

Lab #	Depth (cm)	C^{14} yr BP*	Calibrated Age**
Beta-95496	42	500 ± 60	1315–45 AD
			1390–45 AD
Beta-95497	76	1020 ± 60	895–1170 AD
Beta-95499	88	1790 ± 60	120–390 AD

* C^{13} adjusted; **2 σ (95% probability)

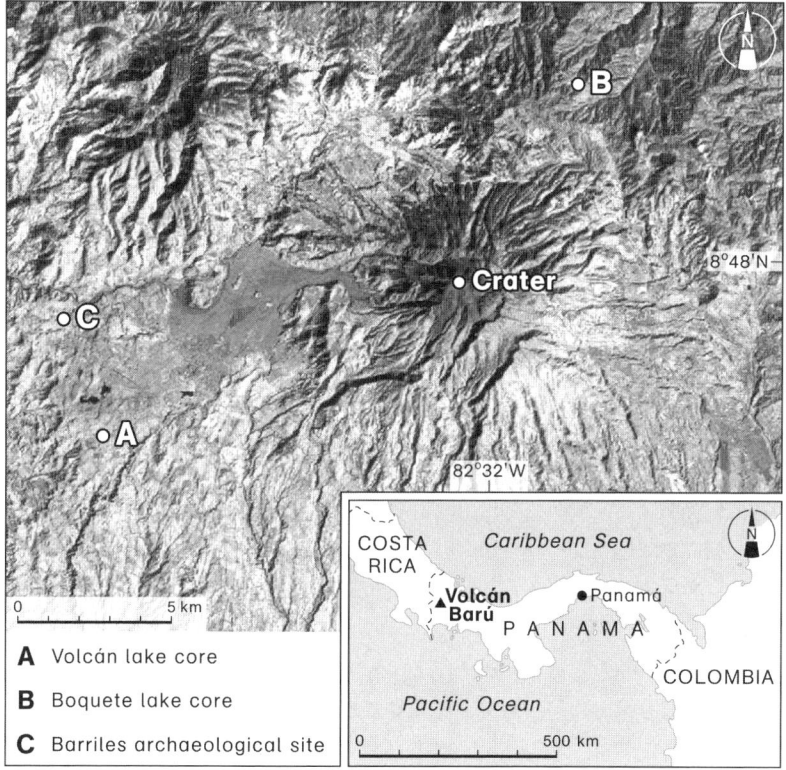

Figure 13.1 Map of the tephra sample locations in relation to the Barú crater.

by a Smithsonian Tropical Research Institute team. Both core samples are currently held in cold storage at the Smithsonian Tropical Research Institute in Panama City.

In contrast, sample set (C) is from the exposed stratigraphy of the Barriles archaeological site, which was excavated in 2001 (Beilke-Voigt *et al* 2004) and sampled by me in 2004. This site is roughly 16 km west of the Barú crater. The Barriles site has only been minimally excavated (Ichon 1968; Künne and Beilke-Voigt in preparation; Rosenthal 1980; Stirling 1950), but is interpreted as the most important known regional centre in prehistoric highland Chiriqui because of the presence of monumental statues and elaborate tomb shafts. The site is thought to have been occupied beginning at some point after the region was settled c 300 BC until c 800 AD. Linares *et al* (1975: 141) state that a 600 AD eruption of Barú 'was responsible for the almost complete depopulation' of the area near the volcano, although Barriles itself does not seem to have been abandoned at that time (Linares and Ranere 1980: 93). Such an eruption, if large enough to prompt abandonment of

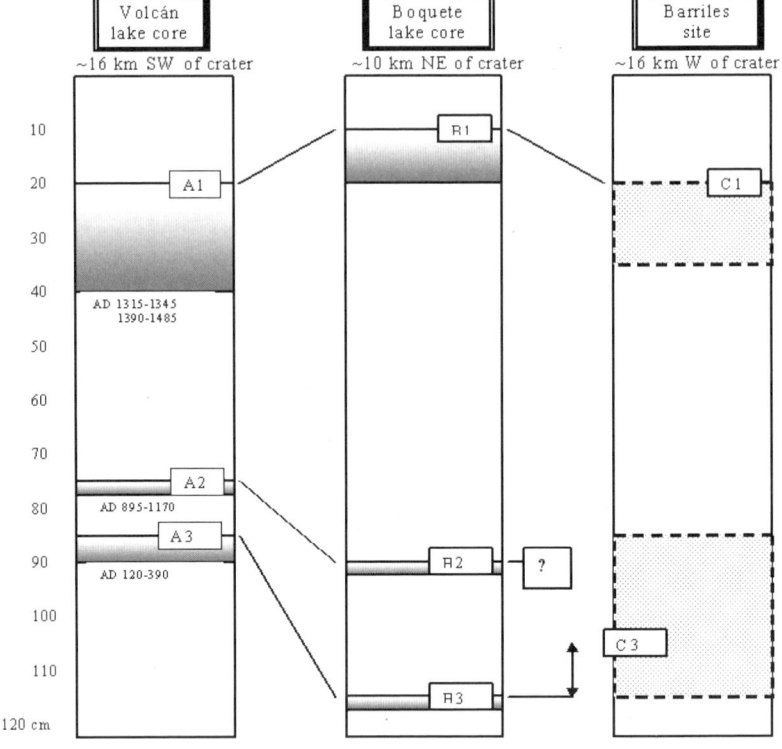

Figure 13.2   Stratigraphic location of the tephra samples.

neighbouring sites, should certainly also be apparent in the Barriles stratigraphy or at least in the more protected lacustrine environment of the Laguna Volcán lake core samples, yet it seems to be absent.

All the samples come from the proximal end of the medial distance range that Wallace (2003: 108) suggests for studies to determine significant coarse tephra falls from stratovolcano eruptions. According to Wallace, the 10–50 km distance is significant as it will only contain material from major eruptions. Due to the nature of an eruption plume, which widens and spreads, minor eruptions will generally fall in the proximal range of 3–8 km and will not be included in samples from the medial range. The samples from this range should show all significant eruptions.

Using feldspar to hornblende ratios present in juvenile clasts in the samples (Table 13.3), the uppermost tephra layer (C1) at the Barriles site and the uppermost tephra layer in the Laguna Volcán coring (A1) appear quite similar. Although the uppermost tephra layer in the Boquete lake core (B1) does not appear strongly correlated to the A1 and C1 samples through the crystal ratios of the clasts, the B1 sample

Table 13.3 Significant characteristics of the tephra layers

	Depth of Tephra Deposit (cm)	Feldspar to Hornblende Ratio in Juvenile Clasts	Vesicularity of Juvenile Clasts
Volcán lake core			
A1	21	2.14	25%
A2	1.5	1.6	15%
A3	4	0.7	15%
Boquete lake core			
B1	10	0.20	20%
B2	1	0.40	15%
B3	1	0.29	10%
Barriles site			
C1	15	2.00	20%
C2	absent	–	–
C3	30*	0.20	10%

*Tephra is not found in a concise layer, but is heavily mixed and interspersed with soil.

does correlate with the higher vesicularity percentages in the A1 and C1 samples, which differentiate them from tephra layers lower in the stratigraphic record. This is likely a result of weathering and reworking over time. The second tephra layer in each lake core (A2 and B2) was dissimilar to the tephra taken from the Barriles archaeological site, and samples from this layer appear to indicate an eruption that eluded detection by previous archaeologists as well as my own archaeological survey and one that is not present or clear in the Barriles stratigraphy.

Interpretations of the Tephra Samples

A definitive eruption history of Barú will require careful geochemical analysis and considerably more work than has been carried out in this project. Some basic conclusions, however, regarding the Barriles archaeological stratigraphy and the tephra record preserved in the lake cores are becoming clear. The most important results concern the interpretation of the impact of the last eruption of Barú and when that eruption occurred. The acceptance of the entrenched 600 AD date leads to confusion when researchers try to reconcile it with ceramic styles assumed to post-date 600 AD that are found below the most recent tephra layer at Barriles (Beilke-Voigt *et al* 2004: 12–13; Künne and Beilke-Voigt in preparation: 9). According to the results of both Behling (2000) and my 2004 fieldwork, a 600 AD eruption seems unlikely. If the similarities between the A1 and C1 samples accurately denote them to be from the same eruption, the AMS determination from organic material directly below the A1 tephra dates the most recent eruption to the 1315–45/1390–45 AD range rather than to 600 AD.

The use of the last eruption of Barú as a catastrophic prime mover needs to be reconsidered. One of the key theories of Linares and Ranere (1980) is that a 600 AD volcanic eruption caused refugees from the Barú area to populate the Caribbean coast and sites such as Bocas del Toro for the first time and that the coast was unsettled before that resettlement. Recent fieldwork at the site of Boca del Drago, however, has recovered Bugaba-style ceramics that pre-date 600 AD by several hundred years (Thomas Wake, personal communication). This casts further doubts on the explanatory need for the volcano to prompt the colonisation of the Caribbean coast, at least at the time of the 600 AD date entrenched in the literature.

A 14th-century date for the last eruption of Barú, as opposed to a 600 AD date, potentially corroborates an historic eruption cited by a number of sources (Montessus de Ballore 1884; Sapper 1913; Simkin and Siebert 1994). Although these sources record an eruption of Barú in the 16th century, these seem to be based on the presence of a single historical maritime document from 1687 and the authors who cite it do so with caution. This document, stored at the *Archivo General de Indias* in Seville, Spain, purportedly shows Barú in eruption. The sketch was made by sailors who did not actually disembark from the Caribbean coast and so never actually saw Barú (Mario Molina, personal communication). It is possible that this sketch is intended only to show that the volcano was active within recent ethnographic and cultural memory and that it was, in fact, a volcano rather than just a mountain. If that is the case and the eruption depicted in the sketch is only a cartographic symbol rather than a record of eruption occurring at the time of the sketch, then the historical record and the palynological and tephrochronological records reported by Behling (2000) mesh well to support a 14th-century eruption.

More investigation is required before the dates can be securely revised, but the combined data may demand a tantalising shift in time scope for the occupation of the region and the interaction that prehistoric people had with volcanic eruption. Regardless of the exact date of Barú's eruptions, it is clear that the overweighting of importance of a proposed 600 AD eruption to explain settlement changes obscures focus on the frequency and number of eruptions that have occurred during human occupation of the area and how they affected prehistoric life.

The presence of only two tephra layers in the Barriles archaeological stratigraphy and three in the lake cores indicates an interesting dilemma for archaeologists accustomed to viewing volcanic strata as secure chronometric markers. Lacustrine environments protect the tephra falls from the wind and heavy rainfall that erode tephra evidence in archaeological contexts, particularly in this region of steep topography. Rather than a concise layer of angular tephra, as found in the lake core samples, the tephra at all archaeological sites surveyed in my 2004 field season were moderately to heavily reworked from erosion. One possibility is that the missing C2 tephra at the Barriles site is simply merged into the very 'dirty' lens in which the C3 layer is found. As juvenile clasts of coarse tephra were heavily interspersed within a 30 cm layer, it is possible that the moderately small (as inferred by the depths of the A2 and C2 deposits) amount of tephra distributed by the eruption leached and sifted down and became incorporated within that of the C3 tephra. It appears that an entire layer of eruptive material is simply 'missing' in the archaeological context. Instead of providing a reliable stratochronological control (cf Stewart 1978: 17), the Barú tephra layers are instead a source of lacunae and mystery. Some eruptive events suffer from archaeological invisibility in the stratigraphic record or are stratigraphically reworked and diffused.

The occurrence of more frequent and repeated volcanic events in the Barú area opens further questions about how past people experienced the volcanic landscape. At the site of Barriles, for example, at least two and likely three eruptions were witnessed by residents over the past two thousand years. From the modern ethnographic examples discussed, it is clear that the social memory of eruption can be incredibly durable. If one uses the Behling (2000) eruption dates as the most reliable baseline to date of Barú's known eruptive activity, a gap of only about 500–700 years likely separated memory of eruption from actual experience of it. This, of course, assumes that Barriles populations did not abandon the site between eruptions. If this was the case, how would the sudden confrontation of 'mythical' or fabled volcanic events that occurred to ancestors interdigitate with physical experience of an actual eruption? Is it possible that the event was met with more terror because of its association with an ancient sublimity, or that it was met with less consternation because ancestors had weathered such events before?

If abandonment occurred at the Barriles site – or at nearby sites as Linares and Ranere (1980) propose – what association to the landscape did people hold as they moved to new locations following eruptions? If terror caused wedding celebrants in Banda Aceh to avoid eating fish after the tsunami, what sorts of social and psychological impacts would forced migration have on refugees?

In a third possible scenario, what is the possibility and impact of a population turnover as opposed to full-scale abandonment? Archaeologists still have a tendency, despite our best efforts, to equate 'pots with people' and measure continuity in populations by a stylistic uniformity over time in ceramics, construction styles, or other diagnostics. Different individuals from the same cultural group, therefore with similar material culture, could have filled the vacated homes and communities of those who left, but without leaving discernible clues for any archaeologists who might enter the scene far in the future.

Forcing the volcano into the catastrophism-inspired role of an entity examined only in the context of *event: affect*, as Barú is by archaeological interpretations, ignores the varied and personal relationship that Cronin *et al* (2004) found the Vanuatu population to have with the volcanic landscape or was evident in post-eruption Congo. The revised dating of the Barú eruption history draws on data previously unavailable to earlier archaeologists, but also on data that were undetected because of the disciplinary constraints of research structures that placed the volcano into a sterile prime mover role. Although by no means implying that perceptions, memories, or individual experience can be excavated with a trowel, I do propose that an incorporation of aids such as ethnohistory, ethnography, and experiential elements of the volcanic landscape are crucial to a more inclusive archaeological language when discussing volcanic contexts.

Barú's Ethnographic Presence

Barú is actively utilised in a number of ways in the modern context. In a utilitarian sense, the volcano is used to support a tangle of communication towers. Volcanic hot springs provide a lure for both locals and tourists for their purported health benefits. More symbolically, as it is the highest point on the isthmus, the volcano is frequently cited in the context of national pride and identity.

The Volcán Barú is most deeply rooted and utilised in a symbolic sense within the province of Chiriqui. It forms a basis and a symbol of local and community identity in the Chiriqui highlands that can be seen in a number of everyday examples. The names of towns and villages in the area give a strong indication of identification with the volcanic landscape, with names like Caldera, Volcán, and Volcancito. The Chiriquí portion of the InterAmerican Highway has recently been dotted with fresh new road signs (Figure 13.3). Each of these pictures an outline of the Volcán Barú as background for whatever road information is listed. A heart is placed within the volcano, whilst a cow on the left of the volcano represents the farming basis of the region and the provincial flag of Chiriquí is placed on the right. The volcano is, in essence, the heart of the region.

A significant number of local businesses also incorporate the volcano in to their logos and names (Figure 13.4). It makes sense for coffee and the volcano to be interlinked in the minds of local residents, such

Figure 13.3 Road signs near the volcano. Barú is heavily referenced as part of the local sense of place (photographs by Karen Holmberg).

Figure 13.4 The Volcán Barú is frequently incorporated in advertising and business signs, such as a hotel mural that place the volcano in a wine glass, the Pizzeria Volcánica, Barú ice machines, and a small neighbourhood store (photographs by Karen Holmberg).

as at Café Barú; the rich soil of Barú's slopes and the coffee that grows in them form the backbone of the agricultural communities surrounding the volcano. Other examples, however, are less predictable. The pizzeria, La Volcánica, is marked by a sign with a handlebar-mustached Italian winking in front of an image of Barú's crater. A local nightclub is called simply Erupción 81, marking the year 1981, when it first 'erupted' into the local nightlife scene. Although the nightclub may want to connote an image of being 'hot', hence a potential link to the volcano in eruption, the linkage of Barú to ice is less obvious. The ice machines found at any gas station in the local area proclaim themselves as containing *hielo Barú* (Barú ice), and, like the pizzeria, also picture a cloud-topped image of the volcano.

The supermarket, too, incorporates the volcano into both its stylised neon logo and its name, the Super Barú. The t-shirts worn by grocery employees combine a logo of the volcano, with the statement in Spanish that the store is 'Chiricano like you'; the store claims that it, like the volcano and the customer, is a basis of the Chiriqui community. A book by a local resident, titled *El Volcán Barú* (Landau 2000), is available for sale inside the Super Barú grocery. The book mixes poetry about the volcano with citations and quotes from various scientific studies, although most of the pages are dedicated to the author's suggestions of names for the more interesting rock formations on the volcano's slopes.

There is quite obviously an imaginative and very human element interwoven within the current Barú landscape. In each of the cited examples, it is clear that a symbolic connection is made between the volcanic landscape and the people who inhabit it. Unlike the examples of Nyiragongo or Vanuatu, however, Barú has neither erupted within memory of the current residents nor within generational memory for the majority of local residents, who are predominantly there as a result of European immigration in the 1920s–40s. The volcano is still, however, an active element of the daily life for those around it. The perception and experience of the volcanic landscape includes, but also extends beyond, the physical impact of eruption.

Barú's Ethnohistoric and Archaeological Presence

The use of ethnohistoric stories and ethnographic attachments or emotional associations to the landscape are gaining refreshing relevance and serious discussion within recent natural disaster literature (eg, Crittenden and Rodolfo 2002; Davies 2002; Malotki and Lomatuway'ma 1987; Shimoyama 2002, and Chapters 6, 10, and 12) and can richly aid the investigation of prehistoric volcanic contexts. Though little is known of the early Barú area inhabitants' beliefs, the neighbouring Bribri people of the Valle de General in the Costa Rican portion of the Chiriqui Province believe that mountains and boulders are inhabited by invisible beings that move with the wind through the upper and lower levels of the world (Garcia and Jaén 1996, Künne *et al* 2000: 137).

The role of the Volcán Barú in the social landscape of the past is hinted at by a series of ethnohistoric stories from the now extinct Doraz language, which repeatedly place Barú as a central focus and reference point within the community (Miranda de Cabal 1974: 9, 27, 28, 35, 46) so as to set the stage for the action of the story. One Doraz story additionally recounts rivers of snakes that ran down from the volcano, which is a potential reference to lava flows or other eruption phenomenon witnessed by residents in the past.

An ethnohistoric story from the Barú region arguably links the largest petroglyph boulder in the area to the act of eruption. The rock-art boulder, found in the small town of Caldera, is 6 m in length and 3 m high and is covered in dozens of geometric and anthropomorphic designs. In the Doraz story, fire and earthquake under the rock-art boulder will one day cause it to shoot into the sky and explode (Miranda de Cabal 1974), an act very much like the volcanic eruptions residents of the area likely witnessed during the period of human occupation in the area. The Caldera rock-art boulder is provocatively located in a direct line-of-sight to Barú and one of the other distinctive geodesic ridges that is said to 'speak' to the volcano in local stories collected during my 2004 fieldwork (Figure 13.5). A boulder that is split in half to form a passageway is located to the immediate west of the rock-art , and the split boulder aligns the rock-art with the solstice sun (Holmberg 2005: 192–93). In the Caldera example, one can see the interlinking of the tangible, marked landscape, the volcano, and ethnohistoric stories.

As Bradley (2000: 36) states, the marking of the landscape is frequently done in association to natural elements like mountains or the course of a river. Rock-art, particularly, is placed in associatively important locations (Nash 2000). I do not suggest that all rock-art or social engagement with the landscape in the area is related to the volcano but rather that the volcano is importantly drawn into the overall web of meaning within the landscape and cannot be viewed as impactive only in its physical aspect of eruption.

Artefactual remains recovered in the Barú area speak to an engagement with the volcanic landscape that is deliberate. Of course, there is an intimate link between the artefacts and the volcano in the expected use of volcanic stone for lithics and tephra as ceramic temper. Beyond this expected and obvious use, however, there are less predictable uses of volcanic materials. As found in archaeological surveys in the early 1970s on the western side of the volcano (Linares 1980), I found a number of crude sculptures or possible pendants made from tephra chunks in my 2004 archaeological field season on the eastern side of Barú. I found a football-sized volcanic bomb in context with a stone tool–manufacturing site. The cylindrical bomb was either carried there and deliberately placed in the midst of the work area or was specifically not moved out of the way when shot into the manufacturing area. A tomb site 6 km from a lava flow that created columnar jointing incorporated columns ripped and transported from the jointing to create

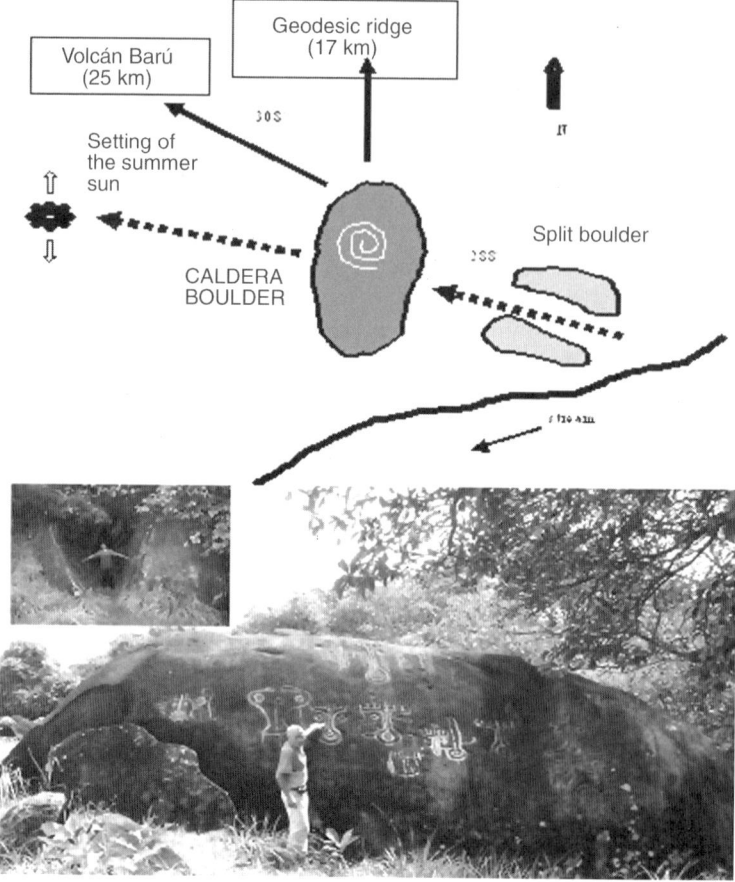

Figure 13.5 The caldera rock-art boulder is provocatively situated in relation to the setting of the solstice sun, a split boulder (insert), the volcano, and a geodesic ridge that 'speaks' to the volcano according to ethnohistoric stories (photograph by Karen Holmberg).

elaborate tombs that were filled with grave goods, including a 'ritually killed' anthropomorphic pot. The effort to transport the heavy stone columns that distance obviously indicates intentional incorporation of the distinctly volcanic landscape, but for what purpose?

Archaeologists in the Barú region will never encounter a prehistoric Pizzaria Volcánica sign and cannot design research structures focused on the hopeful recovery of a Sunset Crater type 'corn rock' or Tetimpa style volcano effigy that puffs smoke. Still, however, the artefactual record strongly indicates a multi-layered interaction that prehistoric residents had with the volcano. When examined in tandem, the linkage of the ethnohistoric stories, the volcano, and the marked landscape represented by the Doraz stories and the volcanic artefacts merge the rigid barriers between natural and cultural elements in the social landscape.

CONCLUSIONS

The interaction between people and volcanic regions proves to be a locus of more complexity than often afforded by many archaeological investigations. It is not simply through the event of eruption that a volcano achieves a presence in human life; the interstices that separate eruptions are a rich component of the long-term social relationship between people and the volcanic landscape. The incorporation of ethnography, ethnohistory, and investigation of the physical landscape and its linkage with past perceptions of it are one way to flesh out the currently unpeopled approach taken to investigating past volcanic contexts over the long term. In doing so, in contrast to the catastrophism implicit in much archaeological interpretation of volcanic regions, nature and culture are seen to be deeply intertwined spheres rather than separate entities.

Archaeology is an interpretive science. Although our task is often to create maps of settlement patterns and population movements, we are also simultaneously mapping our conceptual perspectives of a given region or topic. It is my hope that our conceptions of volcanic regions can broaden to explore and map socialised elements of the volcanic landscape. The incorporation of ethnohistory and examination of the holistic volcanic landscape is one way to attempt an inclusion of such elements in archaeological contexts. This is not an effort to connect living peoples to some ancestral past in the Volcán Barú study region – an impossible task with the mixture of indigenous and European influences – but rather a search for traces of the past landscape relations and physical remainders of them.

The modern contexts of Nyiragongo and Barú discussed in this chapter show a deep texture to the landscape relations in volcanic regions that are not simply environmental or catastrophic. Similarly, the contemporary ethnographic context of the Volcán Barú demonstrates the strong perceptual and lived experience with the volcano, despite the fact that the most recent eruption seems, from my own research and that of Behling (2000), to have occurred at least 500–700 years ago. The volcano, as a highly distinctive and emotive element of its landscape, is interwoven within daily life experience and expression in the present context.

Ethnohistoric stories and artefacts that reference the volcano or volcanic activity hint at a more vibrant and less environmentally constrained past presence of the Volcán Barú than it has been afforded by archaeological assessment to date. Prehistoric interactions with volcanic landscapes, like contemporary ones, were complexly intertwined combinations of material and social interactions that were deeply imbued with memory and sense of place. The volcano cannot be seen simply as attaining prominence only through eruption; a strong experiential component is intimately woven throughout volcanic landscapes and is palpable in both eruptive and non-eruptive time periods. The impacts of a volcano on people who live around it are intimately tangled with what it is to be human and not limited solely to scientifically reducible statements of geologic fact. They instead entail a panoply of expressions and experiential elements, hence requiring a broad range of studies and research structures to elucidate.

ACKNOWLEDGMENTS

Many grateful thanks to the Wenner-Gren Foundation and to the Fulbright Program, which funded the research from which this chapter is produced. I owe much respect and gratitude to Olga Linares, Tony Ranere, Payson Sheets, and the entire team they were a part of for their pioneering work in the Barú area. Hermann Behling graciously provided permission to sample the Laguna Volcán lake core, and Enrique Moreno was integral in helping me sample it and the lake core from Boquete in his lab. As always, I thank Lynn Meskell for her unflagging guidance.

REFERENCES

Basso, K (1996) *Wisdom Sits in Places*, Albuquerque: University of New Mexico Press
Bawden, G and Reycraft, RM (eds) (2000) *Environmental Disaster and the Archaeology of Human Response*, Anthropological Papers Maxwell Museum of Anthropology 7, Albuquerque: Maxwell Museum of Anthropology
Beatty, A (2005) 'Aid in faraway places: the context of an earthquake', *Anthropology Today* 21, 5–7
Beget, J (2000) 'Volcanic tsunamis', in Sigurdsson, H (ed), *Encyclopedia of Volcanoes*, pp 1005–13, San Diego, CA: Academic Press
Behling, H (2000) 'A 2860-year high-resolution pollen and charcoal record from the Cordillera de Talamanca in Panama: a history of human and volcanic forest disturbance', *The Holocene* 10, 387–93
Beilke-Voigt, I, Joly, LG and Künne, M (2004) 'Fechas por radiocarbon de la excavación arqueológica en el Sitio Barriles Bajo (BU-24-I), Chiriquí, Panamá', Universidad Autónoma de Chiriquí – [[http://unachi,ac.pa/publicaciones/Fechamiento_por_radiocarbono.pdf
Bender, B (ed) (1993) *Landscape: Politics and Perspective*, Oxford: Berg
Benjamin, W (1999) 'Excavation and memory', in Benjamin W (ed), *Walter Benjamin: Selected Writings*, vol 2, part 1, *1927–1939*, p 576, Cambridge: Belknap Press
Bradley, R (1984) 'Studying monuments', in Bradley, R and Gardiner, J (eds), *Neolithic Studies*, pp 61–66, Oxford: British Archaeological Reports

Bradley, R (1998) *The Significance of Monuments*, London: Routledge

Bradley, R (2000) *An Archaeology of Natural Places*, London: Routledge

Bradley, R (2002) *The Past in Prehistoric Societies*, London: Routledge

Brandt, P-A (2004) 'What's new? 50,000 years of modernism', in Brandt, P-A (ed), *Spaces, Domains, and Meanings: Essays in Cognitive Semiotics*, pp 245–67, New York: Peter Lang

Colton, H (1932) *A Survey of Prehistoric Sites in the Region of Flagstaff, Arizona*, Bureau of American Ethnology Bulletin 104, Washington, DC: Smithsonian Institution

Connolly, TJ (1999) *Newberry Crater: A Ten-Thousand Year Old Record of Human Occupation and Environmental Change in the Basin-Plateau Borderlands*, Salt Lake City: University of Utah Press

Cordova, C, Martin del Pozzo, AL and Camacho, JL (1994) 'Paleolandforms and volcanic impact on the environment of prehistoric Cuicuilco, southern Mexico City', *Journal of Archaeological Science* 21, 585–96

Crittenden, KS and Rodolfo, KS (2002) 'Bacolor town and Pinatubo volcano, Philippines: coping with recurrent lahar disaster', in Torrence, R and Grattan, J (eds), *Natural Disasters and Cultural Change*, pp 43–65, London: Routledge

Cronin, S, Gaylord, D, Charley, D, Alloway, B, Wallez, S and Esau, J (2004) 'Participatory methods of incorporating scientific with traditional knowledge for volcanic hazard management on Ambae Island, Vanuatu', *Bulletin of Volcanology* 66, 652–68

Dahlin, B (1980) 'Surveying the Volcan region with the posthole digger', in Linares, O and Ranere, A (eds), *Adaptive Radiations in Prehistoric Panama*, Peabody Museum Monographs 5, pp 276–79, Cambridge, MA: Harvard University Press

Davies, H (2002) 'Tsunamis and the coastal communities of Papua New Guinea', in Torrence, R and Grattan, J (eds), *Natural Disasters and Cultural Change*, pp 28–42, London: Routledge

De Boer, JZ, Defant, MJ, Stewart, RH, Restrepo, JF, Clark, LF and Ramirez, AH (1988) 'Quaternary calc-alkaline volcanism in western Panama: regional variation and implication for the plate tectonic framework', *Journal of South American Earth Science* 1, 275–93

Edmonds, M (1993) 'Interpreting causewayed enclosures in the past and the present', in Tilley, C (ed), *Interpretive Archaeology*, pp 99–142, Providence, RI: Berg

Elson, M, Ort, M, Hesse, J and Duffield, M (2002) 'Lava, corn, and ritual in the northern Southwest', *American Antiquity* 67, 119–35

Fullager, R and Head, L (1999) 'Exploring the prehistory of hunter-gatherer attachments to place: an example from the Keep River area, Northern Territory, Australia', in Ucko, P and Layton, R (eds), *The Archaeology and Anthropology of Landscape*, pp 322–35, London: Routledge

Garcia, A and Jaén, A (1996) *Es Sa' Yilìte: Historias Bribris*, San José, Costa Rica: Cooperación Espanola

Himmelheber, H (1993) *Eskimo Artists*, Fairbanks: University of Alaska Press

Hobsbawm, E (1983) 'Introduction', in Gobsbawm, E and Ranger, T (eds),*The Invention of Tradition*, pp 1–14, Cambridge: Cambridge University Press

Hoffman, S and Oliver-Smith, A (eds) (2002) *Catastrophe and Culture: The Anthropology of Disaster*, School of American Research Advanced Seminar Series, Santa Fe, NM: School of American Research Press

Holmberg, K (2005) 'The voices of stones: unthinkable materiality in the volcanic context of western Panamá', in Meskell, L (ed), *Archaeologies of Materiality*, pp 190–211, Oxford: Blackwell

Ichon, A (1968) 'Le probleme de la ceramique de Barriles', *Boletin del Museo Chiricano* 6, 15–24

IHRE (1987) Reconnaissance study of geothermal resources in the Republic of Panama. Unpublished report available at the Institute of Hydraulic Resources and Electrification, Interamerican Development Bank, Latin American Energy

Organization (IRHE-IDB-OLADE0), Panama City, Panama and Quito, Ecuador (IRHE - IDB - OLADE)

Isaacson, J and Zeidler, J (1998) 'Accidental history: volcanic activity and the end of the formative in northwestern Ecuador', in Mothes, P (ed), *Actividad Volcanica y Pueblos Precolombinos en el Ecuador*, pp 41–72, Quito: Abya-Yala

Künne, M and Beilke-Voigt, I (in preparation) 'Mito y realidad: una excavación arqueológica en el sitio Barriles (Panamá) y sus consecuencias sociales', *Revista de Historia de Nicaragua y Centroamérica* (Managua) 19

Künne, M, Beilke-Voigt, I and Voigt, K-U (2000) 'Petroglyphs in the northern part of the general valley in Costa Rica (Central America): their situation in different landscapes', in Nash, G (ed), *Signifying Space and Place*, BAR International Series 902, pp 131–41, Oxford: Archaeopress

Lacey, M (2002) 'Under active Congo volcano, rebel city trembles but makes the most of its lava', in *New York Times*, late edition, final edition, section 1, p 8, column 4

Landau, C (2000) *El Volcan Baru*, Boque, Panama: Parque Natural de Panamá

Lantis, M (1938) 'The mythology of Kodiak Island, Alaska', *Journal of American Folklore* 51, 123–72

Latour, B (1991) *We Have Never Been Modern*, Cambridge, MA: Harvard University Press

Linares, O (1980) 'Miscellaneous artifacts of special use', in Linares, O and Ranere, A (eds), *Adaptive Radiations in Prehistoric Panama*, Peabody Museum Monographs 5, pp 139–45, Cambridge, MA: Harvard University Press

Linares, O and Ranere, A (eds) (1980) *Adaptive Radiations in Prehistoric Panama*, Peabody Museum Monographs 5, Cambridge, MA: Harvard University

Linares, O, Sheets, P and Rosenthal, J (1975) 'Prehistoric agriculture in tropical highlands', *Science* 187, 137–45

Malotki, E and Lomatuway'ma, M (1987) *Earth Fire: A Hopi Legend of the Sunset Crater Eruption*, Flagstaff, AZ: Northland Press

Matsuda, M (1996) *The Memory of the Modern*, New York: Oxford University Press

Miranda de Cabal, B (1974) *Un Pueblo Visto a Través de Su Lenguaje*, David, Panama: private press

Montessus de Ballore, F, comte de (1884) *Temblores y Erupciones Volcanicas en Centro-America*, San Salvador, El Salvador: F Sagrini

Montessus de Ballore, F, comte de (1888) *Tremblements de Terre et Eruptions Volcaniques au Centre-Amérique depuis la Conquête Espagnole Jusqu'à nos Jours*, Dijon, France: Impr E Jobard

Mothes, P (ed) (1998) *Actividad Volcanica y Pueblos Precolombinos en el Ecuador*, Quito: Abya-Yala, Ducotech

Nash, G (ed) (2000) *Signifying Place and Space*, BAR International Series 902, Oxford: Archaeopress

Plunket, P and Uruñuela, G (1998a) 'The impact of Popocatépetl volcano on Preclassic settlement in central Mexico', *Quaternaire* 9, 53–59

Plunket, P and Uruñuela, G (1998b) 'Preclassic household patterns preserved under volcanic ash at Tetimpa, Puebla, Mexico', *Latin American Antiquity* 9, 287–309

Rosenthal, J (1980) 'Excavations at Barriles (BU-24): a small testing program', in Linares, O and Ranere, A (eds), *Adaptive Radiations in Prehistoric Panama*, Peabody Museum Monographs 5, pp 288–91, Cambridge, MA: Harvard University Press

Sanders, W, Parson, J and Santley, R (1979) *The Basin of Mexico: Ecological Processes in the Evolution of a Civilization*, New York: Academic Press

Santley, R, Nelson, S, Reinhardt, B, Pool, C and Arnold, P (2000) 'Environmental disaster and the archaeology of human response', in Bawden, G and Reycraft, RM (eds), *Environmental Disaster and the Archaeology of Human Response*, Anthropological Papers Maxwell Museum of Anthropology 7, pp 143–62, Albuquerque: Maxwell Museum of Anthropology

Sapper, K (1913) *Die Mittelamerikanischen Vulkane*, Petermanns Geographische Mitteilungen, Erganzungsheft 178, Gotha, Germany: Justus Perthes

Sheets, P (ed) (1983) *Archaeology and Volcanism in Central America*, Austin: University of Texas Press

Sheets, P (1992) *The Ceren Site: A Prehistoric Village Buried by Volcanic Ash in Central America*, Ft Worth, TX: Harcourt Brace

Sheets, P and Grayson, D (eds) (1979) *Volcanic Activity and Human Ecology*, New York: Academic Press

Sheets, P and McKee, B (eds) (1994) *Archaeology, Volcanism, and Remote Sensing in the Arenal Region, Costa Rica*, Austin: University of Texas Press

Shimoyama, S (2002) 'Volcanic disasters and archaeological sites in Southern Kyushu, Japan', in Torrence, R and Grattan, J (eds), *Natural Disasters and Cultural Change*, pp 326–42, London: Routledge

Simkin, T and Siebert, L (1994) *Volcanoes of the World*, 2nd ed, Tucson, AZ: Geoscience Press

Simkin, T and Siebert, L (2005) 'Global Volcanism Program: worldwide Holocene volcano and eruption information (www.volcano.si.edu)', Washington DC: Smithsonian Institution

Soetjipto, T (2005) 'Aceh couple finally holds tsunami-delayed wedding', in Reuters Newswire – World News, 30 January, Banda Aceh, Indonesia

Soja, E (1989) *Postmodern Geographies: The Reassertion of Space in Critical Social Theory*, London: Verso

Steffian, A, Beget, J and Saltonstall, P (1996) 'Prehistoric Alutiiq artifact from Kodiak Island provides oldest documentary record of ancient volcanic eruptions in Alaska', *Alaska Volcano Observatory Bimonthly Report* 8, 13–14

Stewart, R (1978) 'Preliminary geology: el Volcan region, province of Chiriqui, Republic of Panama', unpublished manuscript, Smithsonian Tropical Research Institute, Panama

Stirling, M (1950) 'Exploring ancient Panama by helicopter', *National Geographic Magazine* 97, 227–46

Torrence, R and Grattan, J (eds) (2002a) *Natural Disasters and Cultural Change*, London: Routledge

Torrence, R and Grattan, J (2002b) 'The archaeology of disasters: past and future trends', in Torrence, R and Grattan, J (eds), *Natural Disasters and Cultural Change*, pp 1–18, London: Routledge

Wallace, K (2003) 'Characterization and discrimination of Holocene tephra-fall deposits, Mount Spurr volcano, Alaska', unpublished MS thesis, Department of Geology, Flagstaff, AZ, Northern Arizona University

INDEX

About the Authors

EDITORS

JOHN GRATTAN is a Reader in the Institute of Geography and Earth Sciences, University of Wales, Aberystwyth.

ROBIN TORRENCE is a Principal Research Scientist, Australian Museum, Sydney.

CONTRIBUTORS

KIRK A ANDERSON, Bilby Research Center, Northern Arizona University

MICHAEL CARROLL, Dipartimento di Scienze della Terra, Università di Camerino, Italy

KATHARINE V CASHMAN, Department of Geological Sciences, University of Oregon

SHANE J CRONIN, Institute of Natural Resources, Massey University, New Zealand

DAVID K CHESTER, Department of Geography, University of Liverpool

FG DELFIN JR, School of Policy, Planning & Development, University of Southern California, Los Angeles

CAROLYN D DILLIAN, Department of Anthropology, Rutgers University

EZ DIZON, Archaeology Division, National Museum of the Philippines

TRUDY DOELMAN, Department of Archaeology, University of Sydney

ANGUS M DUNCAN, Institute of Applied Natural Science, University of Luton, United Kingdom

MARK D ELSON, Desert Archaeology, Inc, Tucson

FRANCISCO G FEDELE, Laboratorio di Antropologia, Dipartimento delle Scienze Biologiche, Università di Napoli

J-C GAILLARD, Department of Geography, University of the Philippines, Diliman Campus

BIAGIO GIACCIO, Istituto di Geologia Ambientale e Geoingegneria, CNR, Rome

SILVIA GONZALEZ, School of Biological and Earth Sciences, Liverpool John Moores University

JAMES M HEIDKE, Desert Archaeology, Inc, Tucson

DAVID HUDDART, Centre for Outdoor and Environmental Education, Liverpool John Moores University

ROBERTO ISAIA, Osservatorio Vesuviano, INGV, Naples

SABINA MICHNOWICZ, CITES Researcher, Conventions and Policy Section, Royal Botanic Gardens, United Kingdom

ROBERT NELSON, Department of Geology, Colby College, Maine

GIOVANNI ORSI, Osservatorio Vesuviano, INGV, Naples

MICHAEL H ORT, Environmental Sciences and Geology, Northern Arizona University

VJ PAZ, Archaeological Studies Program, University of the Philippines, Diliman Campus

ROLAND RABARTIN, 145, Rue des Branles, 44560, St Denis-en-Val, France

EG RAMOS, Philippine Institute of Volcanology and Seismology, University of the Philippines, Diliman Campus

CT REMOTIGUE, National Institute of Geological Sciences, University of the Philippines, Diliman Campus

KS RODOLFO, Department of Earth and Environmental Sciences, University of Illinois at Chicago

BRUNO SCAILLET, ISTO, CNRS, Orléans, France

PAYSON D SHEETS, Department of Anthropology, University of Colorado at Boulder

FP SIRINGAN, National Institute of Geological Sciences, University of the Philippines, Diliman Campus

JLA SORIA, National Institute of Geological Sciences, University of the Philippines, Diliman Campus

JV UMBAL, BMP Environment and Community Care Inc, Guadalupe, Makati City, Philippines

RICHARD VANDERHOEK, Department of Anthropology, University of Illinois at Urbana-Champaign and Alaska Department of Natural Resources, Division of Parks and Outdoor Recreation